W9-ABI-128

SEPTIC TANK SYSTEM EFFECTS ON GROUND WATER QUALITY

By Larry W. Canter and Robert C. Knox

Library of Congress Cataloging in Publication Data

Canter, Larry W.
Septic tank system effects on ground water quality

Bibliography: p.
Includes index.
1. Water, Underground—Pollution—United States.
2. Septic tanks—Environmental aspects—United States.
3. Water Quality—United States. I. Knox, R.C.
II. Title.
TD223.C26 1985 628.1′682 84-23280
ISBN 0-87371-012-6

Third Printing 1988

Second Printing 1986

LEWIS PUBLISHERS, INC.
121 South Main Street, Chelsea, Michigan 48118

PRINTED IN THE UNITED STATES OF AMERICA

Larry W. Canter

LARRY W. CANTER, P.E., is the Sun Company Professor of Ground Water Hydrology, and Director, Environmental and Ground Water Institute, at the University of Oklahoma, Norman, Oklahoma, in the USA. Dr. Canter received his Ph.D. in Environmental Health Engineering from the University of Texas in 1967, MS in Sanitary Engineering from the University of Illinois in 1962, and BE in Civil Engineering from Vanderbilt University in 1961. Before joining the faculty of the University of Oklahoma in 1969, he was on the faculty at Tulane University and was a sanitary engineer in the U.S. Public Health Service. He served as Director of the School of Civil Engineering and Environmental Science at the University of Oklahoma from 1971 to 1979.

Dr. Canter has published several books and has written chapters in other books; he is also the author or co-author of numerous papers and research reports. His research interests include environmental impact assessment and ground water pollution control. In 1982 he received the Outstanding Faculty Achievement in Research Award from the College of Engineering, and in 1983 the Regent's Award for Superior Accomplishment in Research.

Dr. Canter currently serves on the U.S. Army Corps of Engineers Environmental Advisory Board. He has conducted research, presented short courses, or served as advisor to institutions in Mexico, Panama, Colombia, Venezuela, Peru, Scotland, The Netherlands, France, Germany, Italy, Greece, Turkey, Kuwait, Thailand, and the People's Republic of China.

Robert C. Knox

ROBERT C. KNOX is an Assistant Professor of Civil Engineering at McNeese State University in Lake Charles, Louisiana. Dr. Knox received his BS, MS and Ph.D. degrees in Civil Engineering from the University of Oklahoma. Prior to joining the faculty at McNeese, he was a research engineer at the Environmental and Ground Water Institute at the University of Oklahoma.

Dr. Knox's research interests include ground water contamination and pollution control, environmental impact assessment, and wastewater treatment. Dr. Knox's dissertation research involved one of the first assessments of a ground water pollution control technology focusing on subsurface impermeable barriers. Dr. Knox has published several technical reports and articles concerning ground water pollution control.

PREFACE

Approximately 1/3 of all housing units in the United States dispose of domestic wastewater through septic tank systems, and about 25 percent of all new homes being constructed are including them. Septic tank systems have been frequently identified as sources of localized and regional ground water pollution. Historical concerns have focused on bacterial and nitrate pollution; more recently, synthetic organic chemicals from septic tank cleaners have been identified. These systems represent a significant source of ground water pollution in the United States since many existing systems are exceeding their design life by several-fold, the usage of synthetic organic chemicals in the household and for system cleaning is increasing, and larger-scale systems are being designed and used. A key issue in siting new septic tank systems is related to evaluating their ground water pollution potential in the locality. This book summarizes existing literature relative to the types and mechanisms of ground water pollution from septic tank systems, and provides information on technical methodologies for evaluating the ground water pollution potential of such systems.

The book is organized into five chapters, with Chapter 1 including background information on the historical and current usage of septic tank systems, and Chapter 5 summarizing the key points of the book. Chapter 2 is related to the engineering design, placement, and operation and maintenance procedures for these systems.

Chapter 3 summarizes the types of pollutants and mechanisms of contamination via the unsaturated zone into the ground water system. The transport and fate of bacteria and viruses in soils and ground water are addressed along with inorganic contaminants such as phosphorus, nitrogen, chlorides, and metals, and contaminants such as cleaning agents and pesticides.

Chapter 4 is focused on the evaluation of septic tank system effects on ground water quality. Information on the Surface Impoundment Assessment methodology and the Soil-Waste Interaction Matrix methodology as applied to 13 septic tank system areas is described. Usage of the Hantush Analytical Model is demonstrated by example calculations for a system serving one household unit. Finally, the advantages and limitations of the Konikow and Bredehoeft Numerical Model are demonstrated by a case study wherein it was applied to a geographical area served by septic tank systems.

The authors wish to express their appreciation to several persons instrumental in the assemblage of this book. First, James Kreissl and Dick Scalf of the U.S. Environmental Protection Agency served as project officers on a research study of the effects of septic tank systems on ground water quality. Second, Debby Fairchild of the Environmental and Ground Water Institute at the University of Oklahoma conducted the computer-based literature searches basic to this effort. Finally, and most important, the authors are indebted to Ms. Leslie Rard of the Environmental and Ground Water Institute for her typing skills and dedication to the preparation of this manuscript.

The authors also wish to express their appreciation to the University of Oklahoma College of Engineering for its basic support of faculty writing endeavors, and to their families for their understanding and patience.

Larry Canter
Sun Company Professor
of Ground Water Hydrology
University of Oklahoma
Norman, Oklahoma

Robert Knox
Assistant Professor of
Civil Engineering
McNeese State University
Lake Charles, Louisiana

January, 1985

To Donna, Doug, Steve, and Greg

To Ruth

CONTENTS

Chapter

Chapter

LIST OF TABLES

Table

Table

Table

LIST OF FIGURES

Figure

Figure

CHAPTER 1

INTRODUCTION

The first reported use of a septic tank for serving the wastewater disposal needs of a household was in France about 1870. Septic tanks were introduced in the United States in 1884 through the use of a two-chamber tank utilizing an automatic siphon for intermittent effluent disposal (Cotteral and Norris, 1969). Since their introduction in the United States, septic tanks systems have become the most widely used method of on-site sewage disposal, with over 70 million people depending on them (Hershaft, 1976). Approximately 17 million housing units, or 1/3 of all housing units, dispose of domestic wastewater through the use of septic tank systems. About 25 percent of all new homes being constructed in the United States use septic tank systems for treatment prior to disposal of the home-generated wastewater (U.S. Environmental Protection Agency, October 1980). Figure 1 is a summary of the approximate populations in the United States utilizing septic tank systems. As can be seen, the greatest densities of usage occur in the east and southeast as well as the northern tier and northwest portions of the country.

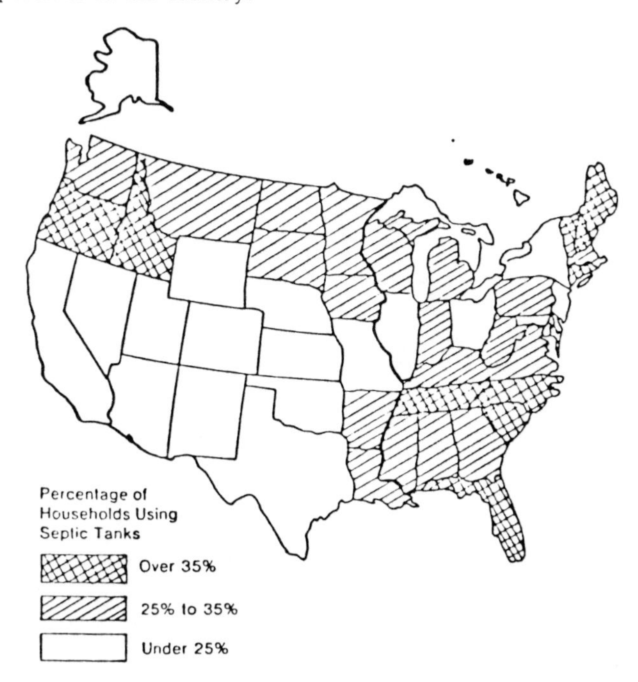

Figure 1: Approximate Populations Using Septic Tanks

A septic tank system includes both the septic tank and the subsurface soil absorption system. Approximately 800 billion gallons of wastewater is discharged annually to the soil via tile fields following the 17 million septic tanks (Scalf, Dunlap and Kreissl, 1977). Of all ground water pollution sources, septic tank systems and cesspools rank highest in total volume of wastewater discharged directly to soils overlying ground water, and they are the most frequently reported sources of contamination (U.S. Environmental Protection Agency, 1977). Figure 2 displays the components of the septic tank system and indicates the general relationship between the soil absorption system and underlying ground water (Bouma, 1979). In sparsely populated urban and rural areas, septic tank systems that have been properly designed, constructed, and maintained are efficient and economical alternatives to public sewage disposal systems. However, due to poor locations for many septic tank systems, as well as poor designs and construction and maintenance practices, septic tank systems have polluted, or have the potential to pollute, underlying ground waters. It is estimated that only 40 percent of existing septic tanks function in a proper manner. A major concern in many locations is that the density of the septic tanks is greater than the natural ability of the subsurface environment to receive and purify system effluents prior to their movement into ground water. A related issue is that the design life of many septic tank systems is in the order of 10-15 years. Due to the rapid rate of placement of septic tank systems in the 1960's, the usable life of many of the systems is being exceeded, and ground water contamination is beginning to occur.

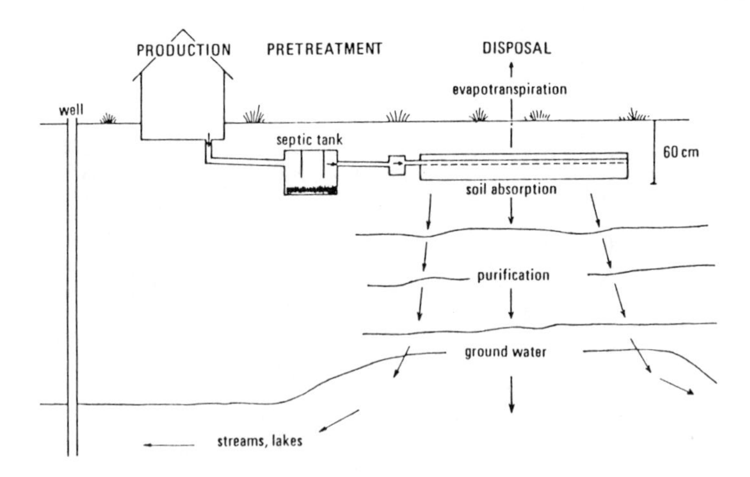

Figure 2: Schematic Cross-Section Through a Conventional Septic Tank Soil Disposal System for On-Site Disposal and Treatment of Domestic Liquid Waste (Bouma, 1979)

A type of ground water pollution of historical as well as current concern is associated with bacterial contamination. Contamination of drinking supplies by malfunctioning septic tank systems has caused outbreaks of waterborne communicable diseases. Documented cases of infectious hepatitis (Hepatitis a) have been traced to contaminated water. In central Appalachia, where few people are served by sewers and septic tank systems often malfunction, the occurrence of infectious hepatitis is high. Many other pathogens, such as typhoid, cholera, streptococci, salmonella, poliomyelitis, and protozoans are also transmitted by septic tank system overflows. Many of these pathogenic organisms have a slow die-off rate in the subsurface environment.

While localized incidents of ground water pollution from septic tank systems are of concern, regional problems have also been recognized in areas of high septic tank system density. Within the United States there are four counties (Nassau and Suffolk, New York; Dade, Florida; and Los Angeles, California) with more than 100,000 housing units served by septic tank systems and cesspools. In addition, there are 23 counties with more than 50,000 housing units served by septic tank systems and cesspools (U.S. Environmental Protection Agency, 1977). Table 1 summarizes relevant county statistics and the density (number per square mile) of septic tank systems and cesspools (U.S. Department of Commerce, 1980; and Newspaper Enterprise Association, Inc., 1982). Densities range from as low as 2 to greater than 346 per square mile. It should be noted that the densities were calculated based on assuming an even distribution of the septic tank systems and cesspools throughout the county. If they are localized in segments of the county the actual densities could be several times greater than those shown in Table 1. Density ranges can be considered as low (less than 10 per square mile or 3.8 per square kilometer), intermediate (between 10 and 40 per square mile, or 3.8 and 15 per square kilometer), and high (greater than 40 per square mile or 15 per square kilometer). Areas with more than 40 per square mile can be considered to have potential contamination problems. Actual densities in areas with documented problems have considerably exceeded the arbitrary 40 per square mile indicator (U.S. Environmental Protection Agency, 1977). Another means of expressing density is by the number per acre, with 40 per square mile equalling 0.062 per acre. The maximum density shown in Table 1 is 346 per square mile, or 0.54 per acre. Considering septic tank system localization within a county, or nonuniform distribution, it would be possible for several counties listed in Table 1 to have densities of greater than 1 septic tank system per acre.

Table 1: Densities of Septic Tank Systems and Cesspools for Counties with More than 50,000 Housing Units Served by These Systems

County	County Statistics			Housing Units With Septic Tank Systems or Cesspools (%)	Density of Septic Tank Systems or Cesspools (No./mi^2)
	1980 Population (x10^3)	1980 Housing Units (x10^3)	Area (sq mi)		
Jefferson, Alabama (1)	671	260	1,115	19-38 (2)	45- 90 (2)
Riverside, California	660	295	7,176	17-34	7- 14
San Bernardino, California	878	368	20,117	14-27	2- 4
Fairfield, Connecticut	807	328 (3)	632	15-30	79-158
Hartford, Connecticut	808	328 (2)	739	15-30	68-135

Table 1: (continued)

County	County Statistics			Housing Unity With Septic Tank Systems or (%) Cesspools	Density of Septic Tank Systems or Cesspools (No./mi^2)
	1980 Population (x10^3)	1980 Housing Units (x10^3)	Area (sq mi)		
New Haven, Connecticut	761	309 (3)	610	16-32	82-164
Broward, Florida	1,006	482	1,219	10-21	41- 82
Duval, Florida	571	227	766	22-44	65-130
Hillsborough, Florida	641	264	1,038	19-38	48- 96
Jefferson, Kentucky	685	266	375	19-38	133-266
Bristol, Massachusetts	475	193 (3)	554	26-52	90-180
Middlesex, Massachusetts	1,367	556 (3)	825	9-18	61-122
Norfolk, Massachusetts	607	247 (3)	394	20-40	127-254
Plymouth, Massachusetts	405	165 (3)	654	30-61	76-153
Worcester, Massachusetts	646	263 (3)	1,509	19-38	33- 66
Genesee, Michigan	450	163	642	31-61	78-156
Oakland, Michigan	1,012	373	867	13-27	58-115
Monmouth, New Jersey	503	186	476	27-54	105-210
Multnomah, Oregon	563	229 (3)	423	22-44	118-236
Westmoreland, Pennsylvania	392	148	1,024	34-68	49- 98
Davidson, Tennessee	478	194 (3)	501	26-52	100-200
King, Washington	1,270	526	2,128	10-19	23- 47
Pierce, Washington	486	187	1,676	27-53	30- 60
Los Angeles, California (4)	7,745	2,854	4,069	>4	> 25
Dade, Florida	1,574	657	2,042	>15	> 49
Nassau, New York	1,322	434	289	>23	>346
Suffolk, New York	1,284	432	929	>23	>108

(1) Counties from Jefferson, Alabama through Pierce, Washington have more than 50,000 housing units, but less than 100,000 housing units, served by septic tank systems.

(2) First number is based on 50,000 housing units served by septic tank systems, and second number by 100,000 housing units served by septic tank systems.

(3) Calculated based on 2.46 persons per housing unit; this value based on reported data for counties of Jefferson, Alabama; Riverside and San Bernardino, California; Broward, Duval, and Hillsborough, Florida; Jefferson, Kentucky; Genesee and Oakland, Michigan; Monmouth, New Jersey; Westmoreland, Pennsylvania; and King and Pierce, Washington.

(4) Counties from Los Angeles, California through Suffolk, New York have more than 100,000 housing units served by septic tank systems.

SEPTIC TANK SYSTEM REGULATION

Several types of institutional arrangements have been developed for regulating septic tank system design and installation, operation and maintenance, and failure detection and correction. Most of the regulatory activities are conducted by state and local governments. Design and siting regulations exist in most states for both individual housing unit systems as well as systems serving clusters of up to several hundred housing units (U.S. Environmental Protection Agency, 1977). Site inspection and installation permit issuance is handled either by the state, regional authority, county, or town, or by a joint effort by two or more of these entities. A state or local governmental entity may regulate all domestic and industrial septic tank system installations; or it may regulate only systems serving multiple housing units and/or industries; or it may regulate only installations in certain critical areas. Where regulations exist, the associated inspections may range from minimal checking to comprehensive evaluations. State regulation and inspection of septic tank installation is generally considered to be more effective than local regulation (U.S. Environmental Protection Agency, 1977).

Operation and maintenance of single housing unit septic tank systems is largely not regulated and is left to the judgment of the system owner. Systems serving multiple housing units or industries may be subject to routine inspections and reporting requirements. Failure detection and correction is difficult to regulate and is typically handled on an individual complaint basis or when a health hazard arises (U.S. Environmental Protection Agency, 1977).

In terms of protection of ground water quality, this is best accomplished by system design, site selection, and installation regulations. Consideration should also be given to the septic tank system density in an area. The U.S. Environmental Protection Agency can become a participant in the regulatory process based on the provision of funding for septic tank systems. Sections 201(h) and (j) of the Clean Water Act of 1977 (P.L. 95-217) authorized construction grants funding of privately owned treatment works serving individual housing units or groups of housing units (or small commercial establishments), provided that a public entity (which will ensure proper operation and maintenance) apply on behalf of a number of such individual systems (Bauer, Conrad and Sherman, 1979). One of the major concerns related to funding applications is to evaluate the ground water pollution potential of the proposed system or systems. This issue becomes even more important for larger systems serving several hundred housing units. To serve as an illustration of possible system size, the U.S. Environmental Protection Agency has funded a system located in the northeastern United States with a design flow of 100,000 gpd (Thomas, 1982).

To provide a basis for evaluation of the ground water pollution potential of septic tank systems, the U.S. Environmental Protection Agency requires that the ground water quality resulting from land utilization practices (septic tank systems) meet the standards for chemical quality (inorganic chemicals) and pesticides (organic chemicals) specified in the

EPA Manual for Evaluating Public Drinking Water Supplies in the case of ground water which potentially can be used for drinking water supply. In addition to the standards for chemical quality and pesticides, the bacteriological standards (microbiological contaminants) specified in the EPA Manual for Evaluating Drinking Water Supplies are required in the case of ground water which is presently being used as a drinking water supply (U.S. Environmental Protection Agency, 1976). Tables 2, 3, and 4 summarize the inorganic, organic, and bacteriological standards, respectively, which should be used in the evaluation process. Current and potential ground water usage should be considered in the evaluation of septic tank systems. The U.S. Environmental Protection Agency requirements have been stated in terms of three cases (U.S. Environmental Protection Agency, 1976):

Case 1: The ground water can potentially be used for drinking water supply.

(1) The maximum contaminant levels for inorganic chemicals and organic chemicals specified for drinking water supply systems as shown in Tables 2 and 3 should not be exceeded.

(2) If the existing concentration of a parameter exceeds the maximum contaminant levels for inorganic chemicals or organic chemicals, there should not be an increase in the concentration of that parameter due to land utilization practices.

Case 2: The ground water is used for drinking water supply.

(1) The criteria for Case 1 should be met.

(2) The maximum microbiological contaminant levels specified for drinking water supply systems as shown in Table 4 should not be exceeded in cases where the ground water is used without disinfection.

Case 3: Uses other than drinking water supply.

(1) Ground water criteria should be established by the EPA Regional Administrator based on the present or potential use of the ground water.

The EPA Regional Administrator in conjunction with the appropriate State officials and the grantee shall determine on a site-by-site basis the areas in the vicinity of a specific land utilization site where the criteria in Case 1, 2 and 3 shall apply. Specifically determined shall be the monitoring requirements appropriate for the project site. This determination shall be made with the objective of protecting the ground water for use as a drinking water supply and/or other designated uses as appropriate and preventing irrevocable damage to ground water. Requirements shall include provisions for monitoring the effect on the native ground water (U.S. Environmental Protection Agency, 1976).

Table 2: Maximum Contaminant Levels for Inorganic Chemicals (U.S. Environmental Protection Agency, 1976)

Contaminant	Level (mg/l)
Arsenic	0.05
Barium	1.
Cadmium	0.010
Chromium	0.05
Lead	0.05
Mercury	0.002
Nitrate (as N)	10.
Selenium	0.01
Silver	0.05

The maximum contaminant levels for fluoride are:

Temperature Degrees Fahrenheit[1]	Degrees Celsius	Level (mg/l)
53.7 and below	12 and below	2.4
53.8 to 58.3	12.1 to 14.6	2.2
58.4 to 63.8	14.7 to 17.6	2.0
63.9 to 70.6	17.7 to 21.4	1.8
70.7 to 79.2	21.5 to 26.2	1.6
79.3 to 90.5	26.3 to 32.5	1.4

[1]Annual average of the maximum daily air temperature.

Table 3: Maximum Contaminant Levels for Organic Chemicals (U.S. Environmental Protection Agency, 1976)

Chemical	Level (mg/l)
Chlorinated hydrocarbons	
Endrin (1,2,3,4,10,10-Hexachloro-6,7 - epoxy - 1,4,4a,5,6,7,8,8a-octahydro-1,4-endo, endo - 5,8,-dimethano naphthalene)	0.0002
Lindane (1,2,3,4,5,6 - Hexachlorocyclohexane, gamma isomer)	0.004

Table 3: (continued)

Chemical	Level (mg/l)
Methoxychlor (1,1,1-Trichloro-2, 2-bis (p-methoxyphenyl) ethane)	0.1
Toxaphene ($C_{10}H_{10}Cl_8$ - Technical chlorinated camphene, 67 to 69 percent chlorine)	0.005
Chlorophenoxys	
2,4-D (2,4-Dichlorophenoxyacetic acid)	0.1
2,4,5-TP Silvex (2,4,5-Trichlorophenoxypropionic acid)	0.01

Table 4: Maximum Bacteriological Contaminant Levels (U.S. Environmental Protection Agency, 1976)

The maximum contaminant levels for coliform bacteria, applicable to community water systems and noncommunity water systems are as follows:

1. When the membrane filter technique is used, the number of coliform bacteria shall not exceed any of the following:

   (a) One per 100 milliliters as the arithmetic mean of all samples examined per month.

   (b) Four per 100 milliliters in more than one sample when less than 20 or more are examined per month.

   (c) Four per 100 milliliters in more than five percent of the samples when 20 or more are examined per month.

2. (a) When the fermentation tube method and 10 milliliter standard portions are used, coliform bacteria shall not be present in any of the following:

   (1) More than 10 percent of the portions in any month;

Table 4: (continued)

---

(2) Three or more portions in more than one sample when less than 20 samples are examined per month; or

(3) Three or more portions in more than five percent of the samples when 20 or more samples are examined per month.

(b) When the fermentation tube method and 100 milliliter standard portions are used, coliform bacteria shall not be present in any of the following:

(1) More than 60 percent of the portions in any month;

(2) Five portions in more than one sample when less than five samples are examined per month; or

(3) Five portions in more than 20 percent of the samples when five or more samples are examined per month.

3. For community or noncommunity systems that are required to sample at a rate of less than 4 per month, compliance with Paragraphs 1, 2(a), or (b) shall be based upon sampling during a 3 month period, except that, at the discretion of the State, compliance may be based upon sampling during a one-month period.

Tables 2 through 4 are based on the National Interim Primary Drinking Water Regulations (40 CFR 141). Any amendments of the National Interim Primary Drinking Water Regulations and any National Revised Primary Drinking Water Regulations hereafter issued by EPA prescribing standards for public water system relating to inorganic chemicals, organic chemicals or microbiological contamination shall automatically apply in the same manner as the National Interim Primary Drinking Water Regulations (U.S. Environmental Protection Agency, 1976).

ORGANIZATION OF BOOK

The objective of this book is to summarize the types and mechanisms of ground water pollution from septic tank systems, and to provide information on technical methodologies for evaluating the ground water pollution potential of septic tank systems. Chapter 1 provides an introduction to the book and includes background information on the use of septic tank systems. Chapter 2 summarizes septic tank system design practices, site selection and evaluation criteria, and operation and maintenance procedures for minimizing ground water pollution concerns. Chapter 3 includes information on the types of pollutants and mechanisms of contamination via migration of pollutants through the unsaturated zone into the ground water system. The transport and fate of bacteria and viruses in soils and ground water are addressed along with similar information on inorganic contaminants such as phosphorus, nitrogen, chlorides, and metals. Information is also included on the transport and fate of organic contaminants.

Chapter 4 represents the focal chapter in terms of the evaluation of septic tank system effects on ground water quality. Information on the Surface Impoundment Assessment methodology and the Soil-Waste Interaction Matrix methodology are included along with descriptions of the Hantush Analytical Model and the Konikow and Bredehoeft Numerical Model. Applications of the two methodologies to 13 septic tank system areas are described. Usage of the Hantush Analytical Model is demonstrated by example calculations for a system serving one household unit. Finally, the advantages and limitations of the Konikow and Bredehoeft Numerical Model are demonstrated through its application to a geographical area served by septic tank systems.

Chapter 5 contains the summary of the key points of the book. Appendix A is an annotated bibliography of published reference materials on septic tank systems and ground water modeling. Appendix B provides information on the characteristics of 13 septic tank system areas located in the central Oklahoma study area, while Appendix C provides specific information on the use of the matrix empirical assessment methodology for these 13 areas. Appendix D contains the error function used in the Hantush Analytical Model. Finally, Appendix E has the Fortran IV program for the Konikow and Bredehoeft Numerical Model.

SELECTED REFERENCES

Bauer, D.H., Conrad, E.T. and Sherman, D.G., "Evaluation of On-Site Wastewater Treatment and Disposal Options", Feb. 1979, U.S. Environmental Protection Agency, Cincinnati, Ohio.

Bouma, J., "Subsurface Applications of Sewage Effluent", in Planning the Uses and Management of Land, 1979, ASA-CSSA-SSSA, Madison, Wisconsin, pp. 665-703.

Cotteral, J.A., Jr. and Norris, D.P., "Septic Tank Systems", ASCE Journal Sanitary Engineering Division, Vol. 95, No. SA4, 1969, pp. 715-746.

Hershaft, A., "The Plight and Promise of On-Site Wastewater Treatment", Compost Science, Vol. 17, No. 5, Winter 1976, pp. 6-13.

Newspaper Enterprise Association, Inc., World Almanac and Book of Facts, 1982, New York, New York.

Scalf, M.R., Dunlap, W.J. and Kreissl, J.F., "Environmental Effects of Septic Tank Systems", EPA-600/3-77-096, Aug. 1977, U.S. Environmental Protection Agency, Ada, Oklahoma.

Thomas, R.E., 1982, personal communication.

U.S. Department of Commerce, Bureau of the Census, State and Metropolitan Area Data Book 1979--A Statistical Abstract Supplement, 1980, Washington, D.C.

U.S. Environmental Protection Agency, "Alternative Waste Management Techniques for Best Practicable Waste Treatment", Federal Register, Vol. 41, No. 29, Feb. 11, 1976, pp. 6190-6191.

U.S. Environmental Protection Agency, "The Report to Congress: Waste Disposal Practices and Their Effects on Ground Water", EPA 570/9-77-001, June 1977, Washington, D.C., pp. 294-321.

U.S. Environmental Protection Agency, "Design Manual--Onsite Wastewater Treatment and Disposal Systems", EPA 625/1-80-012, Oct. 1980, Cincinnati, Ohio.

CHAPTER 2

DESIGN OF SEPTIC TANK SYSTEMS

Septic tank systems consist of the septic tank and associated soil absorption system. Sound principles of engineering should be used in designing both system components. In addition, site suitability criteria should be applied for system location, and routine operational checks and maintenance activities should be conducted. The purpose of this chapter is to summarize design factors, locational criteria, and operational and maintenance measures for septic tank systems. The focus will be on those factors, criteria, and measures which will provide appropriate ground water quality protection. The chapter will begin with some general information on septic tank systems and be followed by sections on septic tank design and operation and subsurface disposal system design and operation. The final section will address a variation of the basic system--the septic tank-mound system popularized in Wisconsin.

OVERVIEW OF SEPTIC TANK SYSTEMS

The basic septic tank system consists of a buried tank where waterborne wastes are collected, and scum, grease and settleable solids are removed from the liquid by gravity separation; and a subsurface drain system where clarified effluent percolates into the soil. System performance is essentially a function of the design of the system components, construction techniques employed, characteristics of the wastes, rate of hydraulic loading, climate, areal geology and topography, physical and chemical composition of the soil mantle, and care given to periodic maintenance (Cotteral and Norris, 1969). A typical on-site system is shown in Figure 3 (Scalf, Dunlap and Kreissl, 1977). The system consists of a building sewer, laid to specified grade, which discharges to the inlet of a septic tank. The septic tank effluent discharges to a series of distribution pipes laid in trenches (absorption trenches) or to a single large excavation (seepage) bed.

The system as displayed in Figure 3 is basically for a single housing unit. Similar systems have been applied at industrial plants and for multiple housing units in a given area, and for small communities with wastewater flows as large as 100,000 gpd (Thomas, 1982). The basic components of larger systems are similar to those for the individual home system; namely, a septic tank and a soil absorption system. Primary differences are associated with the size of the components of the system.

The general advantages of septic tank systems include the following:

1. Minimal maintenance is required for the system, with potential pumpage of septage required every three to five years. While

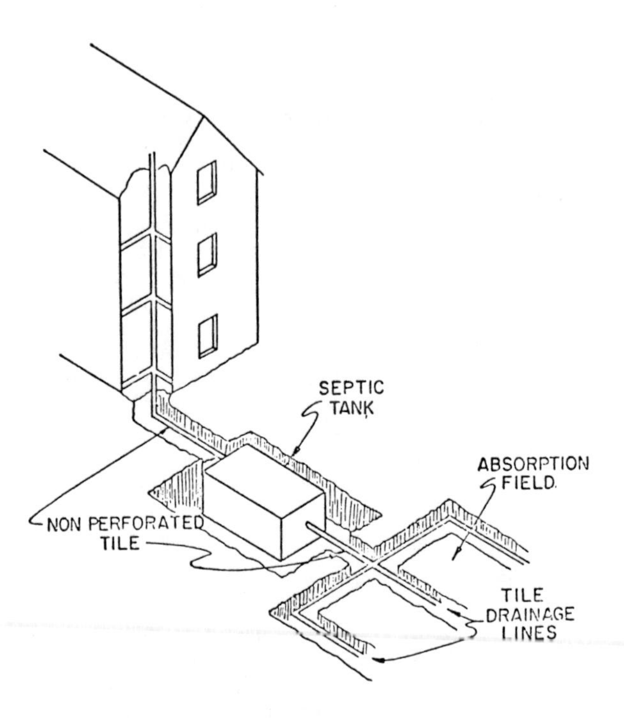

Figure 3: Typical On-Site System (Scalf, Dunlap and Kreissl, 1977)

there are requirements for removal of septage, there is less sludge produced per person through use of a septic tank system than through use of a centralized mechanical plant such as an activated sludge plant.

2. The cost of individual or community septic tank systems is less than the cost of central wastewater collection facilities and treatment plants.

3. The septic tank system represents a low technology system, thus the possibility for long term operation without extensive periods of shutdown is enhanced.

4. The energy requirements of septic tank systems are low in comparison to centralized wastewater treatment facilities.

The general disadvantages of septic tank systems include:

1. The potential for ground water pollution depending upon the soil characteristics and density of systems in a given geographical area.

2. System overflows and pollution of adjacent water wells and surface water courses if the systems are not properly maintained.

3. Cleaners used for maintenance of septic tank systems may create difficulties in terms of ground water pollution, particularly cleaners that have organic solvent bases.

These advantages and disadvantages of septic tank systems must be considered as general statements, with the specific decision to locate a system in a given geographical area based on site suitability and costs relative to other on-site disposal options and central collection and treatment systems.

SEPTIC TANK DESIGN

Septic tanks are buried, water-tight receptacles designed and constructed to receive wastewater from one to multiple housing units or industrial processes. A typical two-compartment septic tank for a housing unit is shown in Figure 4 (Cotteral and Norris, 1969). Heavier sewage solids in the influent settle to the bottom of the tank forming a blanket of sludge. The lighter solids, which includes fats and greases, rise to the surface and form a layer of scum. A considerable portion of the sludge and scum is liquified through decomposition and digestion processes. Gas is liberated from the sludge in this process, carrying some of these solids to the surface where they accumulate with the scum. Further digestion may occur in the scum, and part of the solids may settle again to the sludge layer below. This process may be retarded if there is an excess of grease in the scum layer. The partially clarified liquid between the sludge and scum flows through an outlet located below the scum layer. Proper use of baffles within the septic tank will minimize scum outflow to the soil absorption system. In summary, the septic tank provides for separation of sludges and floatable materials from the wastewater, and an anaerobic environment for decomposition of both retained sludge and nonsettleable materials within the scum layer. Some anaerobic decomposition of the intermediate liquid layer also occurs.

Design considerations related to a septic tank include determination of the appropriate volume, a choice between single and double compartments, selection of the construction material, and placement on the site. The septic tank must be designed to ensure removal of almost all settleable solids in the influent wastewater. Key design considerations basic to this removal from wastewaters from individual housing units include (U.S. Environmental Protection Agency, October 1980):

(1) Liquid volume sufficient for a 24-hr fluid retention time at maximum sludge depth and scum accumulation.

(2) Inlet and outlet devices to prevent the discharge of sludge or scum in the effluent.

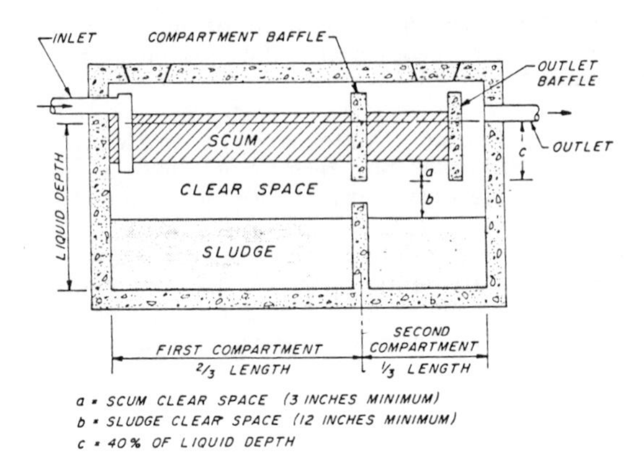

Figure 4: Typical Two Compartment Septic Tank (Cotteral and Norris, 1969)

(3) Sufficient sludge storage space to prevent the discharge of sludge or scum in the effluent.

(4) Venting provisions to allow for the escape of accumulated methane and hydrogen sulfide gases.

Septic Tank Volume

It is important that septic tanks be sized based on the wastewater to be handled. A factor of safety should be provided to allow for variations in wastewater loading and future changes in the character of household wastes. Oversized tanks will not be cost-effective and undersized ones will yield effluent discharges which have not received the level of treatment necessary for optimum usage of the soil absorption system. The first step in selecting the appropriate tank volume is to determine the average daily volume of wastewater produced from the source or sources to be handled. This determination should be based on measurements of actual wastewater flows; however, measurements will not be possible for housing units, commercial establishments, or industrial plants which are under construction. The design volume for septic tanks serving single housing units can be based on the number of bedrooms per home and the average number of persons per bedroom. The average wastewater contribution is about 45 gpcd (170 lpcd). As a safety factor, a value of 75 gpcd (284 lpcd) can be coupled with a potential maximum dwelling density of two persons per bedroom yielding a theoretical design flow of 150 gal/bedroom/day (570 liters/bedroom/day). A theoretical tank volume of 2 to 3 times the design daily flow is common, resulting in a total tank design capacity of 300 to 450 gal per bedroom (1,140 to 1,700 l per bedroom) (U.S. Environmental Protection Agency, October 1980). Single household unit septic tank liquid volume requirements recommended by the Federal Housing Authority, U.S. Public Health Service, and Uniform Plumbing Code are shown in Table 5

(U.S. Environmental Protection Agency, October 1980). State requirements for tank size and minimum water depth are summarized in Table 6 (Senn, 1978). Tank length to width ratios of at least 2 to 1 and not over 3 to 1 have been used for several decades (Senn, 1978).

Table 5: Single Household Unit Septic Tank Liquid Volume Requirements (U.S. Environmental Protection Agency, October 1980)

	Federal Housing Authority	U.S. Public Health Service	Uniform Plumbing Code
Minimum, gal	750	750	750
1-2 bedrooms, gal	750	750	750
3 bedrooms, gal	900	900	1,000
4 bedrooms, gal	1,000	1,000	1,200
5 bedrooms, gal	1,250	1,250	1,500
Additional bedrooms (ea), gal	250	250	150

Table 6: State Requirements for Single Household Unit Septic Tank Size and Water Depth (Senn, 1978)

States	Tank Size in Gallons					Minimum Water Depth (Feet)
	Number of Bedrooms					
	1	2	3	4	5	
Alabama	1000	1000	1000	1200	1400	4
Alaska	750	750	900	1000	1250	4
Arizona	960	960	960	1200	1500	4
Arkansas	-	-	-	-	-	-
California	-	-	-	-	-	-
Colorado	750	750	900	1000	1250	No Minimum
Connecticut	1000	1000	1000	1250	1500	1.5
Delaware	-	-	-	-	-	-
Florida	750	750	900	1000	1200	1.5
Georgia	750	750	900	1000	1250	No Minimum
Hawaii	-	-	-	-	-	-

Table 6: (continued)

States	Tank Size in Gallons					Minimum Water Depth (Feet)
	Number of Bedrooms					
	1	2	3	4	5	
Idaho	750	750	900	1000	1250	4
Illinois	-	-	-	-	-	-
Indiana	750	750	900	1100	1250	
Iowa	750	750	1000	1250	1500	1.5
Kansas	-	-	-	-	-	-
Kentucky	750	750	900	1000	1250	
Louisiana	500	750	900	1150	1400	None
Maine	750	750	900	1000	1250	2
Maryland	-	-	-	-	-	-
Massachusetts	-	-	-	-	-	-
Michigan	-	-	-	-	-	-
Minnesota	-	-	-	-	-	-
Mississippi	-	-	-	-	-	-
Missouri	-	-	-	-	-	-
Montana	750	750	900	1000	1250	4
Nebraska	750	750	900	1000	1250	
Nevada	1000	1000	1000	1000	1250	4
New Hampshire	750	750	900	1000	1250	4
New Jersey	-	-	-	-	-	-
New Mexico	-	-	-	-	-	-
New York	-	-	-	-	-	-
North Carolina	750	750	900	1000	1250	2
North Dakota	-	-	-	-	-	-
Ohio	1000	1000	1500	2000	2000	4
Oklahoma	-	-	-	-	-	-
Oregon	750	750	900	1000	1250	1.5
Pennsylvania	900	900	900	1000	1100	4
Rhode Island	750	750	900	1000	1250	3
South Carolina	-	-	-	-	-	-
South Dakota	1000	1000	1000	1250	1500	4
Tennessee	750	750	900	1000	1250	4
Texas	-	-	-	-	-	-

Table 6: (continued)

States	Tank Size in Gallons Number of Bedrooms 1	2	3	4	5	Minimum Water Depth (Feet)
Utah	750	750	900	1000	1250	1
Vermont	-	-	-	-	-	-
Virginia	30 hour Detention 100 Gallons Per Day					No Minimum
Washington	750	750	900	1000	1250	3
West Virginia	750	750	900	1000	1250	4
Wisconsin	750	750	975	1200	1375	3
Wyoming	750	750	900	1000	1250	4

Septic tanks for commercial, institutional, or industrial sources, or for multiple housing units, must also be sized for daily wastewater flows. For septic tanks being planned for existing sources the flow should be measured to determine average daily flows and peak flows. For multiple housing units, if the total flow cannot be measured, the individually measured or estimated flows based on the expected population and the generation rate of 45 gpcd (170 lpcd) from each unit must be summed to determine the design flow (U.S. Environmental Protection Agency, October 1980). For flows between 750 and 1,500 gpd (2,840 to 5,680 lpd), the capacity of the tank should equal to 1-1/2 days wastewater flow. For flows between 1,500 and 15,000 gpd (5,680 to 56,800 lpd), the minimum effective tank capacity can be calculated as 1,125 gal (4,260 l) plus 75 percent of the daily flow; or

$$V = 1{,}125 + 0.75Q$$

where:

V = net volume of the tank (gal)

Q = daily wastewater flow (gal)

Number of Compartments

Early trends in septic tank design focused on single compartment tanks. More recent trends favor multiple compartment tanks due to resultant improvements in biochemical oxygen demand (BOD) and suspended solids removals. Figure 4 displayed some of the features of a two-compartment tank (Cotteral and Norris, 1969), and Figure 5 shows still others (U.S. Environmental Protection Agency, October 1980). Benefits of the design shown in Figure 5 are due largely to hydraulic isolation, and to the reduction or elimination of intercompartmental mixing. Mixing can occur by two means--water oscillation and true turbulence. Oscillatory mixing can be minimized by making compartments unequal in size (commonly the second compartment is 1/3 to 1/2 the size of the first), reducing the flow-through area, and using an ell to connect compartments. In the first compartment, some mixing of sludge and scum with the liquid always occurs due to induced turbulence from entering wastewater and the digestive process. The second compartment receives the clarified effluent from the first compartment. Most of the time it receives this hydraulic load at a lower rate and with less turbulence than does the first compartment, and thus, better conditions exist for settling low-density solids. These conditions lead to longer working periods before pump-out of solids is necessary, and they improve overall performance.

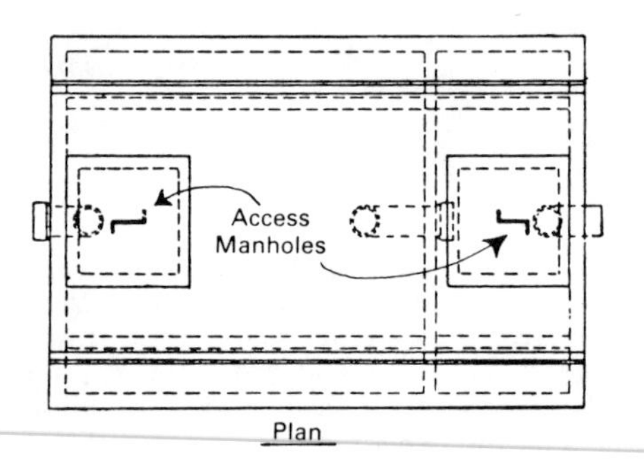

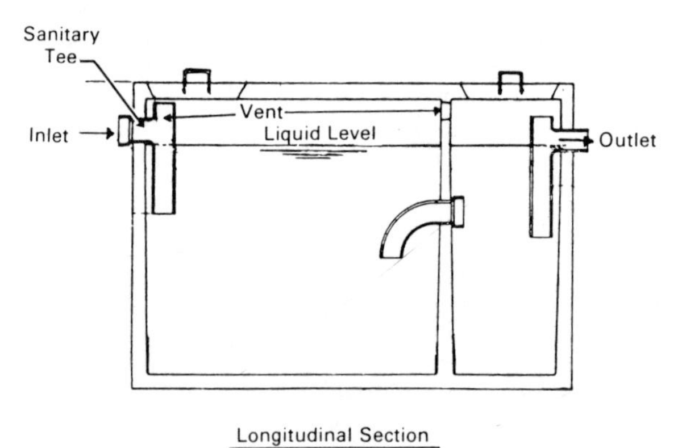

Figure 5: Plan and Section Views of Two Compartment Septic Tank (U.S. Environmental Protection Agency, October 1980)

Construction Material

Septic tanks should be watertight, structurally sound, and reasonably durable. The watertight requirement is to preclude infiltration into the tank which will cause hydraulic overloading and lead to poor quality discharges to soil absorption systems at a rate greater than the system design. In addition, the material of construction should be such that the tank is cost-effective. The Federal Housing Authority and Uniform Plumbing Code indicate that the materials of construction should be durable; the U.S. Public Health Service indicates usage of either concrete or metal (Cotteral and Norris, 1969). Reinforced concrete is the most commonly used septic tank construction material. Most single housing unit septic tanks have been precast for easy installation at the site (U.S. Environmental Protection Agency, October 1980). The walls have thicknesses of 3 to 4 in (8 to 10 cm), and the tanks are sealed for watertightness after installation with two coats of bituminous coating. Care must be taken to seal around the inlet and discharge pipes with a bonding compound that will adhere both to concrete and to the inlet and outlet pipe. When steel is used as the construction material it must be treated to be able to resist corrosion and decay. Such protection includes bituminous coating or other corrosion-resistant treatment. However, despite a corrosion-resistant coating, tanks deteriorate at the liquid level. Past history indicates that steel tanks have a short operational life (less than 10 years) due to corrosion (U.S. Environmental Protection Agency, October 1980). Other construction materials which have been or are being used include redwood and cedar (Cotteral and Norris, 1969) and polyethylene and fiberglass (U.S. Environmental Protection Agency, October 1980). Plastic and fiberglass tanks are very light, easily transported, and resistant to corrosion and decay. While these tanks have not had a good history due to structural failures, some manufacturers are now producing tanks with increased strength. This minimizes the chance of damage during installation or when heavy machinery moves over them after burial. A well-designed and maintained concrete, fiberglass, or plastic tank should last for 50 years. Because of corrosion problems, steel tanks can be expected to last no more than 10 years (U.S. Environmental Protection Agency, October 1980).

Placement on a Site

Placement of the septic tank on a site basically involves consideration of the site slope and minimum setback distances from various natural features or built structures. A typical minimum lot size for a septic tank system serving a single household unit is one acre (Cotteral and Norris, 1969). This lot size takes into account areas unsuitable for drainfields by reason of local variation in soil conditions, and it will normally be sufficient for the required initial drainfield and its replacement, and leave space for standard setbacks and construction of the residence. As noted in Chapter 1, numerous areas within the United States have septic tank densities greater than one per acre. If the site slope is greater than 5 percent, it is desirable to increase the minimum lot size to make allowance for additional construction difficulties on slopes, and to take precautions to prevent slides or downhill surfacing of system effluents. Based upon drainfield area slope, that is, the maximum slope across a minimum 1/2 acre area containing both the required drainfield and the replacement drainfield area, the following minimum lot sizes should be required: 5 to 10 percent--1.25 acres; 10 to 20 percent--1.50 acres; and over 20 percent--2.0 acres (Cotteral and Norris, 1969). Minimum setback distances used by the Federal Housing Authority, Uniform Plumbing Code, and several California counties are listed in

Table 7 (Cotteral and Norris, 1969). Setback information for drainfields will be presented in the next section.

Table 7: Setback Requirements for Septic Tanks (Cotteral and Norris, 1969)

	Setback Requirements (feet)						
	Federal Housing Authority	Uniform Plumbing Code	San Mateo County	Santa Cruz County	Santa Clara County	Contra Costa County	Marin County
	Septic Tanks To:						
Buildings	5	5	5	5	5	10	5
Property lines	10	5	10	5	10	5	5
Wells	50	50	50	50	100	50	50
Creeks or streams	-	50	20	50	-	50	10
Cuts or embankments	-	-	-	15	-	50	15
Pools	-	-	25	-	-	-	10
Water lines	10	5	-	5	-	10	-
Walks and drives	-	-	-	-	-	5	-
Large trees	-	10	-		-	10	-

SOIL ABSORPTION SYSTEM DESIGN

Proper designs of soil absorption systems are critical to the successful operation of septic tank systems. Laak, Healy and Hardisty (1974) identified three aspects in the rational design of soil absorption systems. The first is associated with hydraulic characteristics. This means that the flow regime and the storage and water-carrying capacity of the receiving soil should be measured before design. A soil with a coefficient of permeability of less than 10^{-4} ft/min ($5x10^{-5}$ cm/sec) suggests, for example, that the hydraulic capacity of the system governs the size of the subsurface leaching field. Seasonally high water tables or impervious strata may retard the flow and reduce the quantity of wastewater that can be carried away from the subsurface disposal area. The second consideration concerns the biological mat in leaching fields. Leaching fields can be designed with higher loadings in soils having a greater coefficient of permeability than 10^{-4} ft/min ($5x10^{-5}$ cm/sec) if increased pretreatment is used. A mathematical relationship can be used for reducing the size of leaching fields for effluents with a BOD_5 plus suspended solids less than 250 mg/l. The third design consideration is related to preserving ground water quality. A chief factor is the type of subsurface soil and distance to the top of the water table from the soil absorption system.

Trench and Bed Systems

Soil absorption systems include the design and usage of trenches and beds, seepage pits, mounds, fills and artificially drained systems (U.S. Environmental Protection Agency, October 1980). Trench and bed systems are the most commonly used methods for on-site wastewater treatment and disposal. A typical trench system is shown in Figure 6 (U.S. Environmental Protection Agency, October 1980). Trenches are shallow, level excavations, usually 1 to 5 ft (0.3 to 1.5 m) deep and 1 to 3 ft (0.3 to 0.9 m) wide. The bottom is filled with 6 in (15 cm) or more of washed crushed rock or gravel over which is laid a single line of perforated distribution piping. Additional rock is placed over the pipe and the rock covered with a suitable semipermeable barrier to prevent the backfill from penetrating the rock. Both the bottoms and sidewalls of the trenches are infiltrative surfaces. Additional details on drainfield trench layouts are in Figure 7 (Cotteral and Norris, 1969).

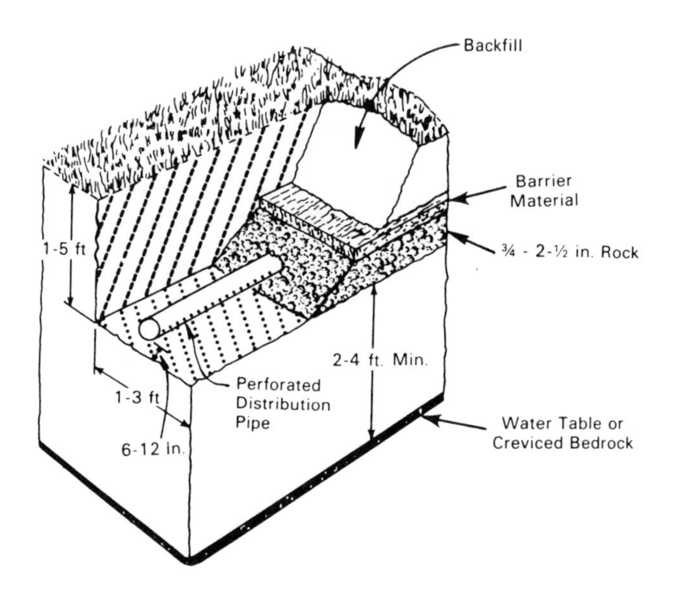

Figure 6: Typical Trench-type Soil Absorption System (U.S. Environmental Protection Agency, October 1980)

A typical bed system is shown in Figure 8 (U.S. Environmental Protection Agency, October 1980). Beds differ from trenches in that they are wider than 3 ft (0.9 m) and may contain more than one line of distribution piping. Thus, the bottoms of the beds are the principal infiltrative surfaces. Site criteria for trench and bed systems are shown in Table 8 (U.S. Environmental Protection Agency, October 1980). These criteria are based upon factors necessary to maintain reasonable infiltration rates and adequate treatment performance over many years of continuous service.

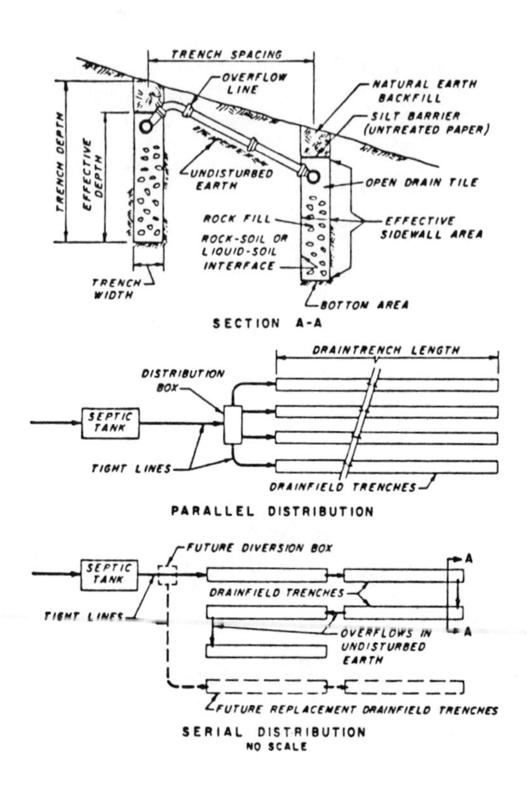

Figure 7: Details of Drainfield Trench Layout (Cotteral and Norris, 1969)

Site Soil Criteria

Specific sites used for soil absorption systems must meet certain basic criteria. A soil is considered suitable for the absorption of septic tank effluent if it has an acceptable percolation rate, without interference from ground water or impervious strata below the level of the absorption system. For a septic tank system to be approved by a local health agency, several criteria normally must be met: a specified percolation rate, as determined by a percolation test; and a minimum 4-ft (1.2 m) separation between the bottom of the seepage system and the maximum seasonal elevation of ground water. In addition, there must be a reasonable thickness, again normally 4 ft, of relatively permeable soil between the seepage system and the top of a clay layer or impervious rock formation (U.S. Environmental Protection Agency, 1977). Bouma (1980) summarized specific limits in a health code for on-site subsurface disposal of septic tank effluent as follows: (1) the percolation rate of the soil should be more than 60 cm per day; (2) bedrock should be at least 90 cm below the surface on 80 percent of the lot area and at least 180 cm below on the remainder; (3) the lot slope should be less than 10 to 20 percent, depending on the percolation rate; (4) the highest ground water level, or depth to the water table, should be at least 90 cm below the bottom of the seepage system; and (5) the lot should not be subject to flooding. Health codes typically list required seepage areas as a function of the soil percolation rate.

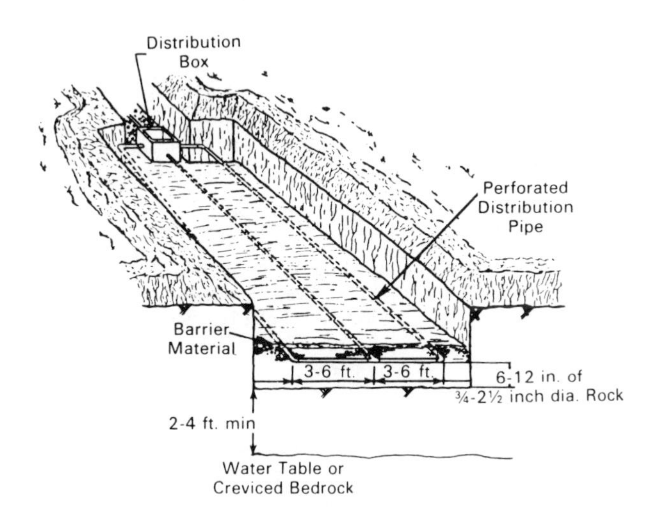

Figure 8: Typical Bed-type Soil Absorption System (U.S. Environmental Protection Agency, October 1980)

Table 8: Site Criteria for Trench and Bed Systems (U.S. Environmental Protection Agency, October 1980)

Item	Criteria
Landscape Position[a]	Level, well drained areas, crests of slopes, convex slopes most desirable. Avoid depressions, bases of slopes and concave slopes unless suitable surface drainage is provided.
Slope[a]	0 to 25 percent. Slopes in excess of 25 percent can be utilized but the use of construction machinery may be limited. Bed systems are limited to 0 to 5 percent.

Table 8: (continued)

Item	Criteria
Typical Horizontal Separation Distances[b]	
Water Supply Wells	50 - 100 ft
Surface Waters, Springs	50 - 100 ft
Escarpments, Manmade Cuts	10 - 20 ft
Boundary of Property	5 - 10 ft
Building Foundations	10 - 20 ft
Soil	
Texture	Soils with sandy or loamy textures are best suited. Gravelly and cobbley soils with open pores and slowly permeable clay soils are less desirable.
Structure	Strong granular, blocky or prismatic structures are desirable. Platy or unstructured massive soils should be avoided.
Color	Bright, uniform colors indicate well-drained, well-aerated soils. Dull, gray or mottled soils indicate continuous or seasonal saturation and are unsuitable.
Layering	Soils exhibiting layers with distinct textural or structural changes should be carefully evaluated to insure water movement will not be severely restricted.
Unsaturated Depth	2 to 4 ft of unsaturated soil should exist between the bottom of the system and the seasonally high water table or bedrock.
Percolation Rate	1 to 60 min/in (average of at least 3 percolation tests).[c] Systems can be constructed in soils with slower percolation rates, but soil damage during construction must be avoided.

Table 8: (continued)

[a]Landscape position and slope are more restrictive for beds because of the depths of cut on the upslope side.

[b]Intended only as a guide. Safe distance varies from site to site, based upon topography, soil permeability, ground water gradients, geology, etc.

[c]Soils with percolation rates less than 1 min/in can be used for trenches and beds if the soil is replaced with a suitably thick (greater than 2 ft) layer of loamy sand or sand.

Cotteral and Norris (1969) identified the four most important factors affecting the performance of a septic tank drainfield as percolative capacity, infiltrative capacity, soil particle size, and drainfield loading rate. Percolative capacity is a measure of the rate at which effluent can be transmitted through the pores or interstices of the soil. Infiltrative capacity is a measure of the rate at which effluent can enter the soil through the surface on which it is applied. Soil particle size refers to a soil characteristic which influences both infiltrative capacity and percolative capacity; the common definition of soil particle size is "effective size," which describes a soil containing 10 percent by weight of particles smaller than the stated size. Loading rate is the rate of application of effluent to the drainfield infiltrative surface; loading rate is normally expressed as cubic feet of liquid per square foot of surface area per day, sometimes shortened to feet per day.

Percolative capacity has traditionally been one of the basic design factors for soil absorption systems. While it is true that the percolative capacity of a soil acts as a limiting factor on the ability of a drainfield to dispose of septic tank effluent, it is the infiltrative capacity of the liquid-soil interface which ultimately determines the life of the drainfield (Cotteral and Norris, 1969). The percolative capacity of the soil must be great enough to transport the effluent away from the system liquid-soil interface at a rate at least equal to the rate at which the liquid enters the soil. The key test used to determine the percolative capacity of the soil is the percolation test which was developed over 50 years ago. The percolation test procedure is described in Table 9 (U.S. Environmental Protection Agency, October 1980). Estimated percolation rates for various soil textures and permeabilities are given in Table 10 (U.S. Environmental Protection Agency, October 1980). Table 11 summarizes the absorption area requirements for single housing units based on measured soil percolation rates (U.S. Public Health Service, 1967). Experience has shown that design hydraulic application rates can sometimes be correlated with soil texture. Table 12 summarizes this experience and is meant only as a guide. Soil texture and measured percolation rates will not always be correlated as indicated, due to differences in structure, clay mineral content, bulk densities, and other factors in various areas of the country (U.S. Environmental Protection Agency, October 1980).

Table 9: Falling Head Percolation Test Procedure (U.S. Environmental Protection Agency, October 1980)

---

1. Number and Location of Tests

Commonly a minimum of three percolation tests are performed within the area proposed for an absorption system. They are spaced uniformly throughout the area. If soil conditions are highly variable, more tests may be required.

2. Preparation of Test Hole

The diameter of each test hole is 6 in, dug or bored to the proposed depths at the absorption systems or to the most limiting soil horizon. To expose a natural soil surface, the sides of the hole are scratched with a sharp pointed instrument and the loose material is removed from the bottom of the test hole. Two inches of 1/2 to 3/4 in gravel are placed in the hole to protect the bottom from scouring action when the water is added.

3. Soaking Period

The hole is carefully filled with at least 12 in of clear water. This depth of water should be maintained for at least 4 hr and preferably overnight if clay soils are present. A funnel with an attached hose or similar device may be used to prevent water from washing down the sides of the hole. Automatic siphons or float valves may be employed to automatically maintain the water level during the soaking period. It is extremely important that the soil be allowed to soak for a sufficiently long period of time to allow the soil to swell if accurate results are to be obtained.

In sandy soils with little or no clay, soaking is not necessary. If, after filling the hole twice with 12 in of water, the water seeps completely away in less than ten minutes, the test can proceed immediately.

4. Measurement of the Percolation Rate

Except for sandy soils, percolation rate measurements are made 15 hr but no more than 30 hr after the soaking period began. Any soil that sloughed into the hole during the soaking period is removed and the water level is adjusted to 6 in above the gravel (or 8 in above the bottom of the hole). At no time during the test is the water level allowed to rise more than 6 in above the gravel.

Table 9: (continued)

Immediately after adjustment, the water level is measured from a fixed reference point to the nearest 1/16 in at 30 min intervals. The test is continued until two successive water level drops do not vary by more than 1/16 in. At least three measurements are made.

After each measurement, the water level is readjusted to the 6 in level. The last water level drop is used to calculate the percolation rate.

In sandy soils or soils in which the first 6 in of water added after the soaking period seeps away in less than 30 min, water level measurements are made at 10 min intervals for a 1 hr period. The last water level drop is used to calculate the percolation rate.

5. Calculation of the Percolation Rate

The percolation rate is calculated for each test hole by dividing the time interval used between measurements by the magnitude of the last water level drop. This calculation results in a percolation rate in terms of min/in. To determine the percolation rate for the area, the rates obtained from each hole are averaged. (If tests in the area vary by more than 20 min/in, variations in soil type are indicated. Under these circumstances, percolation rates should not be averaged.)

Example: If the last measured drop in water level after 30 min is 5/8 in, the percolation rate = (30 min)/(5/8 in) = 48 min/in.

Table 10: Estimated Hydraulic Characteristics of Soil (U.S. Environmental Protection Agency, October 1980)

Soil Texture	Permeability (in/hr)	Percolation (min/in)
Sand	>6.0	<10
Sandy loams Porous silt loams Silty clay loams	0.2-6.0	10-45

Table 10: (continued)

Soil Texture	Permeability (in/hr)	Percolation (min/in)
Clays, compact Silt loams Silty clay loams	<0.2	>45

Table 11: Soil Absorption System Area Requirements for Single Housing Units (U.S. Public Health Service, 1967)

Percolation Rate (Time Required for Water to Fall 1 in) (minutes)	Required Absorption Area[a], in Square Feet per Bedroom[b], for Standard Trench[c] and Seepage Pits[d]
1 or less	70
2	85
3	100
4	115
5	125
10	165
15	190
30[e]	250
45[e]	300
60[e,f]	330

[a]Provides for garbage grinders and automatic-sequence washing machines.

[b]In every case, sufficient area should be provided for at least two bedrooms.

Table 11: (continued)

[c]Absorption area for standard trenches is figured as trench-bottom area.

[d]Absorption area for seepage pits is figured as effective sidewall area beneath the inlet.

[e]Unsuitable for seepage pits if over 30.

[f]Unsuitable for leaching systems if over 60.

Table 12: Recommended Rates of Wastewater Application for Trench and Bed Bottom Areas[a] (U.S. Environmental Protection Agency, October 1980)

Soil Texture	Percolation Rate (min/in)	Application Rate[b] (gpd/ft^2)
Gravel, coarse sand	<1	Not suitable[c]
Coarse to medium sand	1 - 5	1.2
Fine sandy, loamy sand	6 - 15	0.8
Sandy loam, loam	16 - 30	0.6
Loam, porous silt loam	31 - 60	0.45
Silty clay loam, clay loam[d]	61 - 120	0.2[e]

[a]May be suitable estimates for sidewall infiltration rates.

[b]Rates based on septic tank effluent from a domestic waste source. A factor of safety may be desirable for wastes of significantly different character.

[c]Soils with percolation rates <1 min/in can be used if the soil is replaced with a suitably thick (>2 ft) layer of loamy sand or sand.

[d]Soils without expandable clays.

[e]These soils may be easily damaged during construction.

It is recognized that the percolation test has a high degree of variability in terms of measuring the saturated conductivity of the soil (Otis, Plews and Patterson, 1978). In one series of tests, variability was as high as 90 percent, therefore, if the percolation rate is the only criterion used for sizing the soil absorption system, failures may occur due to the high variability of the test results and their inappropriate usage in system design. Saturated hydraulic conductivity does not reveal how the soil will conduct wastewater under prolonged use, because once the surface mat is formed, the liquid movement is through unsaturated soil below. As noted in Table 10, the percolation rate is related to soil texture and permeability (hydraulic conductivity). Classes of soil permeability have been defined, with the class limit values representing saturated or maximum permeability (Otis et al., 1977). If the moisture content of the soil decreases the permeability also decreases; therefore, soils can have an infinite number of permeabilities. Sand has relatively large pores that drain abruptly at relatively low tensions, whereas clay releases only a small volume of water over a wide tension range due to its very fine pores. Figure 9 shows soil moisture retention for four different soil materials (U.S. Environmental Protection Agency, September 1978). Figure 10 shows hydraulic conductivity as a function of soil moisture tension (U.S. Environmental Protection Agency, September 1978). A field method has been developed to directly measure the unsaturated hydraulic conductivity of soils. However, it is a complex technique that requires both time and trained technicians. Therefore, the short term traditional percolation test is still used for most soil evaluations prior to soil absorption system design and installation (Otis, Plews and Patterson, 1978).

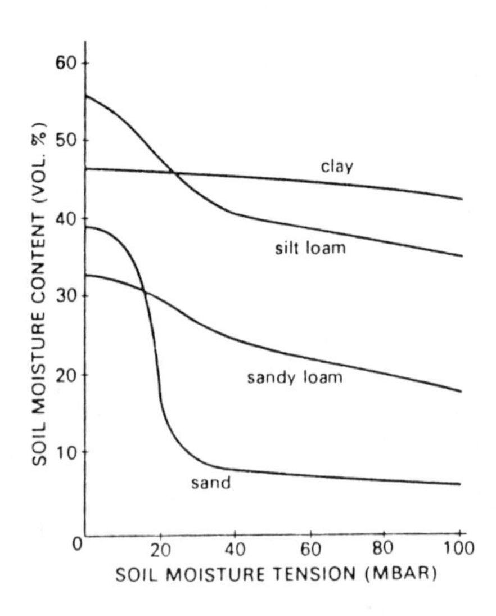

Figure 9: Soil Moisture Content (U.S. Environmental Protection Agency, September 1978)

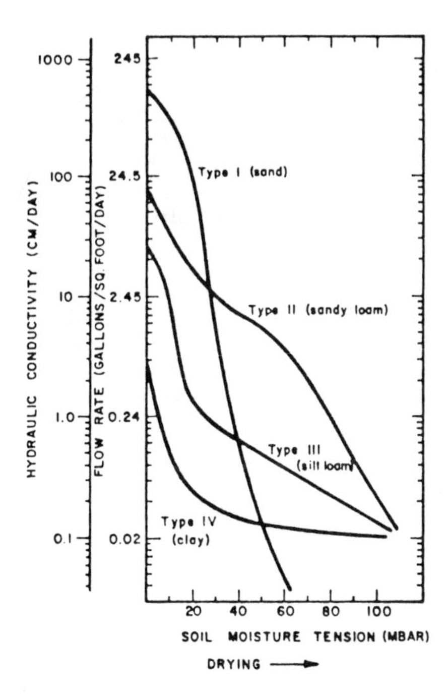

Figure 10: Hydraulic Conductivity of Soils (U.S. Environmental Protection Agency, September 1978)

It has been well demonstrated that the infiltrative capacity of the liquid-soil interface controls the long term capacity of the drainfield system, and that this infiltration rate, due to the various clogging effects which occur, will always be less than the percolation rate of the soil (Cotteral and Norris, 1969). A typical time-rate infiltration curve representing three distinct phases in the infiltration process can be defined as shown in Figure 11. Phase 1 reflects the initial loss of infiltration rate due to slaking of the soil caused by the affinity of the internal soil surface for water and the breakdown of cohesive forces which hold the soil together. Phase 2 represents an increase due to the removal of entrapped air by solution in the percolation water. Phase 3 represents a long term decrease due primarily to microbial action in the soil. Although in reality no equilibrium rate is ever reached, a reasonably stable rate may be expected after a period of from 5 months to 1 year. It has further been shown that the long term infiltration rate of sewage into permeable soils eventually declines to about the same negligible quantity regardless of the difference in soil permeabilities in the beginning.

The process of soil clogging and resultant loss of infiltrative capacity as shown in Figure 11 occurs as a result of combined physical, chemical and biological factors. Physical factors include the compaction of the soils by ponded water or equipment, smearing of soil surfaces by excavating equipment, and physical movement of fines into the voids of the infiltrative surface. The most important chemical factor is the deflocculation of soils when they are irrigated with high sodium-percentage waters. A high percentage of sodium in the wastes, either naturally occurring or resulting from water softener regeneration, may in effect preclude the use of soil absorption systems. Biological factors are the most important factors influencing the clogging phenomenon. The major reduction of infiltrative capacity shown as Phase 3 in Figure 11 results from the formation of an organic mat about 5 mm thick at the liquid-soil interface. It is within this zone that the major biological activity occurs, and it is here that the processes of deposition of suspended materials, bacterial build-up, and decomposition of organic material by bacterial action continually modify the infiltrative capacity. Anaerobic conditions within the clogging zone will lead to further clogging through the growth of slimes and the deposition of ferrous sulfide in an even deeper zone of 2 in to 3 in beyond the surface (Cotteral and Norris, 1969).

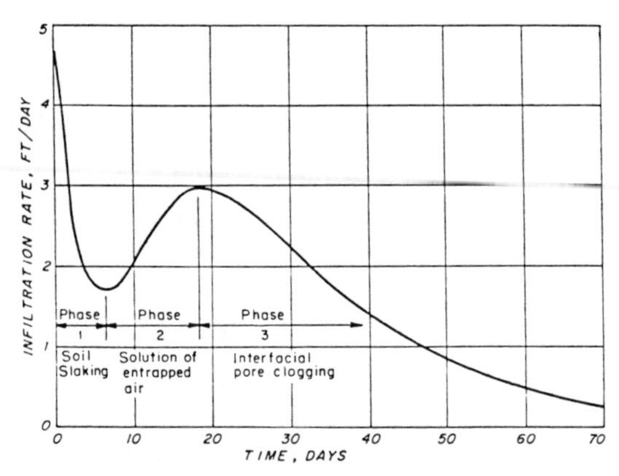

Figure 11: Typical Time-Rate Infiltration Curve for Soil Absorption System (Cotteral and Norris, 1969)

Healy and Laak (1974) concluded that there is a long term acceptance rate, which is a function of the soil permeability, at which septic tank effluent can be absorbed indefinitely. This acceptance rate is independent of whether the soil is continuously or intermittently flooded, and varies from approximately 0.3 gpd/sq ft for a soil with a permeability of 0.0002 ft/min (6×10^{-5} m/min) to approximately 3.0 gpd/sq ft (12 cm/day) for a soil with a permeability of 0.1 ft/min (0.03 m/min).

In a related matter to general soil absorption system clogging, it has been determined that the sidewall area of a trench is the major infiltrative surface, and that the bottom area is of minor importance. The effects of slaking of the soil particles and the more rapid clogging of bottom surfaces by sedimentation contribute to this phenomenon. The finding leads to the conclusion that deep, narrow trenches are the most effective. The minimum desirable width of drainfield trenches appears to be determined primarily by practical considerations of construction (Cotteral and Norris, 1969).

The percolative and infiltrative capacity of a soil cannot be predicted by comparing its particle size characteristics with those of other soils for which these capacities have been established. Nevertheless, there are several broad observations which can be made concerning the effect of soil particle size on absorption field performance. Where small soil grain size makes the effects of surface tension and capillarity a major consideration, a minimum length of soil column becomes necessary to produce complete draining of the soil adjacent to infiltrative surface areas. Failure to provide this minimum soil column length, by construction, for example, in an area with a high ground water table, will lead to continued inundation and failure. If drainage is to take place at all, the infiltrative surface must be sufficiently above the ground water surface to overcome capillary forces. The critical distance varies from soil to soil, but is on the order of 3 ft in many fine-grained soils having good percolative capacity (Cotteral and Norris, 1969).

The topographic and geological characteristics of the soil absorption system site can affect performance. Examples of factors which should be considered in determining the suitability of a given site are the ground slope, ground water level, depth of the soil mantle, and location in relation to surface water streams and wells. The ground slope can affect the stability of hillside trenches, the cost of construction of the soil absorption system, and the distance that the effluent from the system will travel through the soil without surfacing. The ground water level should be low enough to assure that the minimum soil depth necessary to prevent inundation of the drainfield is maintained throughout the year. As inundation may lead to irreversible clogging of the infiltrative surface, suitability should be determined on the basis of the maximum height of the ground water table during the wet season. The depth of the soil mantle at the site should be sufficient to receive and transport the effluent from the piping system. Consideration should be given to the depth and profile of the underlying rock formation to ensure that it will not contribute to the inundation of the drainfield nor short-circuiting and subsequent surfacing of the effluent from the system piping (Cotteral and Norris, 1969).

Placement on a Site

As was the case for septic tank placement, location of the soil absorption system on the site involves consideration of minimum setback distances from various natural features or built structures. Minimum drainfield setback distances used by the Federal Housing Authority, Uniform Plumbing Code, and several California counties are listed in Table 13 (Cotteral and Norris, 1969).

Drainfield Hydraulic Loading

The pollutional strength and volume of the wastewater should be

Table 13: Setback Requirements for Drainfields (Cotteral and Norris, 1969)

	Setback Requirements (feet)						
	Federal Housing Authority	Uniform Plumbing Code	San Mateo County	Santa Cruz County	Santa Clara County	Contra Costa County	Marin County
	Drainfields To:						
Buildings	5	8	5	5	10	10	10
Property lines	5	5	10	5	10	5	5
Wells	100	50	75	100	100	50	100
Creeks or streams	-	50	20	100	-	50	25
Cuts or embankments	-	-	20	15	-	50	25
Pools	-	-	25	-	-	-	25
Water lines	10	5	-	5	-	10	-
Walks and drives	-	-	-	-	-	5	-
Large trees	-	10	-	-	-	10	-

considered prior to designing the soil absorption system. Three types of drainfield loading can be utilized, including continuous ponding, dosing and resting, and uniform application without ponding. In the continuous ponding method the infiltrative surface is covered at all times with wastewater. This method has the advantage of increasing the effective infiltrative area by submerging the sidewalls of the drainfield trenches. It also increases the hydraulic gradient across the infiltrative surface and this in turn may increase the infiltration rate. However, since clogging occurs at the infiltrative surface there will be no aeration at the surface and this may cause subsequent problems in terms of both hydraulic flow and biological decomposition.

To overcome some of the concerns related to continuous ponding, dosing and resting can be utilized. This approach provides reaeration in that periods of loading are followed by periods of resting. The resting phase allows the soil to drain and reaerate, thus encouraging degradation of the clogging mat which may build up at the infiltrative surface.

The process of alternate dosing and resting of a drainfield can therefore markedly prolong the effective life of the system. For practical purposes the resting period required for restoration appears to be on the order of several months (Cotteral and Norris, 1969). In uniform application without ponding the liquid is distributed uniformly over the entire infiltrative surface at a rate lower than that which the soil can accept liquid. Therefore, the soil always remains unsaturated and aerobic conditions prevail at the infiltrative surface. When these aerobic conditions prevail, the resistance of the clogging mat is minimized.

As mentioned earlier, trench and bed systems are the most commonly used designs for soil absorption systems. Bed systems usually require less total land area and are less costly to construct. However, trench systems can provide up to five times more sidewall area than do bed systems for identical bottom areas. Less damage is likely to occur to the soil during construction because the excavation equipment can straddle the trenches so it is not necessary to drive on the infiltrative surface. On sloping sites, trench systems can follow the contours to maintain the infiltrative surfaces in the same soil horizon and keep excavation to a minimum. Bed systems may be acceptable where the site is relatively level and the soils are sandy and loamy sands. Trench system design factors utilized by the Federal Housing Authority, U.S. Public Health Service, and Uniform Plumbing Code are summarized in Table 14 (Cotteral and Norris, 1969). Table 15 summarizes state setback requirements and design factors for trench systems (Senn, 1978).

Table 14: Design Factors for Trench System Drainfields (Cotteral and Norris, 1969)

Data	Federal Housing Authority	U.S. Public Health Service	Uniform Plumbing Code
Percolation test used	yes	yes	as req'd[a]
Surface used for design	bottom	bottom	bottom
Trench width required, in inches	12-36	12-36	18-36
Gravel depth below tile, in inches	6	6	12
Minimum sidewall area,[b] in square feet	85	140	200
Sidewall area,[b] in square feet per bedroom			
15 min per inch percolation rate	190	190	80
30 min per inch percolation rate	250	250	120
60 min per inch percolation rate	330	330	as req'd
Minimum trench spacing, in feet	6	6	6

Table 14: (continued)

Data	Federal Housing Authority	U.S. Public Health Service	Uniform Plumbing Code
Replacement area req'd, as a percentage	--	--	50

[a]As required by the Health Department.

[b]Where design is based on bottom area, sidewall area was calculated based on minimum depth and width requirements.

Design of trench and bed soil absorption systems for small institutions, commercial establishments, and clusters of dwellings generally follows the same design principles as for single dwellings. In cluster systems serving more than about five homes, however, peak flow estimates can be reduced because of flow attenuation, but contributions from infiltration through the collection system must be included. Flexibility in operation should also be incorporated into systems serving larger flows since a failure can create a significant problem. Alternating bed systems should be considered. A three-field system can be constructed in which each field contains 50 percent of the required absorption area. This design allows flexibility in operation. Two beds are always in operation, providing 100 percent of the needed infiltrative surface. The third field is alternated in service on a semi-annual or annual schedule. Thus, each field is in service for one or two years and "rested" for six months to one year to rejuvenate. The third field also acts as a standby unit in case one field fails. The idle field can be put into service immediately while a failed field is rehabilitated. Larger systems should utilize some dosing or uniform application to assure proper performance (U.S. Environmental Protection Agency, October 1980).

OVERVIEW OF SEPTIC TANK-MOUND SYSTEMS

Many soils in the United States are not suited for on-site disposal of wastewaters using conventional septic tank systems. Examples are slowly permeable soils (defined as having a percolation rate slower than 60 min/in), thin layered permeable soils over very permeable creviced bedrock, and soils with periodic or permanent high ground water. A general requirement for on-site waste disposal is the availability of at least 3 ft of sufficiently permeable unsaturated soil below the bottom of the seepage bed. An estimated 50 percent of the soils of Wisconsin do not meet this requirement and are thus unsuitable for on-site disposal. An engineering modification of the conventional septic tank-soil absorption system can be used to overcome

Table 15: Summary of State Setback Requirements and Design Factors for Trench Systems (Senn, 1978)

State	Setbacks (feet) Well	Setbacks (feet) Surface Water	Minimum Spacing (Feet)	Minimum Cover (Inches)
Alabama	50-75		6	6
Alaska	50-100	50-100	6	12
Arizona	50-100	100	6	12
Arkansas	-	-	-	-
California	-	-	-	-
Colorado	100	50	6	12
Connecticut	75	50	6-9	6
Delaware	-	-	-	-
Florida	75-100	50	6-8	12
Georgia	100	50	10	12
Hawaii	-	-	-	-
Idaho	100	100-300	6	12
Illinois	-	-	-	-
Indiana	50-100	50	6-7.5	12
Iowa	100-200	25	7.5	12
Kansas	-	-	-	-
Kentucky				None
Louisiana	100			6-12
Maine	100-300	50-100	10	2-6
Maryland	-	-	-	-
Massachusetts	-	-	-	-
Michigan	-	-	-	-
Minnesota	-	-	-	-
Mississippi	-	-	-	-
Missouri	-	-	-	-
Montana	100	100	6	12
Nebraska	100	50	6	6
Nevada	100	100	6	4-6
New Hampshire	75	75	6-7.5	6
New Jersey	-	-	-	-
New Mexico	-	-	-	-
New York	-	-	-	-
North Carolina	-	-	-	-
North Dakota	-	-	-	-
Ohio	50		6	6
Oklahoma	-	-	-	-
Oregon	50-100	50-100	10	6
Pennsylvania	100	50	6	12
Rhode Island	100	50	6	12
South Carolina	-	-	-	-
South Dakota	100	100	6	
Tennessee	50	50	6	12
Texas	-	-	-	-
Utah	100	100	6-7.5	12
Vermont	-	-	-	-
Virginia	35-100	50-100	6-9	None
Washington	75-100	100	6	6
West Virginia	100	100	6	12
Wisconsin	50-100	50	10	12
Wyoming	100	50	6-7.5	6-12

Table 15: (continued)

State	Minimum Percolation Restrictions	Trench Widths (Inches)	Sizing
Alabama	None	18-36	Perc
Alaska	None	12-36	Perc & Soils
Arizona	None	12-18	Perc
Arkansas	-	-	-
California	-	-	-
Colorado	None	18-36	Perc
Connecticut	None	18-36	Perc
Delaware	-	-	-
Florida	None	18-24	Perc & Soils
Georgia	None	18-36	Perc & Soils
Hawaii	-	-	-
Idaho	None	12-36	Perc & Soils
Illinois	-	-	-
Indiana	None	18-36	Perc
Iowa	None	18	Perc
Kansas	-	-	-
Kentucky			Perc
Louisiana	None	12-18	Perc
Maine	None	24	Soils
Maryland	-	-	-
Massachusetts	-	-	-
Michigan	-	-	-
Minnesota	-	-	-
Mississippi	-	-	-
Missouri	-	-	-
Montana	Yes	12-36	Perc & Soils
Nebraska	No	10-36	Perc
Nevada	Yes	12-24	Perc
New Hampshire	None	12-36	Perc
New Jersey	-	-	-
New Mexico	-	-	-
New York	-	-	-
North Carolina	-	-	-
North Dakota	-	-	-
Ohio	None	8-30	Soils
Oklahoma	-	-	-
Oregon	None	24	Soils
Pennsylvania	Yes	12-36	Perc
Rhode Island	None	18	Perc
South Carolina	-	-	-
South Dakota	Yes		Perc
Tennessee	None	18-36	Perc & Soils
Texas	-	-	-
Utah	None	12-36	Perc
Vermont	-	-	-
Virginia	None	18-36	Perc & Soils
Washington	Yes	18-36	Perc & Soils
West Virginia	None	12-36	Perc
Wisconsin	None	18-36	Perc & Soils
Wyoming	None	12-36	Perc

natural soil limitations, with the modification known as the septic tank-mound system. The mound system is essentially an elevated soil absorption system. The main components of a septic tank-mound system include the septic tank, a pumping chamber, and the mound itself. Most of the developmental work on septic tank-mound systems has been done in Wisconsin.

The septic tank in a septic tank-mound system is sized in the same manner as the septic tank in the conventional septic tank system. A pump then elevates the effluent from the tank and pressurizes the distribution within the mound. A siphon may replace the pump if the mound is located downslope (Converse, 1978). Figure 12 depicts a typical Wisconsin mound system (Harkin et al., 1979). The mound is comprised of a fill material (usually medium-textured sand), an absorption area, a distribution system, a cap, and topsoil. The effluent flows through the fill material where it is purified before entering the natural soil. The cap, which is usually comprised of topsoil or subsoil, provides frost protection, serves as a barrier to infiltration, retains moisture for vegetation, and promotes runoff of precipitation. A minimum of 24 inches of unsaturated natural soil under the mound is recommended (Converse, 1978). This natural soil provides additional purification and acts as a buffer protecting ground water from potential contamination.

There are several advantages and disadvantages of a septic tank-mound system. The advantages are as follows:

1. The topsoil can be selected to be more permeable than the subsoil.

2. There is less chance of changing the hydraulic characteristics of the soil during compaction as construction is eliminated in the wetter subsoil.

3. Slimes that develop in the bottom of the absorption area do not clog the fill as readily as they do in less permeable natural soil.

4. A smaller absorption area is required for a given quantity of wastewater when compared to the traditional septic tank system.

5. There is a reduction in the nitrate nitrogen leaving the system due to denitrification processes.

In contrast to the advantages enumerated above, the septic tank-mound system has certain disadvantages as follows:

1. Septic tank mound systems exhibit increased cost over traditional septic tank systems due to the cost of the fill material and its placement.

2. Construction of the mound will change the landscape and possibly the visual quality of the environmental setting.

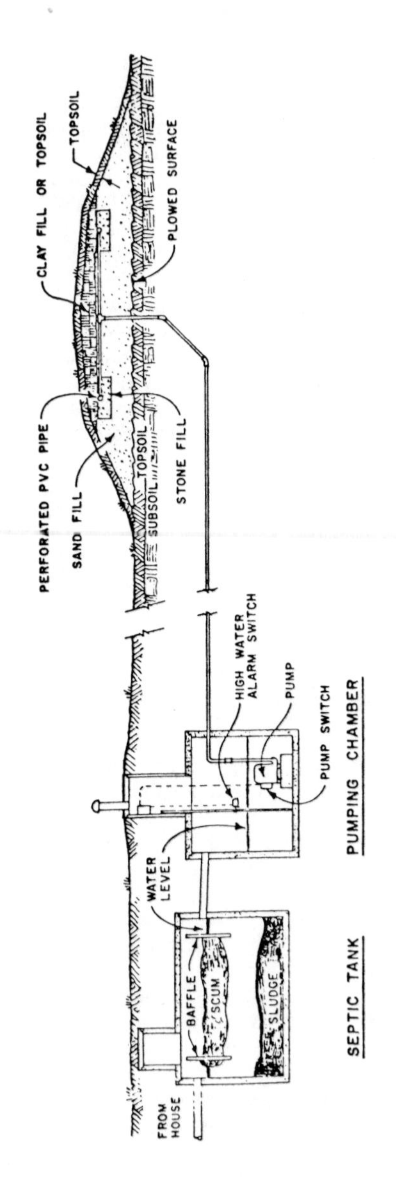

Figure 12: Typical Wisconsin Mound System (Harkin et al., 1979)

3. Even though the absorption area within the mound is smaller, the mound itself may comprise a larger area than the soil absorption system would encompass in the conventional septic tank system.

In summary relative to septic tank-mound systems, they represent an alternative approach to the traditional septic tank system in areas where the soil characteristics are insufficient for use of the conventional system. Care must be taken to appropriately evaluate the features and characteristics of the septic tank-mound system and determine its applicability in a given geographical area.

SELECTED REFERENCES

Bouma, J., "Use of Soil Survey Data for Preliminary Design of Land Treatment Systems and Regional Planning", Ch. 22 in Simulating Nutrient Transformation and Transport During Land Treatment of Wastewater, Aug. 1980, John Wiley and Sons, Inc., New York, New York.

Converse, J.C., Design and Construction Manual for Wisconsin Ponds, University of Wisconsin, Madison, Wisconsin, 1978.

Cotteral, J.A., Jr. and Norris, D.P., "Septic Tank Systems", ASCE Journal Sanitary Engineering Division, Vol. 95, No. SA4, 1969, pp. 715-746.

Harkin, J.M. et al., "Evaluation of Mound Systems for Purification of Septic Tank Effluent", Technical Report WIS WRC 79-05, 1979, Madison, Wisconsin.

Healy, K.A. and Laak, R., "Site Evaluation and Design of Seepage Fields", ASCE Journal of Environmental Engineering Division, Vol. 100, No. 5, Oct. 1974, pp. 1133-1146.

Laak, R., Healy, K.A. and Hardisty, D.M., "Rational Basis for Septic Tank System Design", Ground Water, Vol. 12, No. 6, Nov.-Dec. 1974, pp. 348-351.

Otis, R.J. et al., "Effluent Distribution", Proceedings of the Second National Home Sewage Treatment Symposium, American Society of Agricultural Engineers, 1977, pp. 61-85.

Otis, R.J., Plews, G.D. and Patterson, D.H., "Design of Conventional Soil Adsorption Trenches and Beds", Third Annual Illinois Private Sewage Disposal Symposium, Toledo Area Council of Governments, Toledo, Ohio, 1978, pp. 52-66.

Scalf, M.R., Dunlap, W.J. and Kreissl, J.F., "Environmental Effects of Septic Tank Systems", EPA-600/3-77-096, Aug. 1977, U.S. Environmental Protection Agency, Ada, Oklahoma.

Senn, C.L., "Current Status of On-Site Wastewater Management", Journal of Environmental Health, Vol. 40, No. 5, Mar.-Apr. 1978, pp. 279-284.

Thomas, R.E., 1982, personal communication.

U.S. Environmental Protection Agency, "The Report to Congress: Waste Disposal Practices and Their Effects on Ground Water", EPA 570/9-77-001, June 1977, Washington, D.C., pp. 294-321.

U.S. Environmental Protection Agency, "Management of Small Waste Flows", Report No. EPA-600/2-78-173, Sept. 1978, Cincinnati, Ohio.

U.S. Environmental Protection Agency, "Design Manual--Onsite Wastewater Treatment and Disposal Systems", EPA 625/1-80-012, Oct. 1980, Cincinnati, Ohio.

U.S. Public Health Service, "Manual of Septic Tank Practice", Pub. No. 526, 1967, Washington, D.C.

CHAPTER 3

GROUND WATER POLLUTION FROM SEPTIC TANK SYSTEMS

One of the key concerns associated with the design and usage of septic tank systems is the potential for inadvertently polluting ground water. This concern is increased when considerating systems serving multiple housing units. This chapter begins with the identification of constituents of potential concern in the effluents from septic tank systems. Mechanisms of ground water contamination from septic tank systems are addressed, including the migration of pollutants through soil and ground water systems. The transport and fate of biological contaminants are described, with information included on both bacteria and viruses. The transport and fate of inorganic chemicals are also described, with emphasis given to phosphates and nitrogen compounds as well as chlorides, metals, and other inorganic contaminants. Brief information is included on the transport and fate of organic contaminants; the brief coverage is due to the lack of extensive information in the published literature. Some information on pollution control measures is presented. Ground water monitoring for septic tank systems is also addressed. Finally, a special section is included on the handling of septage.

POTENTIAL POLLUTANTS FROM SYSTEM EFFLUENTS

Potential ground water pollutants from septic tank systems are primarily those associated with domestic wastewater. Contaminants originating from system cleaning can also contribute to the ground water pollution potential of septic tank systems. The volume of wastewater introduced to a septic tank system from a typical household unit ranges from 40 to 45 gpd/person (150 to 170 liters/day/person) (U.S. Environmental Protection Agency, 1977). Typical sources of household wastewater, expressed on a percentage basis, are: toilet(s)--22 to 45 percent; laundry--4 to 26 percent; bath(s)--18 to 37 percent; kitchen--6 to 13 percent; and other--0 to 14 percent.

Influent Wastewater Characteristics

The quality characteristics of wastewater entering septic tank systems are summarized in Table 16 (Bauer, Conrad and Sherman, 1979). The data in Table 16 excludes contributions from garbage disposal units and home water softeners. Garbage disposal contributions are excluded since they can contribute substantial quantities of pollutants which can be effectively disposed of as solid wastes without entering septic tank systems.

To provide a basis for comparison of quality characteristics, Table 17 summarizes the typical composition of community domestic wastewater in

Table 16: Characteristics of Influent Wastewaters to Septic Tank Systems* (Bauer, Conrad and Sherman, 1979)

Constituent (g/cap/d)+	Investigator						Constituent (mg/ℓ)
	Olson Karlgren, and Tullander	Laak	Bennett and Lindstedt	Witt, Siegrist and Boyle	SSWMP	Weighted Value	
BOD_5	45	48.7	34.8	49.5	49.5	48	300
BOD_5 filtered	-	-	-	30.4	30.4	30	188
COD	120	119.4	121.5	-	-	120	750
TOC	-	-	-	32.1	32.1	32	200
TOC filtered	-	-	-	22.0	22.0	22	138
TS	130	-	146.3	113.4	113.4	125	781
TVS	83	-	74.6	63.1	63.1	70	438
SS	48	-	47.3	35.4	35.4	40	250
VSS	40	-	41.6	26.6	26.6	31	194
TKN	12.1	-	6.5	6.1	6.1	6	38
NH_3-N	-	3.2	-	1.3	1.3	2	12
NO_3-N	-	0.1	-	0.1	0.1	0.1	0.6
NO_2-N	-	-	-	-	-	-	-
TP	3.8	-	-	4.0	4.0	4	25
PO_4-P	-	4.0	3.7	1.4	1.4	1.4	8.8
Oil & Grease	-	-	-	14.6	-	15	94
MBAS	-	-	-	-	-	3	19
flow (lpcd)	131.5	156.7	165.3	119.4	161.2	160	

* Also excludes garbage disposal contributions

\+ Data have been rounded

accordance with weak, medium, and strong concentrations of a variety of constituents (Council on Environmental Quality, 1974). Table 18 displays the average characteristics of septic tank influents in relation to medium strength domestic wastewater. The physical and chemical constituents are reasonably comparable in their concentrations, although individual studies of septic tank influents may indicate that the organic strength of household wastewater (septic tank influent) is greater than the organic strength of medium community wastewater (Viraraghavan, 1976). Bacterial counts in household wastewater tend to be lower than in community wastewater, with a possible cause being a shorter incubation time from the house (source) to the septic tank in comparison with the time from the source to the community treatment plant.

Table 17: Typical Characteristics of Domestic Sewage in the United States (Council on Environmental Quality, 1974)

Constituent	Weak	Medium	Strong
Physical Characteristics			
Color (nonseptic)	Gray	Gray	Gray
Color (septic)	Gray-Black	Blackish	Blackish
Odor (nonseptic)	Musty	Musty	Musty
Odor (septic)	Musty-H_2S	H_2S	H_2S
Temperature- F (average)	55°-90°	55°-90°	55°-90°
Total solids* (mg/l)	450	800	1200
Total volatile solids (mg/l)	250	425	800
Suspended solids (mg/l)	100	200	375
Volatile suspended solids (mg/l)	75	130	200
Settleable solids-$\frac{(ML)}{L}$	2	5	7
Chemical Characteristics			
pH (units)	6.5	7.5	8.0
Cl, SO_4, Ca, Mg, etc.*			
Total nitrogen (mg/l)	15	40	60
Organic nitrogen (mg/l)	5	14.5	19
Ammonia nitrogen (mg/l)	10	25	40
Nitrate nitrogen (mg/l)	-	0.5	1.0
Total phosphate-PO_4 (mg/l)	5	15	30
Biological Characteristics			
Total bacteria $\frac{(counts)}{100\ ml}$	1×10^8	30×10^8	100×10^8
Total coliform $\frac{(MPN)}{100\ ml}$	1×10^6	30×10^6	100×10^6
Biochemical oxygen demand	100	200	450

* Quite variable depending on natural water quality of region.

Table 18: Comparison of Septic Tank Influent Wastewater with Community Domestic Wastewater

Constituent	Community Wastewater (1)	Septic Tank Influent Wastewater (2)
Total solids (mg/l)	800	781
Total dissolved solids (mg/l)	600	531
Total suspended solids (mg/l)	200	250
5-day BOD (mg/l)	200	300
Total organic carbon (mg/l)	200	200
Total nitrogen (as N, mg/l)	40	50

Table 18: (continued)

Constituent	Community Wastewater (1)	Septic Tank Influent Wastewater (2)
Organic	14.5	38
Ammonia	25	12
Nitrate	0.5	0.6
Total phosphorus (as P, mg/l)	15	25
Total bacteria (counts/100 ml)	30×10^8	$5.6\text{-}8 \times 10^7$ (3)
Total coliform (MPN/100 ml)	30×10^6	2×10^6 (3)
Fecal coliform (MPN/100 ml)	n.d.	3×10^4 (3)
Fecal streptococci (MPN/100 ml)	n.d.	3×10^4 (3)
Enteric virus (PFU/l)	7000 (4)	32 - 7000 (5)

(1) Based on medium strength wastewater as shown in Table 17.

(2) Based on averages shown in Table 16.

(3) Viraraghavan (1976)

(4) Vilker (1978)

(5) Siegrist (1977)

In regard to the virological characteristics of individual household wastewater, very little characterization work has been conducted (Siegrist, 1977). Viruses are generally not part of the normal microbial flora of healthy individuals and unlike the bacteria, appear to be shed in significant concentrations only as a result of an infection. Assuming a viral concentration of 10^6 PFU/wet gram of feces for a typical individual experiencing an intestinal viral infection, it can be estimated that the concentration in the individual's wastewater could reach 10^7 PFU/liter. As expected, investigations of raw municipal wastewater have demonstrated considerably lower levels. One investigator estimated the level of virus to be about 7000 PFU/liter, while another reported recoveries of only 32 to 107 PFU/liter (Siegrist, 1977).

Septic Tank Treatment Efficiency

Of concern in terms of ground water pollution is the quality of the effluent from the septic tank portion of the system, and the efficiency of constituent removal in the soil underlying the soil absorption system. Those constituents which pass through the septic tank and the unsaturated soil beneath the drainfield represent ones of concern relative to ground water pollution. Numerous studies have been made of the treatment efficiencies and effluent qualities from septic tanks, with fewer reported studies related to soil absorption system efficiencies.

The septic tank serves several important functions such as solid-liquid separation, storage of solids and floatable materials, and anaerobic treatment of both stored solids as well as nonsettleable materials. Viraraghavan (1976) reported on a study of a household two-compartment septic tank serving 12 persons. The tank volume was 200 ft^3, with 148 ft^3 in the first compartment and 52 ft^3 in the second. The average flow rate was 327 gal/day (27.3 gal/person/day), thus the theoretical detention time was 4.6 days (not accounting for sludge accumulation). Table 19 summarizes the statistics associated with the treatment efficiency of the septic tank. The BOD and COD removal efficiencies were in the order of 50 percent on the average, with the TSS removal less than 25 percent.

Lawrence (1973) reported on the efficiencies of two single chamber septic tanks each having a liquid capacity of 740 gal. Tank 1 received domestic and laundry wastes from a family of six. The household water supply was a hard, high-iron water taken from a private well and treated by ion exchange. Waste brine from regeneration was not allowed to enter the system. At the time of this study, the tank had been in continuous operation for five years and, with the exception of scum removed after the first two years of operation, had not been cleaned prior to this investigation. Tank 2 served a family of five and received only domestic wastes without laundry discharge. Household water was from a municipal supply softened by the cold lime process. At the time of this study the tank had been in service five years and had not been serviced since its installation. Observations on water consumption by each household revealed an average daily flow of 186 gpd (31 gpd/capita) for the first and 245 gpd (49 gpd/capita) for the second. This indicates theoretical detention times of four and three days, respectively; however, due to the sludge and scum volume and the fact that quite frequently, two-thirds of the daily flow of wastewater was generated in less than four hours, the effective detention time for both tanks was a matter of a few hours. Table 20 summarizes the measured efficiencies. The wide range in influent and effluent quality and efficiencies may be attributed to the difference in water consumption (both total and per capita) between the two households and the fact that household laundry wastes were excluded from tank 2 but not tank 1. The suspended solids removals were in the order of 35 to 45 percent, with the BOD removals being 15 percent or less.

Table 19: Summary of Treatment Efficiency of a Septic Tank (Viraraghavan, 1976)

	Influent			Effluent			Efficiency		
	Time Equal to or Less Than:			Time Equal to or Less Than:			Time Equal to or Less Than:		
Characteristics (1)	15%	50%	85%	15%	50%	85%	15%	50%	85%
pH (units)	6.45	7.60	8.7	6.65	6.90	7.15			
TSS	80	200	320	80	165	250	nil	18	22
BOD	362	520	670	170	280	350	53	46	48
COD	350	1000	1650	300	550	800	14	45	52
SOC	70	280	470	35	70	105	50	75	78
PO_4-P	0.0	14.0	32.0	6.5	10.5	14.0			
NH_3-N	17.0	47.0	75.0	80	92	105			
NO_3-N	0.0	0.10	0.19	0.0	0.02	0.04			
Total soluble iron	0.0	1.50	3.0	0.0	2.25	4.75			
Chlorides	0.0	130.0	260	35	50	67			

Total coliform/100 ml	$2.2x10^5$	$2x10^6$	$1.6x10^7$	$3.7x10^5$	$2.3x10^6$	$1.45x10^7$			
Fecal coliform/100 ml	$1.5x10^3$	$3x10^4$	$6.6x10^5$	$1x10^4$	$1.6x10^5$	$2.6x10^6$			
Fecal streptococci/100 ml	$1.5x10^3$	$3x10^4$	$5.6x10^5$	$2.2x10^4$	$1.1x10^5$	$5.1x10^5$			
SPC per milliliter ($35^\circ C$)	$1.3x10^5$	$5.6x10^5$	$2.5x10^6$	$5x10^4$	$3x10^5$	$1.7x10^6$	62	46	32
SPC per milliliter ($20^\circ C$)	$1.4x10^5$	$8x10^5$	$4.7x10^6$	$4.2x10^5$	$8.2x10^5$	$1.6x10^6$			
Psuedomonas aeruginosa/100ml	-	150	$4x10^3$	<2	28	520		81	87

(1) mg/1 unless noted otherwise.

Table 20: Summary of Treatment Efficiencies of Two Septic Tanks (Lawrence, 1973)

Tank No.	Parameter	Influent	Effluent	Percentage Reductance
1	Total solids	1128	1034	8
	Volatile solids	483	420	13
	Suspended solids	200	130	35
	Volatile suspended solids	159	107	33
	BOD	241	224	7
	Settleable solids	4.4	0.2	85
	pH	7.5	7.5	--
	Detergents	43	49	0
	Grease	21	26	0
2	Total solids	512	505	1
	Volatile solids	249	239	4
	Suspended solids	126	70	44
	Volatile suspended solids	108	73	32
	BOD	146	124	15
	Settleable solids	0.7	0.06	91
	pH	7.2	7.2	--
	Detergents	3.7	5.0	0
	Grease	16	8.5	47

Septic Tank Effluent Quality

The quality of the effluent from a septic tank is of greater importance in terms of ground water pollution than its treatment efficiency. A summary of the physical and chemical parameter effluent qualities measured for 7 septic tank systems is in Table 21 (University of Wisconsin, 1978). Additional summary information from a study of 34 more tanks is in Table 22 (U.S. Environmental Protection Agency, October 1980). Based on the composite information in Tables 21 and 22, the following represent typical physical and chemical parameter effluent concentrations from septic tanks:

Suspended solids	75 mg/l
BOD_5	140 mg/l
COD	300 mg/l
Total nitrogen	40 mg/l
Total phosphorus	15 mg/l

Table 21: Summary of Effluent Quality from Seven Septic Tanks (University of Wisconsin, 1978)

Parameter and Statistics	Results (1)
Suspended Solids, mg/l	
Mean (# of Samples)	49(148)
Coeff. of Variation	0.16
95% Conf. Int.	44-54
Range	10-695
Volatile Suspended, mg/l	
Mean (# of Samples)	35(148)
Coeff. of Variation	0.18
95% Conf. Int.	32-39
Range	5-320
BOD_5 (Unfiltered) mg/l	
Mean (# of Samples)	138(150)
Coeff. of Variation	0.42
95% Conf. Int.	129-147
Range	7-480
BOD_5 (Filtered), mg/l	
Mean (# of Samples)	190(130)
Coeff. of Variation	0.47
95% Conf. Int.	100-118
Range	7-330
COD, mg/l	
Mean (# of Samples)	327(152)
Coeff. of Variation	0.33

Table 21: (continued)

Parameter and Statistics	Results (1)
95% Conf. Int.	310-344
Range	25-780
Total Phosphorus, mg-P/L	
Mean (# of Samples)	13(99)
Coeff. of Variation	0.34
95% Conf. Int.	12-14
Range	0.7-99
Orthophosphorus, mg-P/L	
Mean (# of Samples)	11(89)
Coeff. of Variation	0.36
95% Conf. Int.	10-12
Range	3-20
Total Nitrogen, mg-N/L	
Mean (# of Samples)	45(99)
Coeff. of Variation	0.40
95% Conf. Int.	41-49
Range	9-125
Ammonia Nitrogen, mg-N/L	
Mean (# of Samples)	31(108)
Coeff. of Variation	0.46
95% Conf. Int.	28-34
Range	0.1-91
Nitrate Nitrogen, mg-N/L	
Mean (# of Samples)	0.4(114)
Coeff. of Variation	6.7
95% Conf. Int.	<0.1-0.9
Range	<0.1-74

(1) Data from seven sites and collected over time period May 1972-December 1976.

Table 22: Summary of Effluent Quality from Various Septic Tank Studies U.S. Environmental Protection Agency, October 1980)

Parameter	7 Tanks	10 Tanks	19 Tanks	4 Tanks	1 Tank	Sample-Weighted Average
Suspended Solids						
Mean, mg/1	49	155[a]	101	95[b]	39	77
Range, mg/1	10-695	43-485	-	48-340	8-270	
No. of Samples	148	55	51	18	47	
BOD_5						
Mean, mg/1	138	138[a]	140	240[b]	120	142
Range, mg/1	7-480	64-256	-	70-385	30-280	
No. of Samples	150	44	51	21	50	
COD						
Mean, mg/1	327	-	-	-	200	296
Range, mg/1	25-780	-	-	-	71-360	
No. of Samples	152	-	-	-	50	
Total Nitrogen						
Mean, mg/1	45	-	36	-	-	42
Range, mg/1	9-125	-	-	-	-	
No. of Samples	99	-	51	-	-	

[a] Calculated from the average values from 10 tanks, 6 series of tests.

[b] Calculated on the basis of a log-normal distribution of data.

Table 23 summarizes the bacteriological character of household septic tank effluent (Siegrist, 1977; and University of Wisconsin, 1978). The quantities of indicator bacteria such as fecal coliform are high, and pathogenic bacteria such as *Pseudomonas aeruginosa*, have commonly been isolated. In addition, results of analyses for *Staphylococcus aureus* and salmonellae have indicated their presence in septic tank effluents, but only infrequently and in much lower concentrations (10 homes, 6 of 63 samples positive at 10-1000/100 ml and 11 homes, 2 of 55 samples positive at 3.4-220/100 ml, respectively) (Siegrist, 1977). Viruses in septic tank effluents are high only if infections have occurred. Salmonellae have been detected in 59 percent of 17 different septic tank pumpout sludges, which shows clearly that septic effluents need to be purified before release to either ground water or surface water (Bouma, 1979).

Table 23: Summary of Bacteriological Character of Household Septic Tank Effluents

Reference	Organism	Number of Samples	Mean (No./100 ml)	95% Confidence Interval (No./100 ml)
1	Total bacteria	88	$3.4x10^8$	2.5 to $4.8x10^8$
1	Total coliform	91	$3.4x10^6$	2.6 to $4.4x10^6$
1	Fecal coliform	94	$4.2x10^5$	2.9 to $6.2x10^5$
2	Fecal coliform	151	$5.0x10^6$	$2.5x10^6$ to $1.0x10^7$
1	Fecal streptococci	97	$3.8x10^3$	2.0 to $7.2x10^3$
2	Fecal streptococci	155	$4.0x10^4$	$8.0x10^3$ to $2.0x10^5$
1	Pseudomonas Aeruginosa	33	$8.6x10^3$	3.8 to $19.0x10^3$

Reference 1 = Siegrist (1977), data from 5 tanks.

Reference 2 = University of Wisconsin (1978), data from 7 tanks.

Soil Absorption System Treatment Efficiency

Viraraghavan and Warnock (1976) conducted a field investigation with the primary objective of determining the efficiency of a soil absorption (drainfield tile) system. By "efficiency" is meant the reduction in concentration of various parameters achieved between the point at which the septic tank effluent was distributed to the tile and the various depths in the soil at which soil water samples were collected. The environmental effects of air and soil temperatures and unsaturated depth of soil on the efficiency of the septic tile were studied as a part of this investigation. The site of the study was near Ottawa, Ontario, Canada. Significant climatic conditions in this region that affected the study are low winter temperatures, usually with snow cover, and a period of melting snow in the spring when ground water levels are usually high. A visual examination of the soil samples taken from various depths at the site indicated that the soil was sandy clay for the initial depth of 2 ft (0.61 m), followed by clay with less sand at depths of 2 to 5 ft (0.61 to 1.53 m). The characteristics of the septic tank effluent applied to the study site are listed in Table 24. The results of the field study conducted from December, 1972, to February, 1974 indicated the following for a 5 ft deep traverse of the underlying soil:

1. The soil had the ability to reduce a high percentage (75 to 90 percent) of the TSS, BOD, COD, and soluble organic carbon present in the septic tank effluent.

2. Reductions of phosphates were usually in the 25 to 50 percent range, much lower than those reported in the literature. This has special significance for lake shore septic tank systems since substantial amounts of phosphorus in the form of phosphates may be added to lakes, thereby causing eutrophication.

3. High ammonia reductions (80 to 90 percent) were observed; with an increase in ammonia reduction, corresponding increases in nitrification were generally observed. Nitrification leads to nitrate build-up in ground water and nearby lakes, thus causing possible health hazards and eutrophication.

4. Efficiency was influenced by seasonal variations. There were greater efficiencies (80 to 90 percent) for the various parameters during the late summer and early fall, extending from September to November when the unsaturated depth of soil was greater. These efficiencies tended to decrease to about 70 to 75 percent with respect to BOD and TSS, and to 20 to 35 percent for ammonia nitrogen, during the winter period when the water levels started to rise. Nitrate nitrogen levels also showed a declining trend during the winter months.

Soil Absorption System Effluent Quality

Based on the above-listed minimum and maximum percentage reductions relative to the average effluent characteristics as shown in

Table 24: Characteristics of Septic Tank Effluent Applied to Study Site (Viraraghavan and Warnock, 1976)

Characteristic	Range	Mean Value
pH	6.53 - 7.45	6.90
TSS	68 - 624	176
BOD	140 - 666	280
COD	240 - 2,026	568
Soluble organic carbon	24 - 190	73
Total phosphates (PO_4-P)	6.25 - 30.0	11.6
Ammonia nitrogen	77 - 111	97
Nitrate-N	0.00 - 0.10	0.026
Total soluble iron	0.00 - 20.0	2.63
Chlorides	37 - 101	53

All values except pH are milligrams per liter.

Table 24, the following concentrations passed the 5 ft depth (if the top of the water table were at the 5 ft depth, these would be the concentrations entering the ground water):

TSS	18 - 53 mg/l
BOD	28 - 84 mg/l
COD	57 - 142 mg/l
SOC	7 - 18 mg/l
Total phosphates	6 - 9 mg/l
Ammonia nitrogen	10 - 78 mg/l

In addition to the above constituents, others of concern relative to ground water pollution from septic tank systems include:

Nitrates--Excessive concentrations of nitrates in drinking water may cause a bitter taste. Water from wells containing more than 45 mg/l of nitrates has been reported to cause methemoglobenemia in infants. Organic and ammonia nitrogen in wastewater can be converted to nitrate nitrogen within the septic tank system.

Organic contaminants--Within recent years there have been several reported instances of organic contamination of ground water, with some cleaning solvents for septic tank systems being identified as potential sources (U.S. Environmental Protection Agency, May 1980). The chief concern relating to organic contaminants is that many of these substances are persistent within the subsurface environment and they are known to be carcinogenic above certain concentration levels. An example of one of these contaminants is trichloroethylene.

Metals (lead, tin, zinc, copper, iron, cadmium, and arsenic)--The concern relating to various metals is associated with their potential toxic effects in excessive concentrations. This is pertinent in terms of ground water usage as drinking water and the movement of gound water into surface waters and subsequent effects on the aquatic ecosystem.

Inorganic contaminants (sodium, chlorides, potassium, calcium, magnesium, and sulfates)--These inorganic constituents in excessive concentrations may cause undesirable health consequences ranging from laxative effects to aggravated cardiovascular or renal disease. These concerns are pertinent in terms of ground water usage as drinking water.

In summary relative to the potential pollutants from septic tank systems, there are a variety of pollutants of concern. Attention is typically given to bacterial and viral contamination along with the introduction of nitrates in the ground water system. Additional pollutants becoming increasingly important include organic contaminants and several metals.

MECHANISMS OF GROUND WATER CONTAMINATION FROM SEPTIC TANK SYSTEMS

Ground water degradation has occurred in many areas having high densities of septic tank systems, with the degradation exemplified by high concentrations of nitrates, bacteria, and other contaminants. Recent studies indicate that significant amounts of organic contaminants have been introduced into ground water through septic tank systems (U.S. Environmental Protection Agency, May 1980). Septic tank problems are magnified by the fact that in many areas, especially rural communities, a substantial reliance on subsurface disposal systems is paralleled by a reliance on private wells for drinking water supplies.

It has been estimated that as many as one-half of all septic tank-soil absorption systems are not operating satisfactorily (Scalf, Dunlap and Kreissl, 1977). One common failure is when the capacity of the soil to absorb effluent from the tank has been exceeded, and the waste added to the system moves to the soil surface above the lateral lines. This type failure results from soil clogging and loss of infiltrative capacity, and is caused by combined physical, chemical, and biological factors. When system failure does occur from soil clogging and wastewaters do seep to the surface, overland flow from rainfall may carry contaminants directly to a stream or lake or into an inadequately sealed well. This transport is shown in Figure 13 (Scalf, Dunlap and Kreissl, 1977).

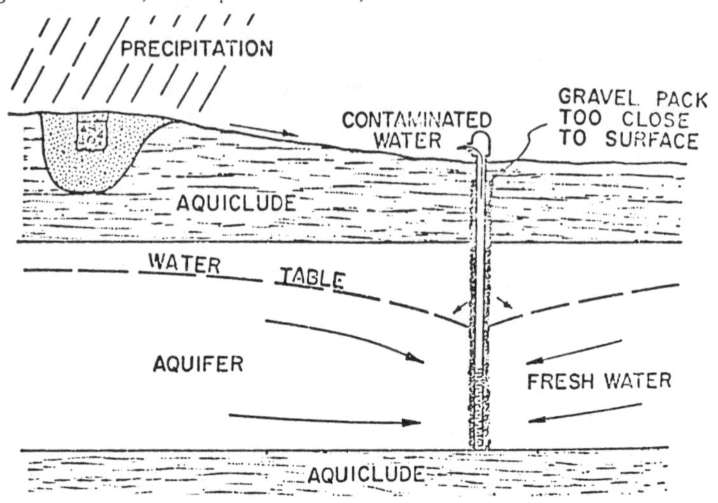

Figure 13: Effect of Clogged Absorption Field on Nearby Well (Scalf, Dunlap and Kreissl, 1977)

Another type of failure which is potentially of more significance is when pollutants move too rapidly through soils. Many soils with high hydraulic absorptive capacity (permeability) can be rapidly overloaded with organic and inorganic chemicals and microorganisms, thus permitting rapid movement of contaminants from the lateral field to the ground water zone. This transport is shown in Figure 14 (Scalf, Dunlap and Kreissl, 1977). This type of system failure has been largely ignored until recent years. The type and thickness of subsurface material is a major determinant for this kind of failure (Scalf, Dunlap and Kreissl, 1977).

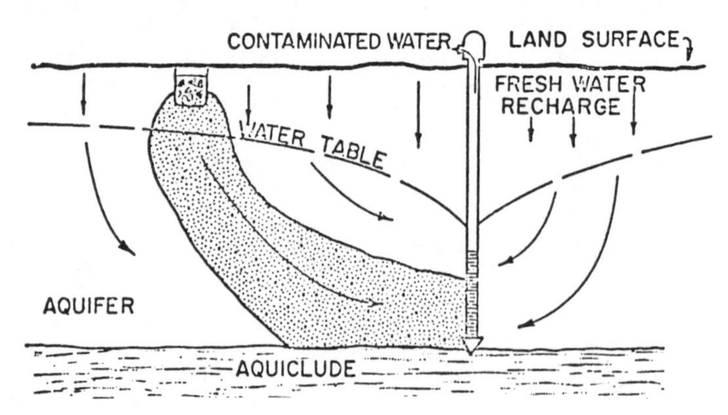

Figure 14: Effect of a Pumping Well on Contaminated Water Movement (Scalf, Dunlap and Kreissl, 1977)

In considering ground water contamination from septic tank systems, attention must be directed to the transport and fate of pollutants from the soil absorption system through underlying soils and into ground water. Physical, chemical and biological removal mechanisms may occur in both the soil and ground water systems (Loehr, 1978). As septic tank system effluent moves through the soil pores, suspended solids are removed by filtration. The depth at which removal occurs varies with the size of the particles, soil texture, and rate of water movement. The larger the hydraulic application rate and the coarser the soil, the greater the distance the particles will move. Adsorption, ion exchange, and chemical precipitation are the most important chemical processes governing the movement of constituents in the septic tank system effluent. A key soil parameter is the cation exchange capacity (CEC); the CEC of soils can range from 2 to 60 meq/100 grams of soil, with most soils having a CEC value between 10 and 30. The differences occur because soils vary widely in their humus and clay content, the components that have the highest CEC. The biological transformations that occur in the soil include organic matter decomposition and nutrient assimilation by plants. Greater biological activity can be anticipated in the upper layers of soil underneath the soil absorption system.

While the focus of this discussion is on ground water contamination from septic tank systems, pertinent information from other sources such as sanitary and chemical landfill leachates will also be included as appropriate. The mechanisms of contamination are independent of pollutant sources and dependent on pollutant types and soil and ground water characteristics in the vicinity of the source.

Soil Systems

Pollutants have been found to interact with natural organic matter, clays, and microorganisms in soils and sediments, and it is these interactions that may render the chemical constituents in landfill leachates more or less mobile (Van Hook, 1978). However, contaminants have been found to travel several hundred meters beyond their point of origin despite the attenuating characteristics of the soil (Lofty et al., 1978). The fate of pollutants in soils can be estimated by knowing the characteristics of different soil types. Chun et al. (1975), conducted a two-year study to determine the removal characteristics of selected Oahu soils with respect to the substances found in landfill leachates. Soil ion exchange capacity was primarily responsible for altering the concentration of inorganic substances, and microbial degradation appeared to be the major mechanism in removal of organic substances. The importance of the cation exchange capacity of soil was first discussed by Bower, Gardner and Goetzen (1957).

A good soil system for receiving septic tank system effluents should absorb all effluent generated, provide a high level of treatment before the effluent reaches the ground water, and have a long, useful life (Otis, Plews and Patterson, 1978). Ideally, a soil should be able to convert a pollutant into an unpolluted state at a rate equal to or greater than the rate at which it is added to the soil (Andrews, 1978). Bradford (1978) studied trace elements in soil-plant-water systems and attempted to determine how trace element concentrations are modified upon passage of water through the soils. Removal of elements in the soil through plant uptake is another potential mechanism of pollutant attenuation. Chemical reactions such as adsorption, fixation, precipitation, and other soil interactions, all influence the transport process. LeGrand (1972) studied the hydrogeological factors controlling pollutant movement. Pertinent factors include the presence of clays to retard movement and facilitate sorption, and the distance to the water table to provide an opportunity for attenuation to occur.

Ground Water Systems

Ground water typically moves in the direction of the slope of the water table, that is, from the area of higher water table to areas of lower water table. Since the water table usually follows the general contours of the ground surface, septic tanks should be located downhill from wells or springs. Information on the mechanisms of pollutant movement in ground water systems can be considered independent of the specific waste source. In other words, nitrates within ground water will move with certain typical characteristics irrespective of whether the nitrates originate from a septic tank system or a landfill leachate. Information in this section will be drawn from a variety of source types, with the principles applicable to the movement of contaminants from septic tank systems within the ground water system.

The transport and fate of pollutants in ground water was discussed by Muir (1977). Pollutants addressed included inorganic and organic nitrogen compounds, bacteria and viruses. A general discussion on the chemistry and movement of ground water was presented by Vandenberg et al. (1977) with reference to a subsurface waste disposal project and its potential for ground water contamination. Sampling strategies for determining pollutant transport and fate in ground water have been discussed by Pimentel et al. (1979).

Changes in ground water geochemistry through the influence of liquid wastes and changes in certain constituents of wastewater as they move through an aquifer have been discussed by Ku, Ragone and Vecchioli (1975). Water quality transformations resulting from the passage of reclaimed water through an aquifer may be due to adsorption, precipitation, ion exchange, dissolution, chemical oxidation, nitrification and denitrification, degradation of organic substances, mechanical dispersion, and filtration (Roberts et al., 1978).

General information on the transport of pollutants and plume migration in ground water are given by Lakshman (1979); Childs and Upchurch (1976); and Childs, Upchurch and Ellis (1974). Specific discussions on the underground movement of chemical and bacterial pollutants are presented by Butler, Orlob and McGaughey (1954) and Roberts et al. (1978). Substances like ammonia, trace metals, and trace organic compounds move slowly in an aquifer when compared with chloride ions. Exler (1972) found that the elements from a garbage deposit could be found up to 3,000 meters away from the source.

TRANSPORT AND FATE OF BIOLOGICAL CONTAMINANTS

The potential for biological contamination of ground water by percolation from such sources as surface spreading of untreated and treated wastewater, sludge landspreading, septic tank systems and landfill leachates is high (Vilker, 1978). Biological contaminants (pathogens) have a wide variety of physical and biological characteristics, including wide ranges in size, shape, surface properties, and die-away rates. This section will address the transport and fate of bacteria and viruses in soils and ground water. Information resulting from specific studies associated with septic tank systems will be presented along with pertinent information from studies of other source types.

Bacteria in Soils

Brown et al. (1979) studied the movement of fecal coliforms and coliphages from a septic tank system through undisturbed soil to ground water. Samples taken 1 and 2 years after system start-up indicated limited mobility and survival of fecal coliforms in the soils. Coliphages were present in the samples in very low concentrations immediately after spiking the applied sewage. At the end of 2 years, the soils below the soil absorption system lines were dissected and sampled in a grid pattern. On only a few occasions were fecal coliforms present in samples collected 120 cm below the soil absorption system lines.

In a study by Reneau and Pettry (1975) of total and fecal coliform bacteria from septic tank system effluents in Virginia coastal plain soils, the most probable number (MPN) of both total and fecal coliforms decreased with horizontal distance and depth from the soil absorption system lines. They concluded that coliform bacteria were unlikely to move into the ground water system. However, extensive movement of coliform bacteria is possible depending upon soil and geological features in a given area. For example, Rahe et al. (1978) found that in a perched water table fecal bacteria moved at a rate of 15 m/hour through a western Oregon hill slope soil. Strains of Escherichia coli survived in large numbers for at least 96 hours in the soils examined.

Table 25 summarizes some information on the movement of bacteria through soil (Gerba, 1975). The distance of travel of bacteria through soil is of considerable significance since contamination of ground or surface water supplies may present a health hazard. A number of environmental factors can influence the transport rate, and certain design considerations can be based on experimental results and studies of removal mechanisms. Environmental factors include rainfall; soil moisture, temperature, and pH; and availability of organic matter. Design considerations are related to soil type and depth as well as the hydraulic loading rate from the soil absorption system.

Table 25: Movement of Bacteria through Soil (Gerba, 1975)

Nature of pollution	Organism	Media	Maximum observed distance of travel (ft)	Time of travel (days)
Canal water on percolation beds	E. coli	sand dunes	10	-
Sewage introduced through a perforated pipe	coliforms	fine-grained sands	6	-
Oxidation pond effluent	coliforms	sand-gravel	2,490	-
Secondary sewage effluent on percolation beds	fecal coliforms	fine loamy sand to gravel	30	-
Diluted settled sewage into injection well	coliforms	sand and pea gravel aquifer	100	
Tertiary treated wastewater	coliforms	fine to medium sand	20	-
Tertiary treated wastewater	fecal coliforms and streptococcus	coarse gravel	1,500	2

Lake water and diluted sewage	B. stearo-thermophilis	crystalline bedrock	94	1.25
Primary and treated sewage effuent	coliforms	fine sandy loam	13	-
Secondary sewage	coliforms	sandy gravels	3	-

Hagedorn, Hanson and Simonson (1978) found that the numbers of bacteria peaked in sampling wells in association with rainfall patterns, and the populations required longer periods to peak in wells farthest from inoculation pits. This study supports the fact that bacterial movement through unsaturated soil is influenced by local infiltration, while bacterial movement in ground water is influenced by local ground water movement rates and direction.

Table 26 (Gerba, 1975) delineates several environmental factors that affect survival of enteric bacteria in soil. Gerba (1975) reported that under adverse conditions survival of enteric bacteria seldom exceeded 10 days; under favorable field conditions survival may extend up to approximately 100 days. The principle factor determining the survival of bacteria in soil is moisture (Peavy, 1978). Temperature, pH, and the availability of organic matter can also influence enteric bacteria survival. Survival in all types of soil tested was found to be greatest during the rainy season. In sand where drying was rapid due to its low moisture retaining power, survival time was short--between 4 days and 7 days during dry weather (Peavy, 1978). In soils that retain a high amount of moisture, such as loam and adobe peat, the organisms persisted longer than 42 days. Temperature changes, the presence of oxygen, a reduction in readily available food supply, and predation by native soil organisms can create unsuitable conditions for bacterial growth. Periodic or partial drying of the soil increases the death rate. Also, bacteria seem to survive longer in cool soils than in warm soils, while low pH, low organics and low moisture content increase the death rate. It was surmised that low pH could not only act to adversely affect the viability of the organisms but also the availability of nutrients; pH could also interfere with the action of inhibiting agents.

Table 26: Factors Affecting Survival of Enteric Bacteria in Soil (Gerba, 1975)

Factor	Remarks
Moisture content	Greater survival time in moist soils and during times of high rainfall.
Moisture holding capacity	Survival time is less in sandy soils than in soils with greater water-holding capacity.
Temperature	Longer survival at low temperatures; longer survival in winter than in summer.
pH	Shorter survival time in acid soils (pH 3-5) than in alkaline soils.
Sunlight	Shorter survival time at soil surface.

Table 26: (continued)

Factor	Remarks
Organic matter	Increased survival and possible regrowth when sufficient amounts of organic matter are present.
Antagonism from soil microflora	Increased survival time in sterile soil.

Several mechanisms combine to remove bacteria from water percolating through the soil. The physical process of straining (chance contact) and the chemical process of adsorption (bonding and chemical interaction) appear to be the most significant. Additional mechanisms include competition for nutrients and the production of antibiotics by high populations of actinomycetes, Pseudomonas, and Bacillus in the aerated zone beneath the clogged layer formed at the soil-trench or soil-bed interface in a soil absorption system. These antibiotics have been suggested as playing an important role in the rapid die-off of fecal coliforms and streptococci (Bouma, 1979).

Physical straining occurs when the bacteria are larger than the pore openings in the soil. Partial clogging of soil pore space by organic particles in the septic tank system effluent increases the efficiency of straining. Finer soil materials such as clay and silt generally function better for bacterial straining due to their small pore spaces (Peavy, 1978). Studies using sandy soils of various effective porosity concluded that removal of bacteria from a liquid percolating through a given depth is inversely proportional to the particle sizes of the soil. The same study also found that the greatest removal occurred on the mat (top 2 to 6 mm) that formed the soil surface (Gerba, 1975). When suspended particles, including bacteria, accumulate on the soil surface, these particles can act as a filter. Such a filter is capable of removing even finer particles, by bridging or sedimentation, before they reach and clog the original soil surface. Removal is accomplished largely by mechanical straining at the soil surface and sedimentation of bacterial clusters.

Adsorption is the other major mechanism in the removal of bacteria by soil. The process of adsorption appears to be significant in soils having pore openings several times larger than typical sizes of bacteria. Since most soils also carry a net negative charge, one might expect rejection rather than attraction of bacteria on soils. This adsorption takes place in spite of the fact that bacteria are hydrophilic colloids which possess a net negative charge at the surface. Adsorption will occur in water with high ionic strength and neutral or slightly acidic pH; these are typical characteristics of septic tank effluents. Cations (Ca^{++}, Mg^{++}, Na^{+}, H^{+}) in water neutralize and sometimes supersaturate the surface of the bacteria, thus making them susceptible to adsorption by negatively charged soil particles (Peavy, 1978).

Both physical straining/filtration and adsorption can be influenced by the flow rate of the septic tank system effluent. Bouma (1979) suggested that the removal of fecal bacteria from percolating effluent is very strongly a function of the flow regime. Rapid movement decreases the travel time and contact between the bacteria, soil, and liquid phase constituents. Better purification was achieved when system effluent was applied at a rate of 5 cm/day to a sand as compared to a rate of 10 cm/day. Laboratory studies have indicated that perhaps 60 cm, but certainly 90 cm, of sand can be effective in removing both pathogenic bacteria and viruses from septic tank effluent if the loading rate does not exceed 5 cm/day and if temperatures do not become too low (Bouma, 1979). Decreased removals at low temperatures suggest biological mechanisms in the removal.

Viruses in Soils

Allen (1978) reported that viruses have been found to migrate in soils to distances greater than 600 ft from their source, and that continued use of septic tank systems and cesspools have resulted in localized pollution. Viruses are more resistant to environmental changes and have a longer life span in the soil than bacteria. Virus survival times of up to 170 days have been reported. It has been shown that viruses attached to clay particles are still infective. Studies show that virus removal, like bacteria removal, is enhanced by low pH and high ionic strength water (Gerba, 1975). Table 27 lists factors that may influence the removal efficiency of viruses by soil (Gerba, 1975).

Table 27: Factors That May Influence Removal Efficiency of Viruses by Soil (Gerba, 1975)

Factors	Remarks
Flow rate	Low flow rates (less than 1/64 gpm/sq ft) result in very efficient removal of viruses (greater than 99%) in clean waters. As flow rate increases, virus retention decreases proportionally.
Cations	Cations, especially divalent cations, can act to neutralize or reduce repulsive electrostatic potential between negatively charged virus and soil particles, allowing adsorption to proceed.
Clays	This is the active fraction of the soil. High virus retention by clays results from

Table 27: (continued)

Factors	Remarks
	their high ion exchange capacity and large surface area per volume.
Soluble organics	Soluble organic matter has been shown to compete with viruses for adsorption sites on the soil particles, resulting in decreased adsorption or elution of an already adsorbed virus.
pH	The hydrogen ion concentration has a strong influence on virus stability as well as adsorption and elution. Generally, a low pH favors virus adsorption while a high pH results in elution of adsorbed virus.
Isoelectric point of virus	The most optimum pH for virus adsorption is expected to occur at or below its isoelectric point, where the virus possesses no charge or a positive charge. A corresponding negative charge on a soil particle at the same pH would be expected to favor adsorption.
Chemical composition of soil	Certain metal complexes such as magnetric iron oxide have been found to readily absorb viruses to their surfaces.

The first factor listed in Table 27 is flow rate, with less than 1/64 gpm/ft^2 corresponding to less than 91 cm/day (Gerba, 1975). The general point is that the lower the hydraulic flow rate, the better the virus removal rate. This point was also found by Bouma (1979) when he indicated that polio-virus type 1 (strain CHAT), when added to septic tank effluent and applied to 60 cm long sand columns, was effectively removed if the hydraulic flow rate did not exceed 5 cm/day. These results are shown in Figure 15 (Bouma, 1979).

The most important mechanism of virus removal in soil is by adsorption of viruses onto soil particles (Drewry and Eliassen, 1968). Virus adsorption is greatly affected by the pH of the soil-water system. This effect is due primarily to the amphoteric nature of the protein shell of the virus particles. At low pH values, below 7.4, virus adsorption by soils is rapid and effective. Burge and Enkiri (1978) noted that coarser soils with higher pH values were less effective in adsorbing viruses. Higher pH values

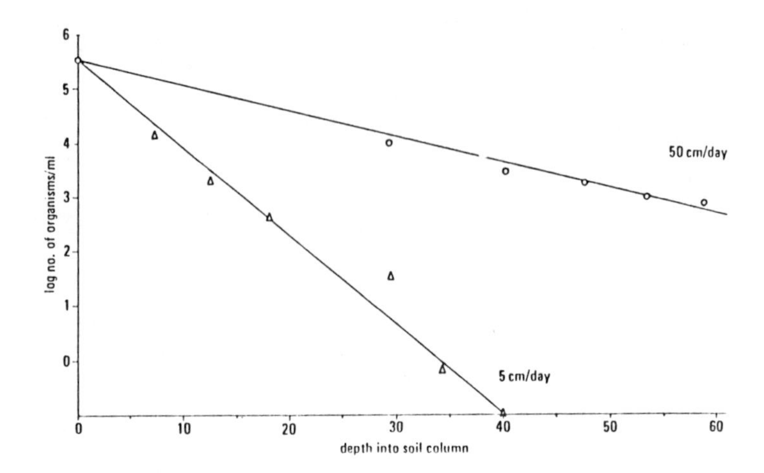

Figure 15: Removal of Poliovirus (added to septic tank effluent) in Sand-columns at Two Different Flow Regimes (Bouma, 1979)

considerably decrease the effectiveness of virus adsorption by soils because of increased ionization of the carboxyl groups of the virus protein and the increasing negative charge on the soil particles. Virus adsorption by some soils is greatly enhanced by increasing the cation concentration of the liquid phase of the soil-water system. The cations in the water neutralize or reduce the repulsive electrostatic potential (the negative charge) on either the virus particles or the soil particles, or both, and allow adsorption to proceed. This study further indicated that adsorption of virus particles by soils increases with increasing clay content, silt content, and ion-exchange capacity (Drewry and Eliassen, 1968). Experiments by Drewry (1969) with different-sized soil particles showed that finer soils are more efficient in removing viruses from water.

As noted above, virus adsorption is influenced by soil pH, liquid phase constituents, and other soil characteristics such as clay content, silt content, ion exchange capacity and particle size. Adsorption also differs as a function of virus type. Gerba and Goyal (1978) conducted a study to determine if poliovirus adsorption to soil truly reflected the behavior of other members of the enterovirus group, including recently isolated strains. It quickly became evident that, while poliovirus to a large extent reflects the behavior of most reference laboratory strains of enteroviruses in adsorption to soil, it was not reflective of many strains recently isolated from sewage-polluted waters. In the initial screening of enteroviruses, the adsorption of laboratory strains to soil from Flushing Meadows, New York, was evaluated. Of the 27 different enterovirus types, only echo 1, echo 12 and echo 29 adsorbed significantly less than the other reference enteroviruses. In addition, the rate of echo 1 adsorption was found to be less than that of polio 1. No difference in adsorption was observed between the laboratory and natural isolates of poliovirus, but a great deal of variability was observed between the natural isolates of echo 1 and coxsackie B4. Adsorption of echo 1 strains to Flushing Meadows soil ranged from 0 to 99 percent. Polioviruses are the enteroviruses most commonly isolated from sewage because of the widespread use of oral poliovirus vaccine. From this work it is apparent that virus adsorption to soil is highly dependent on the strain of virus. Differences in adsorption between different strains of the same virus type might result from variability in the configuration of proteins in the outer capsid of the virus, since this will influence the net charge on the virus. The net charge on the virus would affect the electrostatic potential between virus and soil, and thereby could influence the degree of interaction between the two particles.

Vilker (1978) has conducted experimentation and developed equations for the prediction of breakthrough of low levels of virus from percolating columns under conditions of both adsorption (application of wastewater to uncontaminated beds) and elution (application of clean water to contaminated beds). This breakthrough is described by the ion exchange/adsorption equations with the effects of external mass transfer and nonlinear adsorption isotherms included. Predictions are in qualitative agreement with reported observations from the experiments which measured virus uptake by columns packed with activated carbon or a silty soil.

While the actual mechanism of viral adsorption to soils is not known, two general theories have been proposed. Both are based on the net electronegativity of the interacting particles. Bacteriophage T2 adsorption to natural clay particles is highly dependent on the concentration and type of cations present in solution. It was shown that maximum adsorption of T2 was about 10 times greater for divalent cations than monovalent cations at the same concentration in solution. In addition, no definite relationship between the degree of virus adsorption to clay particles and electrophoretic mobility was evident. This led to the conclusion that a clay-cation-virus bridge was operating to link the two negatively charged particles. Therefore, a reduction in cation concentration results in a breakdown of the bridging effect and desorption of the virus (Gerba, 1975). The second theory of adsorption is that fixation of multivalent cations onto the ionizable groups on the virus particle is accompanied by a reduction of the net charge of the particle. This reduction or elimination of the electric charge on the particle allows the solid and the virus to come close enough for intermolecular van der Waals forces to interact. The predictability of this process, however, is complicated by the existence of considerable variation in the affinity of inorganic cations toward the different functional groups on the virus. In addition, ions that enhance adsorption at low concentration may cause desorption at higher concentrations (Gerba, 1975).

From the two proposed theories it can be concluded that virus adsorption cannot be considered a process of absolute immobilization of the virus from the liquid phase. Any process that results in a breakdown of virus association with solids will result in their further movement through porous media. For example, it has been demonstrated that organic matter in the water phase will compete with viruses for adsorption sites, thus resulting in either decreased virus adsorption or elution of previously adsorbed viruses from clay particles (Gerba, 1975).

Bacteria and Viruses in Ground Water

In a study on the fate of bacteria in ground water, lake water containing Salmonella typhimurium was passed through columns of sand. Results showed that S. typhimurium could not multiply under these conditions and that die-off continued for 44 days, after which no bacteria were detected (Gerba, 1975). However, E. coli bacteria have been found to survive and even multiply on organic matter filtered from lake water during underground recharge projects in Israel. Table 28 shows the survival times of several types of bacteria in ground water (Gerba, 1975).

Table 28: Survival of Bacteria in Ground Water (Gerba, 1975)

Organism	Survival Time	Media
E. coli	63 days	recharge well
Salmonella	44 days	water infiltrating sand columns
Shigella	24 days	water infiltrating sand columns
E. coli	3-3.5 months	ground water in the field
E. coli	4-4.5 months	ground water held in the lab
Coliforms	17 hr/50% reduction	well water
Shigella flexneri	26.8 hr/50% reduction	well water
Vibrio cholerae	7.2 hr/50% reduction	well water

As noted by Gerba (1975), Lefler and Kott studied the survival of f2 bacteriophage and poliovirus 1 in sand saturated with distilled water, distilled water containing cations, tap water, and oxidation pond effluents. The poliovirus titer was lost between 63 days and 91 days after the start in distilled water, while f2 bacteriophage survived longer than 175 days. The viruses survived even longer in distilled water containing cations. When tap water or oxidation pond effluent was used, it was noted that a very high initial kill of virus occurred, but poliovirus particles could still be detected after 91 days. In this media, f2 particles again survived in excess of 175 days (Gerba, 1975). In summary, while this information is not specifically related to ground water, the results can be considered as indicative of potential survival times in ground water.

TRANSPORT AND FATE OF INORGANIC CONTAMINANTS

Potential inorganic contaminants from septic tank systems include phosphorus, nitrogen, chlorides and metals. This section will address the subsurface movement and fate of these contaminants, with the information primarily based on, but not limited to, studies on septic tank systems.

Phosphorus

As shown in Table 16, the total phosphorus in influent wastewaters to septic tank systems serving single household units averages 25 mg/l, with 8.8 mg/l, or 35 percent, being in the inorganic, or orthophosphate form, and 65 percent being in the organic form. The anaerobic digestion process occurring in the septic tank converts most of the influent phosphorus, both organic and condensed phosphate forms, to soluble orthophosphate. Bouma (1979) reported on studies by others who found that more than 85 percent of the total phosphorus in septic tank effluents was in the soluble orthophosphate form. Total phosphorus concentrations in the effluents from seven septic tanks monitored in a field study averaged 13 mg/l, with 85 percent, or 11 mg/l, in the orthophosphate form (University of Wisconsin, 1978). In a separate study of septic tank treatment efficiency, the orthophosphates in the tank effluent averaged 10.5 mg/l (Viraraghavan, 1976). Therefore, the septic tank portions of septic tank systems are not highly efficient in phosphorus removals. As noted based on data from studies of 41 septic tank systems presented in Tables 21 and 22, the typical total phosphorus concentration entering the soil absorption system is 15 mg/l.

While phosphorus can move through soils underlying soil absorption systems and reach ground water, this has not been a major concern since phosphorus can be easily retained in the underlying soils due to chemical changes and adsorption. In a study by Jones et al. (1977) it was confirmed that phosphorus from septic tank wastewater disposal system effluent is not usually transported through the soil to ground water.

Phosphate ions become chemisorbed on the surfaces of Fe and Al minerals in strongly acid to neutral systems, and on Ca minerals in neutral to alkaline systems. As the concentration in the soil solution is raised, there comes a point above which one or more phosphate precipitates may form. In the pH range encountered in septic tank seepage fields, hydroxyapatite is the stable calcium phosphate precipitate. However, at relatively high phosphorus concentrations similar to those found in septic tank effluents, dicalcium phosphate or octacalcium phosphate are formed initially, followed by a slow conversion to hydroxyapatite (Bouma, 1979). Therefore, both chemical precipitation as well as chemisorption is involved in phosphorus retention in soils. Phosphates can be removed at practically all pH ranges (Peavy, 1978). These removal mechanisms have been found by many investigators, including Enfield et al. (1975) who determined that the ability of soil to remove wastewater orthophosphate from a solution passing through a soil matrix is primarily related to the formation of relatively insoluble phosphate compounds of aluminum, iron, and calcium. Studies to confirm phosphorus retention in soils have also been made, with Bouma (1979) reporting on a study in central Wisconsin where it was determined that phosphorus extracted from sandy soils beneath septic tank seepage fields which had operated for several years ranged from about 100 to about 300 ug/gm.

The rate at which phosphorus is sorbed from solution onto the surfaces of soil constituents has been shown to consist of a rapid initial reaction followed by an important, much slower, reaction which appears to follow first order kinetics (Bouma, 1979). Since the removal involves chemisorption, it is possible to exceed the sorptive capacity of the soil based on either long term use of a septic tank system or the application of high hydraulic loading rates such as might occur for a system serving multiple housing units. Sawhrey and Starr (1977) described two laboratory experiments which illustrate how sorptive capacity can be exceeded. Wastewater containing 6 to 9 mg/l phosphorus was applied to 75-cm long soil columns for 240 days at the rate of 20 cm/day for 2 hours a day. The concentration of phosphorus in the effluent solution reached 0.1 mg/l in the column filled with 0.1 to 0.25 mm soil particles, whereas in the effluent from a column containing 0.25 to 5.0 mm particles it reached a concentration of 5.8 mg/l. In another experiment a septic tank system effluent containing 18 mg/l phosphorus was applied at a rate of 8 cm/day to columns containing 60 cm of sandy fill underlain by 30 cm of silt loam. Concentrations in the effluent were very low during the first 20 days and then continued to increase with time. Obviously, movement of phosphorus through soil columns is minimum until "sorption sites" are occupied. Thereafter, movement through the soil continues to increase, depending upon the application rate, percolation rate, and the pH of the soil.

Measurements of phosphorus in ground water underlying septic tank systems have generally confirmed that only minimal concentrations are introduced from these systems. Sawhrey and Starr (1977) described two studies by others of the phosphorus introduced into ground water from septic tank systems. One study of five systems in sandy soils was conducted in August, October, and November, 1971. Dissolved inorganic phosphorus concentrations as high as 1.9 mg/l was observed in the underlying ground water in August. During the two remaining periods, concentrations of soluble phosphorus in the ground water were less than 0.25 mg/l in systems with no perched water table. Another study focused on the phosphorus concentrations in ground water in Varina and Goldsboro soils where plinthic horizons (iron-rich hardpan) were 54 and 132 cm below the drainline and produced a seasonal perched water table. In the Varina soil, a concentration of 1.05 mg/l was observed in the perched water 36 cm below and 15 cm away from the drainline, while only 0.01 mg/l of phosphorus was observed in soil solution within the plinthic horizon. On the other hand, in the Goldsboro soil, the concentration in the perched water was only 0.01 mg/l while in the soil solution from the plinthic horizon, the concentration of phosphorus was 0.91 mg/l. The higher concentration in the plinthic horizon in Goldsboro soil was attributed to the movement of phosphorus in water under saturated flow to the plinthic horizon when the water table was low.

In summary, phosphorus in septic tank system effluents is effectively retained in underlying soils, and only low concentrations will be typically introduced into ground water. There will be exceptions in localized situations based on geohydrology and the soil barrier. For example, Viraraghavan and Warnock (1976) indicated phosphate concentrations ranging from less than one to in excess of 20 mg/l in ground water beneath a tile field, with the higher concentrations corresponding to periods of high ground water. In addition, Peavy (1978) reported on minimal attenuation of phosphate 50 ft from a tile field.

If phosphate contamination of ground water becomes a problem, it is possible to reduce the phosphorus concentration in system effluents through chemical additions to septic tanks (Brandes, 1977). Aluminum sulfate (alum), lime, and ferric chloride have been widely used in municipal wastewater treatment plants in North America and Europe for removal of phosphates, BOD, and suspended solids. Phosphorus can be completely removed from solution when aluminum is present in large excess. Additional benefits of the use of alum is the removal of coliform organisms (about 80 percent) and intestinal parasite ova and protozoa. In accordance with the stoichiometry of the reaction between the alum and the orthophosphates of the domestic wastewater, a solid product ($AlPO_4$) is formed

$$Al_2(S)_4)_3 + 2PO_4 \rightarrow 2AlPO_4 \downarrow + 3SO_4 \quad (1)$$

which precipitates to the bottom of the septic tank. The chemical reaction and the precipitation process are affected by the Al:P ratio in the solution and by the pH. Other reaction products like $Ca_{10}(PO_4)_6\ (OH)_2$ (hydroxyapatite) and $FePO_4 \cdot 2H_2$ (strengite) are also formed as a result of the affinity between multivalent metal ions and orthophosphates. All the above precipitated solids and flocs are components of the sludge that is removed later from the septic tank. Based on the stoichiometry of reaction (1), it would take 0.87 gm of aluminum to precipitate 1.0 gm of phosphorus. Due to its composition, it would take 11.0 gm of alum to get 1 gm of aluminum; therefore, based on stoichiometric considerations, it would take 9.57 gm of alum to precipitate 1.0 gm of phosphorus. In practice, however, the recommended aluminum:phosphorus ratio is in the order from 2 to 3, depending on the phosphorus concentration in the wastewater and on the phosphorus concentration permitted in the final effluent. Using a ratio of 2, it would take 22.0 gm of alum to precipitate 1.0 gm of phosphorus. Periodic usage of alum during periods of high ground water might be justified in order to minimize the phosphorus concentration in the ground water.

Nitrogen

As shown in Table 16, the total Kjeldahl nitrogen (organic plus ammonia) in influent wastewaters serving single household units averages 38 mg/l, with 12 mg/l (32 percent) in the ammonium (NH_4^+) form. Anaerobic conditions prevail in the septic tank and organic nitrogen is converted to the ammonium form. Nitrogen in tank effluents averages about 40 mg/l and consists of about 75 percent in the ammonium form and 25 percent in the organic form. Therefore, the septic tank is ineffective in nitrogen removal, but it does cause conversion of organic nitrogen to ammonium ions. The nitrates concentration in septic tank effluents is low due to the lack of oxygen in septic tanks (anaerobic conditions).

Nitrogen contamination of ground water has occurred as a result of septic tank systems. On Long Island, New York, it was determined that the major sources of nitrogen in the recharge water for the aquifers was lawn fertilizers and septic tank systems (Shoemaker and Porter, 1978). Several recent studies have reported on the extent of nitrate contamination of ground water adjacent to septic tank seepage beds. Nitrogen is a key nutrient of concern because it contributes to eutrophication of surface water, and excess nitrogen reaching ground water can be a health hazard.

The transport and fate of nitrogen in the subsurface underneath a septic tank system is dependent upon the form of the entering nitrogen and various biological conversions which may take place. Figure 16 displays the forms and fate of nitrogen in the subsurface environment (Freeze and Cherry, 1979).

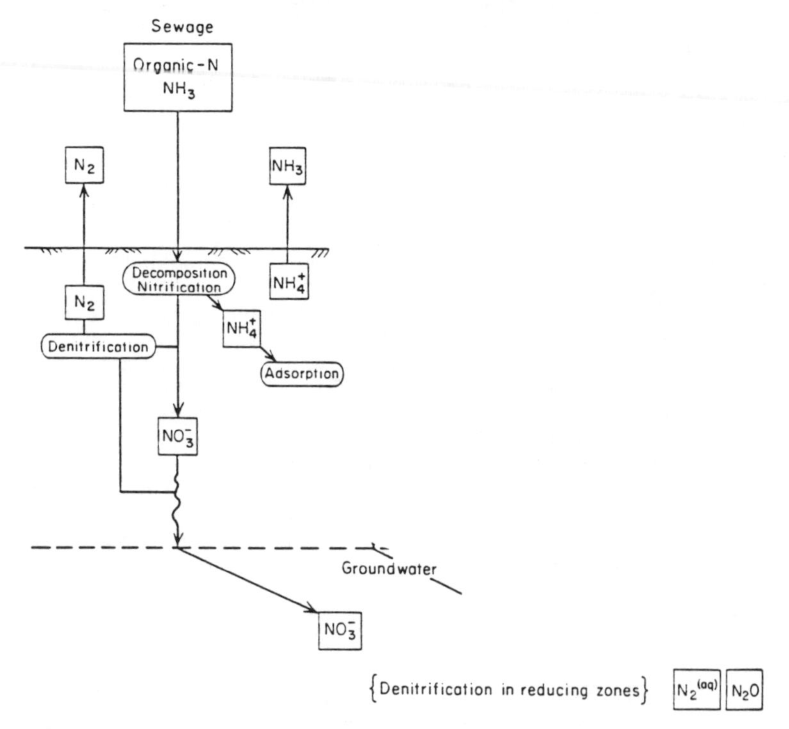

Figure 16: Form and Fate of Nitrogen in the Subsurface Environment (Freeze and Cherry, 1979)

As noted earlier, the predominant nitrogen form entering the soil from the soil absorption system is the ammonium form. Some organic nitrogen will also be introduced. The fate of the introduced nitrogen is dependent upon its initial form as well as biological conversions in the soil and ground water. Nitrates (NO_3^-) can be formed by nitrification involving ammonium ion conversion to nitrites and then to nitrates. Nitrification ($NH_4^+ \rightarrow NO_2^- \rightarrow NO_3^-$) is an aerobic reaction performed primarily by obligate autotrophic organisms and NO_3^- is the predominant end product (Bouma, 1979). Nitrification is dependent on the aeration of the soil, which, in turn, is dependent on soil characteristics, percolation rate, loading rate, distance to impervious strata, and distance to ground water (Peavy, 1978). Effluents from septic tank systems located in sandy soils can be expected to undergo predominantly aerobic reactions; this has been demonstrated to be the case in field systems located in sands and laboratory column studies employing sands. However, incomplete nitrification may occur in more clayey soils, such as silt loams and clays (Bouma, 1979).

Denitrification is another important nitrogen transformation in the subsurface environment (soils and ground water) underlying septic tank systems. It is the only mechanism by which the NO_3^- concentration in the percolating (and oxidized) effluent can be decreased. Denitrification or the reduction of NO_3^- to N_2O or N_2 is a biological process performed primarily by ubiquitous facultative heterotrophs. In the absence of O_2, NO_3^- acts as an acceptor of electrons generated in the microbial decomposition of an energy source. However, in order for the denitrification to occur in soils beneath a home waste disposal system, the nitrogen must usually be in the NO_3^- form and an energy source must be available. Therefore nitrification, an aerobic reaction, must occur before denitrification. Therefore, knowing the aeration conditions beneath a seepage bed will provide information as to the probable nitrogen forms present (Bouma, 1979).

Based upon the forms of nitrogen in septic tank system effluents, and the biological transformations which can occur in the subsurface environment, there are two forms of major concern relative to ground water pollution--ammonium ions (NH_4^+) and nitrates (NO_3^-). Ammonium ions can be discharged directly from soil absorption system drainage tiles into the subsurface environment, or they can be generated within the upper layers of soil from the ammonification process (conversion of organic nitrogen to ammonia nitrogen). The transport and fate of ammonium ions may involve adsorption, cation exchange, incorporation into microbial biomass, or release to the atmosphere in the gaseous form. Adsorption is probably the major mechanism of removal in the subsurface environment.

Anaerobic conditions will normally prevail below the upper layers of soil beneath a soil absorption system. Under these conditions, positively charged ammonium ions (NH_4^+) are readily adsorbed onto negatively charged soil particles. This adsorption is essentially complete in the first few inches of soil. After the adsorption capacity of the first few inches of soil is reached, the ammonia must travel through saturated soil to find unoccupied sites. This movement will go farther if there are still anaerobic conditions. Since anaerobic conditions in soils are usually associated with saturated soils, some movement of ammonia with ground water can occur if the effluent is transmitted through a continuously saturated soil into the aquifer. This movement will be slow, however, since adsorption continues to occur onto soil particles in the aquifer (Peavy, 1978). To illustrate the adsorption process, Viraraghavan and Warnock (1976) conducted an analysis of a soil absorption system located in fine soil (D_{10} ranging from 0.0010 to 0.0062 mm and D_{60} from 0.051 to 0.41 mm in the top 1.5 meters) over high ground water. The ground water fluctuated from near the ground surface to a maximum depth of 3.05 meters. Percolation rates of approximately 1.2 in/hr were found. The predominant species of nitrogen in the ground water under this drainfield was ammonia during periods of operation. During periods when loading ceased for a period of a few weeks, the concentration of ammonia decreased with a corresponding rise in the concentration of nitrate. An attenuation of ammonia with distance away from the system was observed. The concentration dropped from approximately 40 mg/l beneath the tiles to less than 5 mg/l at 10 ft. The adsorption of ammonium ions may be aided by the presence of organic matter; however, the exact chemical nature of this organic-ammonia complex is not well understood.

Cation exchange may be involved along with adsorption in the retention of ammonium ions in soils underneath septic tank systems. However, just as the adsorption capacity of a soil can be exceeded, the cation exchange capacity can also be exhausted. Under these conditions the cation exchange sites in the soil beneath a seepage bed would become equilibrated with the cations in the effluent. The effluent would then move to the ground water with its cation composition essentially unchanged (Bouma, 1979). Ammonia nitrogen can be incorporated into microbial or plant biomass in the subsurface environment; however, this is probably not a major removal mechanism relative to nitrogen in septic tank system effluents. Finally, ammonia gas can be released to the atmosphere as a function of the soil-liquid pH conditions. When the pH is neutral or below, most of the nitrogen is in the ammonium ion form. As the pH become basic the NH_4^+ is transformed into ammonia and can be released from the soil as a gas.

Nitrate ions can also be discharged directly from soil absorption system drainage tiles into the subsurface environment, or they can be generated within the upper layers of soil from the nitrification process. The transport and fate of nitrate ions may involve movement with the water phase, uptake in plants or crops, or denitrification. Since nitrate ions (NO_3^-) have a negative charge, they are not attracted to soils which also possess negative charges. Accordingly, nitrates are more mobile than ammonium ions in both unsaturated and saturated soils.

Immobilization of nitrates by plants in the immediate vicinity of disposal fields can occur as indicated by the characteristic lush growth often seen near septic tank systems. But this amount is minor inasmuch as the amount of nitrogen in system effluents greatly exceeds that which can be utilized by nearby plants (Bouma, 1979). Some of the nitrates could be removed by crops needing nutrient materials. Nitrogen uptake by crops grown on the drainfield site is an effective way to reduce the nutrient content of the system effluent. Removal of nitrates depends on plant roots into the effluent-laden layer of soil. A crop should be selected that has a long growing season and a high nitrogen requirement.

Denitrification can also remove nitrates from soils underlying septic tank systems. Denitrification occurs in soils which contain an abundance of denitrifying bacteria that can use free oxygen or nitrate as a substitute hydrogen acceptor if free oxygen is absent. In the denitrification process bacteria convert nitrates back to nitrites and then to nitrogen gas, N_2. This gas can be released from the soil.

Nitrogen in the form of nitrate usually reaches ground water, and becomes very mobile because of its solubility and anionic form. Nitrates can move with ground water with minimal transformation. They can migrate long distances from input areas if there are highly permeable subsurface materials which contain dissolved oxygen. The only condition which can effect this process is a decline in the redox potential of the ground water. In this case, the denitrification process can occur.

Chlorides

Chlorides are natural constituents in surface and ground water, and they are also found in household and community wastewaters. Both septic tank systems and conventional community wastewater treatment plants are ineffective for chlorides removal. The chlorides concentrations in septic tank system effluents will be variable depending on the natural quality of the water supply. To serve as an example, the concentrations of chlorides in septic tanks, and thus in the effluent discharged to the soil through soil absorption systems, have been reported by Peavy (1978) to range from 37 to 101 mg/l. Due to their anionic form (Cl^-) and mobility with the water phase, chlorides can be useful as a tracer or indicator of septic tank system pollution.

Metals and Other Inorganic Contaminants

Metals in the effluents from septic tank systems may be responsible for contamination of shallow water supply sources. Sandhu, Warren and Nelson (1977), in a random survey of Chesterfield County, South Carolina, showed that metallic contamination was quite common from septic tank systems. The levels of arsenic, iron, lead, mercury and manganese were, in some cases, higher than recommended limits. Lower and more acceptable concentrations of cadmium, copper, and zinc were found in the study. The lead and cadmium found in the ground water may have originated from corrosion of antiquated plumbing in old houses being served by septic tank systems.

A review of the transport and fate of heavy metals in the subsurface environment has been prepared by Bates (1980). The four major reactions that metals may be involved in with soils are adsorption, ion exchange, chemical precipitation and complexation with organic substances. Of these four, adsorption seems to be the most important for the fixation of heavy metals. Ion exchange is thought to provide only a temporary or transitory mechanism for the retention of trace and heavy metals. The competing effects exhibited by more common metal ions such as Ca^{+2}, Na^{+}, H^{+} and K^{+} limit the cation exchange sites available for heavy metal removal (Jenne, 1968). Precipitation reactions as a mechanism of metal fixation in soils have been well documented (Jenne, 1968; Hahne and Kroontje, 1973; Kee and Bloomfield, 1962; and Lindsay, 1972). This type of reaction is greatly influenced by pH and concentration, with precipitation predominantly occurring at neutral to high pH values and in macro-concentrations (Bingham et al., 1964). Organic materials in soils may immobilize metals by complexation reactions or cation exchange. Organic materials have a very high cation exchange capacity, therefore providing more available exchange sites than most clays. Complexation reactions between metals and organic substances, although definitely serving to fix the metals, may only provide for temporary immobilization. If the organic complex is biodegradable, the metal may be subsequently released back to the soil environment. Fixation of heavy metals by soils by either of these four mechanisms is dependent on a number of factors, including soil composition, soil texture, pH and the oxidation-reduction potential of the soil and associated ions (Bates, 1980).

Soil type or composition is a very important factor in all heavy metal fixation reactions. Clays are extremely important in adsorption reactions because of their high cation exchange capacity. In addition, soils high in humus or other organic matter also exhibit good exchange capacity. The type of clay mineral present is, in addition, an important factor. Many sorption reactions take place at the surface of iron and aluminum hydroxides and hydroxy oxides and, therefore, the iron and aluminum content of soils becomes an essential factor governing the ability of a soil for heavy metal immobilization. A number of studies have been conducted on the retention of zinc, copper, cadmium, lead, arsenic, mercury and molybdenum by various soil types (Bates, 1980).

Soil texture or soil particle size is another factor that can influence the fixation of metals by soils. In general, finely textured soils immobilize trace and heavy metals to a greater extent than coarse textured soils. Also, finely textured soils usually have a greater cation exchange capacity which is an important factor in heavy metal fixation. Soil texture has been found to influence the transport of mercury, lead, nickel and zinc (Bates, 1980).

Soil pH plays a very important role in the retention and mobility of metals in soil columns (Korte et al., 1976; Zimdahl and Skogerboe, 1977). The pH is a controlling factor in sorption-desorption reactions and precipitation-solubilization reactions. In addition, the cation exchange capacity of soils generally increases with an increase in pH. Even with a soil that has a high affinity for a specific metal, the degree to which the metal is fixed is a function of pH. Soil pH has been determined to be a major factor along with cation exchange capacity for the fixation of lead by soils. Soil pH also influences the retention of zinc, molybdenum, mercury and copper (Bates, 1980).

The oxidation-reduction or redox potential of a soil is very important in determining which species of an element is available for sorption, precipitation, or complexation. In general, the reduced forms of a metal are more soluble than the oxidized forms. The redox potential of a soil system is usually altered through biological activity and a change in redox potential is many times correlated with changes in pH. Reducing conditions may be associated with a low pH resulting from the formation of CO_2 and organic acids from the microbial degradation of organic matter. A reducing environment typically exists in saturated soils underneath septic tank systems. The anaerobic conditions would enhance the mobility of metals in system effluents. Iron is a good example of a metal which readily undergoes redox reactions. In the oxidized or ferric state, iron may form insoluble compounds of $Fe(OH)_3$ or $FePO_4$. However, when iron is reduced under anoxic conditions, the ferrous form, which is more soluble, predominates (Bates, 1980).

Another factor affecting the retention or mobility of metal ions is competing ions. The presence of phosphate affects the retention of both arsenic and zinc. Arsenic tends to become more mobile in the presence of phosphate and zinc is more highly retained. The effects of chlorides on the mobility of several heavy metals have also been investigated. For example, the presence of chloride decreases the adsorption of mercury (II) and enhances its mobility. Doner (1978) conducted studies on the effect of chlorides on the mobilities of nickel (II), copper (II) and Cd (II) in soils. Cadmium forms stable complexes with chloride while nickel and copper form weak chloride complexes. Using a sandy loam soil, Doner found that chloride increased the rate of mobility of nickel, copper and cadmium through soil. Of the three metals, copper was held more strongly than nickel or cadmium and the mobility of cadmium was increased more than that of nickel or copper.

In summary relative to the transport and fate of metals in septic tank system effluents, a number of mechanisms and influencing conditions are involved. Although generalities can be drawn with respect to the soil types and textures favorable for optimum metal retention, other factors such as pH, redox potential, and the presence of specific associated ions makes the chemistry of each metal ion in the soil column unique. Of particular concern is the influence of anaerobic conditions and associated ions in increasing the mobility of metals in the subsurface environment. These factors can increase the possibility of ground water contamination by heavy metals from system effluents.

TRANSPORT AND FATE OF ORGANIC CONTAMINANTS

Recent evidence indicates that many aquifers have been contaminated by organic chemicals. Some of these chemicals are known to be carcinogenic, and thus they pose a public health threat. Studies have also demonstrated that these contaminants have entered some ground water systems through septic tank systems. For example, a municipal landfill in Jackson Township, New Jersey, was licensed to receive wastewater sludges and septic tank wastes, but it now appears that dumping of chemicals has also occurred at the site. As a result of the variety of potential ground water pollutants, approximately 100 wells surrounding the landfill have been closed due to organic chemical contamination. It is impossible to determine whether the chemicals were from sludges, septic tank wastes, or other indiscriminate dumping. Water analyses revealed the presence of chloroform (33 micrograms per liter), methylene chloride (3000 micrograms per liter), benzene (330 micrograms per liter), toluene (6400 micrograms per liter), trichloroethylene (1000 micrograms per liter), ethylbenzene (2000 micrograms per liter), and acetone (3000 micrograms per liter) (U.S. Environmental Protection Agency, May 1980). One additional example of the movement of organic chemicals into ground water is based on a study of subsurface migration of hazardous chemical constituents at 50 land disposal sites that had received large volumes of industrial wastes (Miller, Braids and Walker, 1977). The facilities included landfills, lagoons, and combinations of the two, both active and abandoned. They were located in 11 states east of the Mississippi River. At 43 of the 50 sites migration of one or more hazardous constituents was confirmed. Migration of heavy metals was confirmed at 40 sites; selenium, arsenic and/or cyanide at 30 sites; and organic chemicals at 27 sites. Eighty-six wells and springs used for monitoring yielded water containing one or more hazardous substances with concentrations above background.

The most frequently found organic contaminant in ground water is trichloroethylene, an industrial solvent and degreaser which is also used as a septic tank cleaner. Other volatile organics include tetrachloroethylene, 1,1,1-trichloroethane, 1,1-dichloroethane, and dichloroethylene (U.S. Environmental Protection Agency, May 1980).

The transport and fate of organic contaminants in the subsurface environment is a relatively new topical area of concern. A variety of possibilities exist for the movement of organics, including transport with the water phase, volatilization and loss from the soil system, retention on the soil due to adsorption, incorporation into microbial or plant biomass, and bacterial degradation. The relative importance of these possibilities in a given situation is dependent upon the characteristics of the organic, the soil types and characteristics, and the subsurface environmental conditions. This very complicated topical area is being actively researched at this time. One study is being conducted at a ground water recharge facility being operated by the Santa Clara Water District in California (Roberts, 1980). Effluent from a 2 mgd advanced waste treatment plant is used in the recharge system. The study objectives are to acquire quantitative data regarding the removal of organic micropollutants (chlorinated and nonchlorinated trace organic compounds) during aquifer passage; obtain evidence that processes such as adsorption and biodegradation influence the transport of such pollutants relative to the velocity of the injected water; estimate the field capacity of the aquifer for retaining specific pollutants; and ascertain to what degree extreme concentration fluctuations are attenuated by aquifer passage.

Griffin and Chow (1980) studied the adsorption, mobility, and degradation of polybrominated biphenyls (PBBs) and hexachlorobenzene (HCB) in soil materials and in a carbonaceous adsorbent. The aqueous solubilities of both materials were low (less than 16 ppb), but solubilities were higher in river water and landfill leachate than in distilled water. The solubilities can be directly correlated with the level of dissolved organics in the waters. The PBBs and HCB were immobile in all soils studied when leached with deionized water and landfill leachate; they were highly mobile in all soil materials when leached with organic solvents. The PBBs and HCB were found to be strongly adsorbed by the carbonaceous adsorbent and by soil materials, with HCB being adsorbed to a greater extent than PBBs. The adsorption capacity and mobility of PBBs and HCB were highly correlated with the organic carbon content of the soil materials. In a soil incubation study, it was found that PBBs and HCB persisted for 6 months in soil with no significant microbial degradation. Because of their low water solubilities, strong adsorption, and persistence in soils, these two compounds are highly resistant to aqueous phase mobility through earth materials; however, they are highly mobile in organic solvents.

Considerable research has been conducted on the transport and fate of organic pesticides and herbicides in soils. It is possible for pesticides to be introduced into septic tank systems through normal household use and disposal practices. A bibliography on the transport, transformation, and soil retention of pesticides is available (Copenhover and Benito, 1979). Based on studies involving the infiltration of aldrin, a chlorinated hydrocarbon insecticide, through columns of Ottawa sand, it was determined that aldrin penetrability through soils is dependent upon the type of formulation applied, frequency of its application, soil conditions, and the frequency and rate of rainfall or irrigation (Robertson and Kahn, 1969).

Several studies have been conducted on the movement and biodegradation of large concentrations of pesticides in soils. Davidson, Ou and Rao (1976) examined the factors affecting pesticide mobility from hazardous waste disposal sites containing high pesticide concentrations. Major consideration was given to the influence of the shape of the adsorption isotherm on pesticide mobility. Equilibrium adsorption of the dimethylamine salt of 2, 4-D ((2, 4-Dichlorophenoxy) acetic acid)) on Webster silty clay loam was measured in the concentration range of 0-5000 ug/ml. The adsorption sites for 2, 4-D on the Webster soil were not saturated even in the presence of 5000 ug/ml of 2, 4-D. The adsorption isotherm was nonlinear in shape with the Freundlich equation exponent being 0.71. The mobility of 2, 4-D in the Webster soil at various 2, 4-D concentrations was simulated with a numerical solution to the solute transport model. These simulations revealed that pesticide mobility increased as solution concentration increased when the Freundlich equation exponent was less than 1.0. However, an increase in solution concentration when the exponent was greater than 1.0 resulted in a decreased mobility. Serious errors may be introduced by assuming a linear adsorption isotherm when predicting pesticide transport under waste disposal sites where high pesticide concentrations exist. A procedure for estimating the arrival time of a selected pesticide concentration at various soil depths below a disposal site was developed by Davidson, Ou and Rao (1976).

Additional studies by Davidson et al. (1980) revealed equilibrium adsorption isotherms of the nonlinear Freundlich type for atrazine, methyl parathion, terbacil, trifluralin, and 2, 4-D and four soils. Pesticide solution concentrations used in the study ranged from zero to the aqueous solubility limit of each pesticide. The mobility of each pesticide increased as the concentration of the pesticide in the soil solution phase increased. These results were in agreement with the equilibrium adsorption isotherm data. Biological degradation of each pesticide was measured by $^{14}CO_2$ evolution resulting from the oxidation of uniformly ^{14}C ring-labeled pesticides, except trifluralin which was labeled at $^{14}CF_3$. Technical grade and formulated forms of each pesticide at concentrations ranging from zero to 20,000 ug/g of soil were used in the biological degradation experiments. Pesticide degradation rates and soil microbial populations generally declined as the pesticide concentration in the soil increased; however, some soils were able to degrade a pesticide at all concentrations studied, some soils degraded a pesticide at a low concentration but not a higher concentration, while others remained essentially sterile throughout the incubation period. Several pesticide metabolites were formed and identified in various soil-pesticide systems.

The movement of 2, 4-D in three soils was studied by Dregne, Gomez and Harris (1969) to determine the extent to which herbicides applied in the field enter ground water systems. Adsorption isotherms, breakthrough curves, leaching studies, and bioassays indicate that 2, 4-D in the acid or salt form, is only slightly adsorbed by soil particles. It is easily leached if the soils are permeable. Virtually 100 percent of applied 2, 4-D was recovered from a sandy loam in 6 1/2 hours of leaching. Only 38 percent was recovered from a slowly permeable silty clay loam over a period of 10 months. Degradation products of 2, 4-D were leached as easily as 2, 4-D itself.

Schneider, Wiese and Jones (1977) conducted a field study of the movement of three herbicides in a fine sand aquifer. Low concentrations of atrazine, picloram and trifluraline, and a $NaNO_3$ tracer were injected into a sand aquifer through a dual-purpose well. Recharge by injection continued for 10 days at an average rate of 81.8 cu m/hour. After a 10-day pause, the well was pumped for 12 days to determine if the herbicides and tracer could be recovered. Water samples were pumped from observation wells located 9, 20, and 45 m from the dual-purpose well. Herbicides were detected in the 9- and 20-m distant wells, but none of the herbicides or the tracer was detected in the 45-m distant well.

GROUND WATER POLLUTION CONTROL MEASURES

Several measures can be identified to minimize the possibility for undesirable ground water pollution to result from septic tank system usage. Examples include proper system design and site selection, establishment of institutional requirements, and consideration of influent wastewater segregation. Siting criteria and design features for septic tanks and soil absorption systems were addressed in Chapter 2. Some control measures can be used for existing septic tank systems experiencing problems with overloaded soil absorption fields. One approach would be to require any existing subdivision subject to septic tank system failures to join sewage districts with specific collection and treatment facilities. Another approach is to require householders to connect to sewers as urban development occurs and sewers are provided.

Table 29 summarizes several positive actions that can be used for new septic tank systems (U.S. Environmental Protection Agency, 1973). One is to require approval of the site and design by competent hydrogeologists, soil scientists and engineers. Another approach is to construct percolation systems by methods which do not compact the infiltrative surface. There are some operational practices which can minimize the potential for ground water pollution. These include alternately loading and resting the percolation system, inspecting and removing scum and grease from septic tanks, and cleaning of septic tanks by withdrawal of only one-half the sludge rather than the entire contents. A final suggestion for control of septic systems is the use of zoning and other land management controls in urban areas to prevent installations in unsuitable soils. Unsuitable soils are those that are too impervious to accept effluents, or too coarse or fractured to maintain the required biological and physical treatment.

Table 29: Ground Water Pollution Control Measures for New Septic Tank Systems (U.S. Environmental Protection Agency, 1973)

Require approval of the site and design by competent hydrogeologists, soil scientists and engineers before septic systems are approved for any subdivision, recognizing that simple percolation tests and standard codes offer only partial criteria for the design of a septic system.

Construct percolation systems by methods which do not compact the infiltrative surface.

Operate septic systems effectively by:

> Alternately loading and resting one-half the percolation system; the cycle to be determined by the onset of ponding in the system at the observation well.
>
> Inspecting and removing scum and grease from septic tanks annually.
>
> Drawing off half of the sludge rather than pumping out the entire contents of tanks.

Use of zoning and other land management controls to prevent septic system installation in unsuitable soils (i.e., soils too impervious to accept effluents, or too coarse or fractured to maintain a biological and physical treatment system).

Another measure to minimize the ground water pollution potential of septic tank systems is to reduce the wastewater strength entering the systems. Segregation of household wastewaters is depicted in Figure 17 (Siegrist, 1977). Various wastewater streams within the household unit can be divided into two major fractions: the toilet wastes, commonly referred to as "black water"; and the other household wastewaters, commonly referred to collectively as "grey water". The characteristics of the black water and grey water streams are summarized in Tables 30 and 31, respectively (Bauer, Conrad and Sherman, 1979). On the average, the black water contributes about 30 percent of the BOD, 50 percent of the suspended solids, 70 percent of the total Kjeldahl nitrogen, 17 percent of the total phosphorus, and 30 percent of the flow from a household unit. Removal of the black water from the household waste stream through use of a nonconventional toilet system (e.g. composting, incinerating, recycle, low volume flush/holding tank), would reduce the wastewater loading to the septic tank and the soil absorption system.

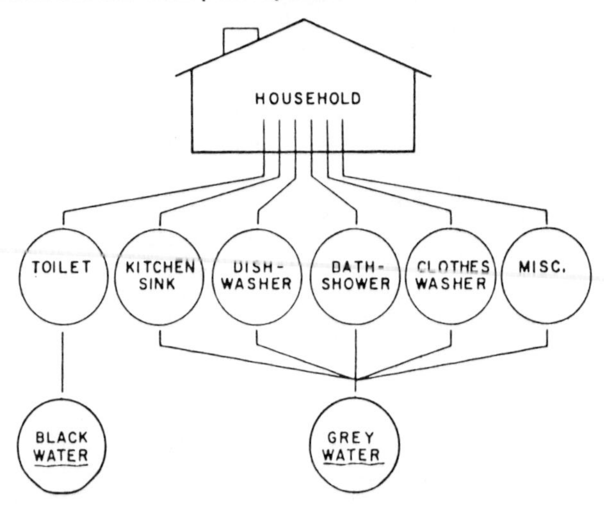

Figure 17: Segregation of Household Wastewater (Siegrist, 1977)

Table 30: Characteristics of Black Water (Bauer, Conrad and Sherman, 1979)

Parameter (g/cap/d)	Investigator					
	Olsson, Karlgren and Tullander	Laak	Bennett and Linstedt	Witt, Siegrist and Boyle	SSWMP	Weighted Value
BOD_5	20	23.5	6.9	10.7	10.7	15
BOD_5 filtered	-	-	-	6.3	6.3	6
COD	72	67.8	65	-	-	68

Table 30: (continued)

Parameter (g/cap/d)	Investigator					
	Olsson, Karlgren and Tullander	Laak	Bennett and Linstedt	Witt, Siegrist and Boyle	SSWMP	Weighted Value
TOC	-	-	-	7.7	7.8	8
TOC filtered	-	-	-	4.8	4.8	5
TS	53	-	76.5	28.5	28.5	45
TVS	39	-	55.8	19.7	19.7	30
SS	30	-	36.5	12.8	12.5	20
VSS	25	-	31	10.2	10.2	16
TKN	11	-	5.2	4.1	4.1	5
NH_3-N	-	2.78	-	1.11	1.11	1
NO_3-N	-	0.02	-	0.03	0.03	0.03
NO_2-N	-	-	-	-	-	-
TP	1.6	-	-	0.55	0.55	0.55
PO_4-P	-	2.16	3.1	0.31	0.31	0.3
Oil and Grease	-	-	-	3.35	-	3
MBAS	-	-	-	-	-	-
pH	8.9	-	5.6	-	-	-
Total Bacteria (#/cap/d)	6.2×10^{10}	-	-	-	-	-
Total coliform (#/cap/d)	4.8×10^{9}	-	-	-	-	-
Fecal coliform (#/cap/d)	3.8×10^{9}	-	-	-	-	-
Fecal strep	-	-	-	-	-	-
Flow (lpcd)	8.5*	74.9	55.6	26.6	34.7	50

* Study households equipped with vacuum toilets.

Table 31: Characteristics of Grey Water (Bauer, Conrad and Sherman, 1979)

Parameter (g/cap/d)	Investigator							
	Olsson, Karlgren, and Tullander	Hypes	Laak	Ligman, Hutzler, and Boyle	Bennett and Linstedt	Witt, Siegrist, and Boyle	SSWMP	Weighted Value
BOD_5	25	-	25.2	24.5	27.9	38.8	38.8	33
BOD_5 filtered	-	-	-	-	-	24.1	24.1	24

Table 31: (continued)

Parameter (g/cap/d)	Investigator							
	Olsson, Karlgren, and Tullander	Hypes	Laak	Ligman, Hutzler, and Boyle	Bennett and Linstedt	Witt, Siegrist, and Boyle	SSWMP	Weighted Value
COD	48	-	51.6	-	56.5	-	-	52
TOC	-	-	-	-	17.8	24.4	24.4	24
TOC filtered	-	-	-	-	-	17.2	17.2	17
TS	77	-	-	70.8	69.8	85	85	80
TVS	44	-	-	-	18.8	43	43	40
SS	18	-	-	15.4	10.8	22.6	22.6	20
VSS	15	-	-	-	10.6	16.5	16.5	15
TKN	1.1	-	-	-	1.3	1.9	1.9	2
NH_3-N	-	-	0.44	-	-	0.16	0.16	0.2
NO_3-N	-	-	0.6	-	-	0.04	0.04	0.05
NO_2-N	trace	-	-	-	-	-	-	-
TP	2.2	-	-	2.7	-	3.43	3.43	3
PO_4-P	-	-	1.8	-	0.6	1.10	1.10	1.1
Oil and Grease	-	-	-	-	-	11.3	-	11
MBAS	-	-	-	-	3.4	-	-	3
pH	-	7.2	-	-	-	-	-	7.2
Total Plate Count (#/cap/d)	7.6×10^{10}#	-	-	-	-	-	-	-
Total coliform (#/cap/d)	1.3×10^{7}#	1.0×10^{7}	-	-	-	6500**	6500**	-
Fecal coliform (#/cap/d)	2.5×10^{9}#	-	-	-	-	550**	550**	-
Fecal strep (#/cap/d)	-	-	-	-	-	94**	94**	-
Flow (lpcd)	121.5*	-	81.8	98.3	109.7	92.8	126.5	110

* Excluding garbage disposal and water softner.
\+ Based on bath/shower, dishwashing, and laundry only.
\# Based on kitchen and bath/shower data only.
**Based on laundry and bath/shower data only.

GROUND WATER MONITORING

As noted in Chapter 1, ground water monitoring may be required for septic tank systems funded or regulated by the U.S. Environmental Protection Agency. Monitoring is of greater importance for geographical areas with high septic tank densities, and for specific systems serving large numbers of housing units. The first requirement for a ground water monitoring program should be the clear delineation of monitoring objectives. Nelson and Ward (1982) suggested two basic objectives based on system failure detection: (1) the detection of temporary overloads of high polllutant concentrations in ground water; and (2) the detection of permanent overloads of high concentration.

The three primary components of a septic tank system monitoring program are: (1) determination of the sampling locations; (2) selection of parameters to be monitored; and (3) selection of the required number of samples (Nelson and Ward, 1982). Since the treatment system consists of the septic tank, the soil absorption system, and the unsaturated soil zone beneath the drainage tiles, the most logical sampling location is in the upper portion of the saturated zone directly beneath the field lines. Sampling at this location should be most representative of the input to the ground water aquifer. Location of the sampling points in the upper portion of the saturated zone entails possible physical difficulties associated with collecting samples at varying depths as the water table fluctuates. This problem can be circumvented by using either a cluster of wells installed at various depths or a ground water profile sampler. This particular sampler consists of a well point filled with sand and divided into sections with partitions made of caulking. A sampling probe is located within each section and tubing extends from each probe to the ground surface. One advantage of using a cluster of wells is that they can be located at various points in the leach field to give more extensive areal coverage. The degree to which more areal coverage is required will depend on the homogeneity of the soil in the leach field and the uniformity of effluent distribution. More sampling wells will be required if the soil is nonhomogeneous and/or the effluent distribution is nonuniform (Nelson and Ward, 1982).

A number of parameters could be measured as a part of a ground water monitoring program for a septic tank system. Tables 16 and 18 illustrate the variety of physical, chemical and biological constituents in septic tank system wastewaters. Parameters of importance in terms of monitoring include those which might be considered health hazards, including bacteria, viruses and nitrates.

Since bacteriological testing is easier and less expensive than virological testing, the former should be given precedence (Nelson and Ward, 1982). Fecal coliforms and fecal streptococci can serve as suitable indicators of bacterial or viral contamination of ground water by septic tank systems. Nitrate monitoring is important since the unsaturated zone is not effective in nitrogen removal. An adequate depth of unsaturated flow, necessary for bacteriological and virological treatment and for phosphorus removal, also establishes conditions which allow for rapid nitrification within the first few centimeters of the unsaturated zone. Nitrate is then transported uninhibited to the ground water. Simple dilution of the nitrate with the ground water provides adequate reduction of the nitrate concentrations if the density of septic tanks in a given area is sufficiently low. However, high densities could result in significant increases in nitrate concentrations in the ground water (Nelson and Ward, 1982).

The number of parameters included in a ground water monitoring program is typically limited by budgetary and time constraints. As noted above, routine monitoring for fecal coliforms, fecal streptococci, and nitrates would be reasonable in most instances. It might be desirable to monitor for total solids, dissolved solids, and chlorides if the background concentrations are low for these constituents. Due to the growing importance of metals and organic constituents in septic tank system effluents, it may become increasingly important to monitor for these constituents, particularly for systems serving multiple household units.

Sampling location and parameter selection is envisioned to be similar for detection of either temporary or permanent overloads of high pollutant concentrations in ground water. Sampling frequency is more dependent on whether the objective is to detect temporary or permanent overloads. Nelson and Ward (1982) used a mathematical model to determine the sampling frequency required to achieve a specific probability of failure detection. The mathematical model was based on detection of nitrates alone. This approach was used since on-site systems are least effective in the removal of nitrate. Also, the modeling of nitrate flow through a porous medium has been well documented. Since most of the bacteria, viruses, and phosphorus will be removed before reaching the ground water, the statistical variability associated with these variables should be much lower than that of nitrate. Therefore, if the sampling frequency for all variables considered in the monitoring program can be taken as the frequency determined for nitrate, then the resulting precision of the estimates for all variables will be at least as good as the precision for nitrates.

The mathematical model used by Nelson and Ward (1982) to determine sampling frequencies consisted of a mass transport model which described the flow of nitrates through the leach field, and a simulation model which was used to establish input concentrations. The convective-dispersion equation was used in the mass transport portion of the model and model parameters were selected to represent different soil types. A number of random components associated with water use patterns in a house were included in the simulation portion of the model. Consequently, varying results were obtained on each simulation run. The output resulting from each simulation was superimposed with a sampling plan in order to determine the effectiveness of that plan. Major emphasis in the modeling was placed on the detection of a permanent system failure. In this case it was assumed that the assimilative capacity of the ground water reservoir is such that a failure of a temporary nature is not severe. Hence, the monitoring objective was to detect a permanent system failure while avoiding the classification of a temporary overload as a permanent failure. Three sampling plans were evaluated by Nelson and Ward (1982) to determine their effectiveness in meeting this objective:

Plan 1 -- Samples were taken at equally spaced intervals at frequencies of 1, 3, 5, 10, 15, and 20 samples per year. If a concentration above the detection limit was found, system failure was assumed and sampling was terminated.

Plan 2 -- Primary samples were taken at the frequencies given in Plan 1. However, if a concentration above the detection limit was found, the primary sample was followed by one secondary sample one week later. If both the primary and secondary samples were above the detection limit, system failure was assumed and sampling was terminated.

Plan 3 -- Primary samples were taken at the frequencies given in Plan 1. However, if a concentration above the detection limit was found, the primary sample was followed by two secondary samples 3 days and 6 days later. If all three samples were above the detection limit, system failure was assumed and sampling was terminated.

From the description of the various sampling plans considered, it is obvious that the primary objective of plans 2 and 3 is to avoid classifying a detection as a permanent failure when it is actually a temporary overload. For each sampling plan and sampling frequency several quantities were determined. For purposes of this discussion the most important are percent permanent failure detection and percent temporary overload detection. As shown in Figure 18, sampling plan 3 is the most effective in detecting system failure. With this plan a sampling frequency of at least 7 samples per year is necessary to detect a system failure 90 percent of the time on the average. Figure 18 also indicates that an increase in sampling frequency beyond this point would not be very beneficial in terms of increased failure detection. It is also noted that sampling plan 1 reaches a maximum percent failure detection at a frequency of 3 or 4 samples per year and higher frequencies actually reduce the effectiveness. The reason for this apparent anomaly is that higher frequencies tend to begin detecting an increasing number of temporary system overloads as indicated in Figure 19 (Nelson and Ward, 1982). The results shown in Figures 18 and 19 indicate that sampling plan 1 would be inadequate in detecting a system failure given the characteristics of the system that was modeled. Sampling plans 2 and 3 provide significant improvement. Plan 3 requires more samples, and consequently, would be more costly. The choice between plan 2 and 3 would be dependent on the value a management agency is willing to accept as the probability of making an error by classifying a detection as failure when it is actually a temporary overload (Nelson and Ward, 1982).

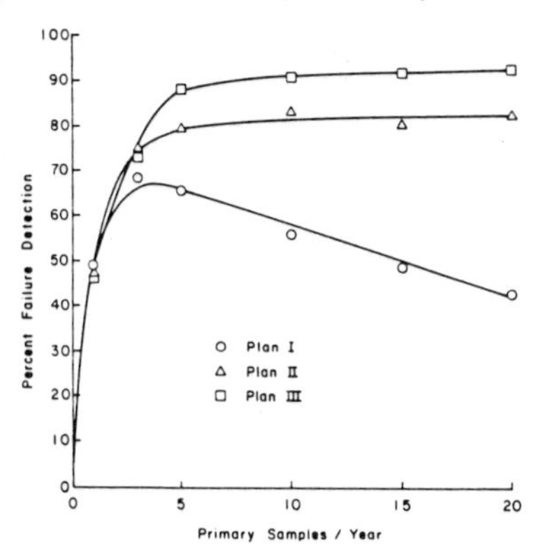

Figure 18: Comparison of the Effectiveness of Sampling Plans in Detecting System Failure (Nelson and Ward, 1982)

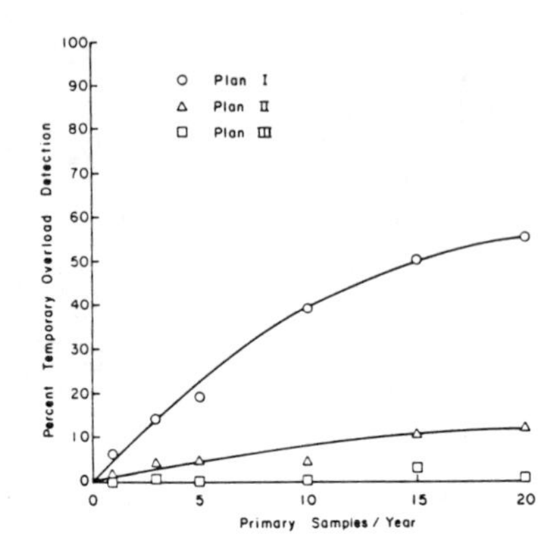

Figure 19: Comparison of the Effectiveness of Sampling Plans as Measured by Temporary Overload Detection (Nelson and Ward, 1982)

SEPTAGE--A SPECIAL CONCERN

Septic tanks serving single or multiple household units must be periodically cleaned to remove septage. Septage refers to the mixture of sludge, fatty materials, and wastewater removed during the pumping of a septic tank (U.S. Environmental Protection Agency, October 1980). Tank clean-out and septage removal may occur every 3 to 5 years or more frequently as needed depending upon wastewater loading. Septage is often highly odoriferous and may contain significant quantities of grit, grease, and hair that may make pumping, screening, or settling difficult. Of particular importance is the high degree of variability of this material, some parameters differing by two or more orders of magnitude. This is reflected to some extent by the variability in mean values from different studies presented in Table 32 (U.S. Environmental Protection Agency, October 1980). In general, the heavy metal content of septage is low relative to municipal wastewater sludge, although the range of values may be high. Table 33 presents typical concentration ranges for indicator organisms and pathogens in septage (U.S. Environmental Protection Agency, October 1980). These values are not unlike those found for raw primary wastewater sludge. It is evident that septage may harbor disease-causing organisms, thus demanding proper management to protect public health.

While it is beyond the scope of this book to address the ground water pollution potential of septage, it should be noted that inappropriate disposal of septage can cause ground water quality programs. Septage is typically disposed into sanitary landfills, thus the issues of concern are associated with leachate formation and transport to ground water. As noted earlier, leachates from the Jackson Township, New Jersey landfill probably contain constituents from septage (U.S. Environmental Protection Agency, May 1980).

Table 32: Characteristics of Domestic Septage (U.S. Environmental Protection Agency, October 1980)

Parameter	Mean Value (mg/l)
Total Solids	22,400
	11,600
	39,500
Total Volatile Solids	15,180
	8,170
	27,600
Suspended Solids	2,350
	9,500
	21,120
	13,060
Volatile Suspended Solids	1,770
	7,650
	12,600
	8,600
BOD	4,790
	5,890
	3,150
COD	26,160
	19,500
	60,580
	24,940
	16,268
pH	6-7 (typical)
Alkalinity ($CaCO_3$)	610
	1,897
TKN	410
	650
	820
	472
NH_3-N	59
	100
	120

Table 32: (continued)

Parameter	Mean Value (mg/l)
	92
	153
Total Phosphorus	190
	214
	172
	351
Grease	3,850
	9,560
Aluminum	48
Arsenic	0.16
Cadmium	0.1
	0.2
	9.1
Chromium	0.6
	1.1
Copper	8.7
	8.3
Iron	210
	160
	190
Mercury	0.02
	0.4
Manganese	5.4
	4.8
Nickel	0.4
	<1.0
	0.7
Lead	2.0
	8.4
Selenium	0.07

Table 32: (continued)

Parameter	Mean Value (mg/l)
Zinc	9.7 .62 .30

Table 33: Indicator Organism and Pathogen Concentrations in Domestic Septage (U.S. Environmental Protection Agency, October 1980)

Parameter	Typical Range (number/100 ml)
Total Coliform	$10^7 - 10^9$
Fecal Coliform	$10^6 - 10^8$
Fecal Streptococci	$10^6 - 10^7$
Ps. aeruginosa	$10^1 - 10^3$
Salmonella sp.	$<1 - 10^2$
Parasites	
Toxacara, Ascaris Lumbricoides, Trichuris trichiura, Trichuris vulpis	Present

SELECTED REFERENCES

Allen, M.J., "Microbiology of Groundwater", Journal of Water Pollution Control Federation, Vol. 50, No. 6, June 1978, pp. 1342-1343.

Andrews, W.F., "Soil as a Media (sic) for Sewage Treatment", Third Annual Illinois Private Sewage Disposal Symposium, Feb. 1978, pp. 18-20.

Bates, M.H., "Fate and Transport of Heavy Metals", Proceedings of the Seminar on Ground Water Quality, July 1980, University of Oklahoma, Norman, Oklahoma, pp. 213-229.

Bauer, D.H., Conrad, E.T. and Sherman, D.G., "Evaluation of On-Site Wastewater Treatment and Disposal Options", Feb. 1979, U.S. Environmental Protection Agency, Cincinnati, Ohio.

Bingham, F.T. et al., "Retention of Cu and Zn by H-Montmorillonite", Soil Sci. Soc. Amer. Proc., Vol. 28, 1964, p. 351.

Bouma, J., "Subsurface Applications of Sewage Effluent", in Planning the Uses and Management of Land, 1979, ASA-CSSA-SSSA, Madison, Wisconsin, pp. 665-703.

Bower, C.A., Gardner, W.P. and Goetzen, J.O., "Dynamics of Cation Exchange in Soil Columns", Soil Science Society of America, Proceedings, 1957.

Bradford, G.R., "Trace Element Study of Soil-Plant-Water Systems", Contract No. 0011628, 1978, U.S. Department of Agriculture, Washington, D.C.

Brandes, M., "Effective Phosphorus Removal by Adding Alum to Septic Tank", Journal of the Water Pollution Control Federation, Vol. 49, No. 11, Nov. 1977, pp. 2285-2296.

Brown, K.W. et al., "Movement of Fecal Coliforms and Coliphages Below Septic Lines", Journal of Environmental Quality, Vol. 8, No. 1, Jan.-Mar. 1979, pp. 121-125.

Burge, W.D. and Enkiri, N.K., "Virus Adsorption by Five Soils", Journal of Environmental Quality, Vol. 7, 1978, pp. 73-76.

Butler, R.G., Orlob, G.T. and McGaughey, P.H., "Underground Movement of Bacterial and Chemical Pollutants", Journal of American Water Works Association, Vol. 46, 1954, p. 97.

Childs, K.E. and Upchurch, S.B., "Documenting Ground Water Contamination from Surface Sources", Presented at NWWA 96th Annual Conference: Water Supply Management Resources and Operations, Vol. 1, June 20-25, 1976, New Orleans, Louisiana.

Childs, K.E., Upchurch, S.B. and Ellis, B., "Sampling of Variable Waste-Migration Patterns in Ground Water", Ground Water, Vol. 12, No. 6, Nov.-Dec. 1974, pp. 369-371.

Chun, M.J. et al., "Groundwater Pollution from Sanitary Landfill Leachate, Oahu, Hawaii", Hawaii University, Honolulu, Water Resources Research Center, OWRT-A-040-HI(1), Apr. 1975, 87 pp.

Copenhover, E.D. and Benito, K.W., "Movement of Hazardous Substances in Soil: A Bibliography. Volume 2. Pesticides", EPA/600/9-79/024B, Aug. 1979, U.S. Environmental Protection Agency, Municipal Environmental Research Laboratory, Cincinnati, Ohio.

Council on Environmental Quality, "Evaluation of Municipal Sewage Treatment Alternatives", Feb. 1974, Washington, D.C., 402 pages.

Davidson, J.M., Ou, L.T. and Rao, P.S.C., "Behavior of High Pesticide Concentration in Soil Water Systems", Hazardous Waste Research Symposium on Residual Management by Land Disposal, University of Arizona, 1976, pp. 206-212.

Davidson, J.M. et al., "Movement and Biological Degradation of Large Concentrations of Selected Pesticides in Soils", Disposal of Hazardous Wastes: Proceedings of the 6th Annual Symposium, Chicago, Illinois, EPA-600/9-80-010, Mar. 1980, U.S. Environmental Protection Agency, Solid and Hazardous Waste Research Division, Washington, D.C., pp. 93-107.

Doner, H.E., "Chloride as a Factor in the Mobilities of Ni (II), Cu (II), and Cd (II) in Soil", Soil Sci. Soc. Am. J., Vol. 42, 1978, pp. 882-885.

Dregne, H.E., Gomez, S. and Harris, W., "Movement of 2,4-D in Soils", New Mexico Agricultural Experiment Station Western Regional Research Project, Progress Report, Nov. 1969, New Mexico State University, Las Cruces, New Mexico.

Drewry, W.A., "Virus Movement in Groundwater Systems", OWRR-A-005-ARK (2), 1969, 85 pp., Water Resources Research Center, University of Fayetteville, Arkansas.

Drewry, W.A. and Eliassen, R.S., "Virus Movement in Ground Water", Journal of Water Pollution Control Federation, Vol. 40, No. 8, 1968, pp. R247-R271.

Enfield, C.G. et al., "Fate of Water Phosphorus in Soil", Journal of Irrigation and Drainage Division, Vol. 101, No. IR3, Sept. 1975, pp. 145-155.

Exler, H.J., "Distribution and Reach of Groundwater Pollution in the Subflow of a Sanitary Landfill", Gar-Wasserfach, Wasser-Absasser, Vol. 113, No. 3, 1972, pp. 101-112.

Freeze, R.A. and Cherry, J.A., Ground Water, Prentice-Hall, Inc., Englewood Cliffs, New Jersey, 1979.

Gerba, C.P., "Fate of Waste Water Bacteria and Viruses in Soil", Journal of the Irrigation and Drainage Division, Vol. 101, No. IR3, Sept. 1975, pp. 157-173.

Gerba, C.P. and Goyal, S.M., "Adsorption of Selected Enteroviruses to Soils", Proceedings of International Symposium on Land Treatment of Wastewater, Vol. 2, 1978, U.S. Army Corps of Engineers Cold Regions Research and Engineering Laboratory, Hanover, New Hampshire, pp. 225-232.

Griffin, R.A. and Chow, S.J., "Disposal and Removal of Halogenated Hydrocarbons in Soils", Disposal of Hazardous Wastes: Proceedings of the

Sixth Annual Symposium, Chicago, Illinois, EPA 600/9-80-010, Mar. 1980, Southwest Research Institute/Solid and Hazardous Waste Research Division, U.S. Environmental Protection Agency, pp. 82-92.

Hagedorn, C., Hansen, D.T. and Simonson, G.H., "Survival and Movement of Fecal Indicator Bacteria in Soil Under Conditions of Saturated Flow", Journal of Environmental Quality, Vol. 7, No. 1, Jan.-Mar. 1978, pp. 55-59.

Hahne, H.C.H. and Kroontje, W., "The Simultaneous Effect of pH and Chloride Concentration Upon Mercury (II) as a Pollutant", Soil Sci. Soc. Amer. Proc., Vol. 37, 1973, p. 838.

Jenne, E.A., "Controls on Mn, Fe, CO, Ni, Cu and Zn Concentrations in Soils and Water: The Significant Role of Hydrous Mn and Fe Oxides", Adv. in Chem. Ser., Vol. 73, 1968, pp. 337-387.

Jones, R.A. et al., "Septic Tank Disposal Systems as Phosphorus Sources for Surface Waters", Report No. EPA-600/3-77-129, Nov. 1977, U.S. Environmental Protection Agency, Ada, Oklahoma.

Kee, N.S. and Bloomfield, C., "The Effect of Flooding and Aeration on the Mobility of Certain Trace Elements in Soils", Plant and Soil, Vol. 16, No. 1, 1962, p. 108.

Korte, N.E. et al., "Trace Element Movement in Soils: Influence of Soil Physical and Chemical Properties", Soil Science, Vol. 122, 1976, pp. 350-359.

Ku, H.F.H., Ragone, S.E. and Vecchioli, J., "Changes in Concentration of Certain Constituents of Treated Wastewater During Movement through the Magothy Aquifer, Bay Park, New York", Journal of Research of the U.S. Geological Survey, Vol. 3, No. 1, Jan.-Feb. 1975, pp. 89-92.

Lakshman, B.T., "Faulty Waste Disposal Practices Endanger Groundwater Quality", Water and Sewage Works, Vol. 126, No. 6, July 1979, pp. 94-98.

Lawrence, C.H., "Septic Tank Performance", Journal of Environmental Health, Vol. 36, No. 3, Nov.-Dec. 1973, pp. 226-228.

LeGrand, H.E., "Hydrogeologic Factors Controlling Pollutant Movement in Shallow Ground", Geological Society of America, Vol. 4, No. 2, 1972, pp. 86-87.

Lindsay, W.L., "Inorganic Phase Equilibria of Micronutrients in Soil", in J.S. Mortuedt, P.M. Giordano, W.L. Lindsay (ed), Micronutrients in Agriculture, Madison, Wisconsin, 1972, p. 41.

Loehr, R.C., "Pretreatment Requirements for Land Application of Wastewater", Proceedings of International Symposium on Land Treatment of Wastewater, Vol. 1, 1978, U.S. Army Corps of Engineers Cold Regions Research and Engineering Laboratory, Hanover, New Hampshire, pp. 283-287.

Lofty, R.J. et al., "Environmental Assessment of Subsurface Disposal of Municipal Wastewater Treatment Sludge", June 1978, 380 pp., SCS Engineers, Long Beach, California, EPA/530/SW-167C.

Miller, D.W., Braids, O.C. and Walker, W.H., "The Prevalence of Subsurface Migration of Hazardous Chemical Substances at Selected Industrial Waste Land Disposal Sites", PB-272 973, Sept. 1977, National Technical Information Service, Springfield, Virginia.

Muir, K.S., "Initial Assessment of the Groundwater Resources in the Monterey Bay Region, California", Report No. USGS/WRD/WRI-77/053, Aug. 1977, U.S. Geological Survey, Menlo Park, California.

Nelson, J.D. and Ward, R.C., "Ground Water Monitoring Strategies for On-Site Sewage Disposal Systems", Proceedings of the Third National Symposium on Individual and Small Community Sewage Treatment, ASAE Publication 1-82, 1982, American Society of Agricultural Engineers, St. Joseph, Michigan, pp. 301-308.

Otis, R.J., Plews, G.D. and Patterson, D.H., "Design of Conventional Soil Adsorption Trenches and Beds", Third Annual Illinois Private Sewage Disposal Symposium, Toledo Area Council of Governments, Toledo, Ohio, 1978, pp. 52-66.

Peavy, H.S., "Groundwater Pollution from Septic Tank Drainfields", June 1978, Montana State University, Montana.

Pimentel, K.D. et al., "Sampling Strategies in Groundwater Transport and Fate Studies for In Situ Shale Retorting", Conf-7903-34-8, June 1979, Lawrence Livermore Lab., California University, Livermore, California.

Rahe, T.M. et al., "Transport of Antibiotic-Resistant Escherichia Coli through Western Oregon Hillslope Soils Under Conditions of Saturated Flow", Journal of Environmental Quality, Vol. 7, No. 4, Oct.-Dec. 1978, pp. 487-494.

Reneau, R.B. and Pettry, D.E., "Movement of Coliform Bacteria from Septic Tank Effluent through Selected Coastal Plain Soils of Virginia", Journal of Environmental Quality, Vol. 4, No. 1, Jan.-Mar. 1975, pp. 41-45.

Roberts, P.V. et al., "Direct Injection of Reclaimed Water into an Aquifer", Journal of American Society of Civil Engineers, Environmental Engineering Division, Vol. 104, No. EE5, Oct. 1978, pp. 933-949.

Roberts, P.V., "Organic Contaminant Behavior During Ground Water Recharge", Journal Water Pollution Control Federation, Vol. 52, No. 1, Jan. 1980, pp. 161-171.

Robertson, J.B. and Kahn, L., "The Infiltration of Aldrin through Ottawa Sand Columns", Professional Paper 650-C, 1969, U.S. Geological Survey, Idaho Falls, Idaho, pp. C219-C223.

Sandhu, S.S., Warren, W.J. and Nelson, P., "Trace Inorganics in Rural Potable Water and Their Correlation to Possible Sources", Water Research, Vol. 12, 1977, pp. 257-261.

Sawhrey, B.L. and Starr, J.L., "Movement of Phosphorus from a Septic System Drainfield", Journal Water Pollution Control Federation, Vol. 49, No. 11, Nov. 1977, pp. 2238-2242.

Scalf, M.R., Dunlap, W.J. and Kreissl, J.F., "Environmental Effects of Septic Tank Systems", EPA-600/3-77-096, Aug. 1977, U.S. Environmental Protection Agency, Ada, Oklahoma.

Schneider, A.D., Wiese, A.F. and Jones, O.R., "Movement of Three Herbicides in a Fine Sand Aquifer", Agronomy Journal, Vol. 69, No. 3, May-June 1977, pp. 432-436.

Shoemaker, C.A. and Porter, K.S., "Recharge and Nitrogen Transport Model for Nassau and Suffolk Counties, New York", NTIS BP-276 906/55, Jan. 1978, Cornell University, Ithaca, New York.

Siegrist, R., "Waste Segregation as a Means of Enhancing Onsite Wastewater Management", Journal of Environmental Health, Vol. 40, No. 1, July-Aug. 1977, pp. 5-8.

U.S. Environmental Protection Agency, "Ground Water Pollution from Subsurface Excavations", EPA-430/9-73-012, 1973, Washington, D.C.

U.S. Environmental Protection Agency, "The Report to Congress: Waste Disposal Practices and Their Effects on Ground Water", EPA 570/9-77-001, June 1977, Washington, D.C., pp. 294-321.

U.S. Environmental Protection Agency, "Planning Workshops to Develop Recommendations for a Ground Water Protection Strategy, Sections I, II and III", May 1980, Washington, D.C.

U.S. Environmental Protection Agency, "Design Manual--Onsite Wastewater Treatment and Disposal Systems", EPA 625/1-80-012, Oct. 1980, Cincinnati, Ohio.

University of Wisconsin, "Management of Small Wastewater Flows", EPA-600/2-78-173, Sept. 1978, U.S. Environmental Protection Agency, Cincinnati, Ohio, pp. 19-20.

Vandenburg, A. et al., "Subsurface Waste Disposal in Lambton County Ontario--Piezometric Head in the Disposal Formation and Ground Water Chemistry of the Shallow Aquifer", Technical Bulletin - Canadian Inland Waters Directorate, No. 90, 1977.

Van Hook, R.I., "Transport and Transportation Pathways of Hazardous Chemicals from Solid Waste Disposal", Environmental Health Perspectives, Vol. 27, Dec. 1978, pp. 295-308.

Vilker, V.L., "An Adsorption Model for Prediction of Virus Breakthrough from Fixed Beds", Proceedings of International Symposium on Land Treatment of Wastewater, Vol. 2, 1978, U.S. Army Corps of Engineers Cold Regions Research and Engineering Laboratory, Hanover, New Hampshire, pp. 381-388.

Viraraghavan, T., "Septic Tank Efficiency", ASCE Journal Environmental Engineering Division, Vol. 102, No. EE2, Apr. 1976, pp. 505-508.

Viraraghavan, T. and Warnock, R.G., "Efficiency of a Septic Tile System", Journal Water Pollution Control Federation, Vol. 48, No. 5, May 1976, pp. 934-944.

Zimdahl, R.L. and Skogerboe, R.N., "Behavior of Lead in Soil, Environmental Science and Technology, Vol. 11, 1977, pp. 1202-1207.

CHAPTER 4

SEPTIC TANK SYSTEM MODELING

There are several technical methodologies for evaluating the ground water pollution potential of septic tank systems. This evaluation is desirable prior to installation of new systems; it is required based on the Section 201 (h) and (j) provisions of the Clean Water Act of 1977 (P.L. 95-217) which authorized construction grant funding of privately owned treatment works serving individual housing units or groups of housing units, provided that a public entity apply on behalf of a number of such individual systems (Bauer, Conrad and Sherman, 1979). As noted earlier, this evaluation is of greater importance for larger systems serving up to several hundred housing units. Evaluations may also be necessary for single systems up to several hundred individual systems in a given geographical area.

Technical methodologies range from empirical index approaches to sophisticated mathematical models. Models can range from analytical approaches addressing ground water flow to numerical approaches which aggregate both flow and solute transport considerations. Technical methodologies vary in input data requirements and specificity of output-oriented calculations. Only minimal work has been done on modeling of the ground water effects of septic tank systems; therefore, the major focus of this chapter will be on the application of existing technical methodologies not previously applied to septic tank systems. The initial section will describe the septic tank system as an area source of ground water pollution. Previous usage of models and selection criteria utilized herein will be summarized. Information will then be presented on two empirical assessment methodologies and their application in a geographical study area; the Hantush analytical model; and the Konikow and Bredehoeft numerical model. Finally, a suggested heirarchical structure for model usage will be presented.

CONCEPT OF AREA SOURCE

An important consideration in selecting technical methodologies is the source type to be modeled, that is, is it a point source, line source, or area source? A point source would be represented by a pipe discharge to the subsurface environment; a very shallow brine disposal well would be an example. A line source would represent discharges to the subsurface along either a horizontal or vertical line. An example of a horizontal line source would be a series of closely spaced recharge wells; a vertical line source example would be a corroded oil production pipeline penetrating the fresh ground water zone. Area sources represent potential ground water pollution sources that range in geographical size from small surface ponds to aquifer recharge areas. Septic tank systems can be considered as area sources of ground water pollution, with the rectangular dimensions of the drainage field representing the source boundaries. Waste stabilization ponds (surface impoundments), and sanitary and chemical landfills also can be considered as potential area sources of ground water pollution.

PREVIOUS USAGE OF TECHNICAL METHODOLOGIES

Technical methodologies which have actual or potential applicability for estimating the ground water pollution potential or single or multiple septic tank systems include empirical assessment approaches, ground water flow models, solute transport models, and specific predictive equations.

Empirical Assessment Approaches

Empirical assessment methodologies refer to simple approaches for development of numerical indices of the ground water pollution potential of man's activities. Several methodologies have been developed for evaluating the ground water pollution potential of wastewater ponds and sanitary and chemical landfills. Table 34 summarizes the general features of empirical assessment methodologies (Canter, 1981). Methodologies typically focus on a numerical index, with larger numbers used to denote greater ground water pollution potential; however, some methodologies encourage the grouping or ranking of pollution potential without extensive usage of numerical indicators. Methodologies typically contain several factors for evaluation, with the number and type, and importance weighting, varying from methodology to methodology. Methodologies also include descriptions of measurement techniques for the factors, with information provided on the scaling of importance weights (points). Final integration of information may involve summation of factor scores or their multiplication followed by summation. Empirical assessment methodologies should be utilized for relative evaluations and not absolute considerations of ground water pollution. Considerable professional judgment is needed in the interpretation of results. However, they do represent approaches which can be used, based on minimal data input, to provide a structured procedure for preliminary source evaluation, site selection, and monitoring planning.

Table 34: Summary Features of Empirical Assessment Methodologies (Canter, 1981)

Numerical Indices of Ground Water Pollution Potential
Multiple Factors and Relative Importance Weighting
Measurement Techniques for Factors and Scaling (Scoring) of Importance Weights
Indices Based on Summation of Factor Scores or Products of Scores
Need for Careful Interpretation with Professional Judgment

Ground Water Flow Models

Ground water models can be classified into flow models and solute transport models. Ground water modeling begins with a conceptual understanding of the physical problem. The next step in modeling is translating the physical system into mathematical terms (Mercer and Faust, 1980). In many cases the equations are simplified, using site-specific assumptions, to form a variety of equation subsets. An understanding of these equations and their associated boundary and initial conditions is necessary before a modeling problem can be formulated. Prickett (1979) identified the following four main groups of flow models: (1) sand tank models which are a scaled down representation of an aquifer, including its boundary configuration and usually its hydraulic conductivity; (2) analytical models, where the behavior of an aquifer is described by differential equations which are derived from basic principles such as the laws of continuity and conservation of energy; (3) analog models, which can be subdivided into the three major categories of viscous fluid models, electrical models, and miscellaneous models and techniques; and (4) numerical models, which can be subdivided into four groups--finite-difference, finite-element variational, finite-element Galerkin, and miscellaneous.

Several studies reviewing the applicability of various ground water models have been conducted. Prickett and Lonnquist (1971) presented information on generalized digital computer program listings that can simulate one-, two-, and three-dimensional nonsteady flow of ground water in heterogeneous aquifers under water table, nonleaky, and leaky artesian conditions. Programming techniques involving time varying pumpage from wells, natural or artificial recharge rates, the relationships of water exchange between surface waters and the ground water reservoir, the process of ground water evapotranspiration, and the mechanism of converting from artesian to water table conditions are also included. The discussion of the digital techniques includes the necessary mathematical background, documented program listings, theoretical versus computer comparisons, and field examples. Also presented are sample computer input data and explanations of job set-up procedures. A finite difference approach is used to formulate the equations of ground water flow.

Appel and Bredehoeft (1976) discussed the types of problems for which models have been, or are being, developed, including ground water flow in saturated or partially unsaturated material, land subsidence resulting from ground water extraction, flow in coupled ground water-stream systems, coupling of rainfall-runoff basin models with soil moisture-accounting aquifer flow models, interaction of economic and hydrologic considerations, predicting the transport of contaminants in an aquifer, and estimating the effects of proposed development schemes for geothermal systems. The status of modeling activity for various models is reported as being in a developmental, verification, operational, or continued improvement phase. Bachmat et al. (1978) assessed the present status of 250 numerical models as a tool for ground water related water resource management. Among the problem areas considered were the accessibility of models to users, communications between managers and technical personnel, inadequacies of data, and inadequacies in modeling. The 250 models were categorized as prediction, management, identification, and data management models.

Solute Transport Models

Prediction of the movement of contaminants in ground water systems through the use of models has been given increased emphasis in recent years because of the growing trend toward subsurface disposal of wastes. Anderson (1979) reviewed the formulation of contaminant transport models, their application to field problems, the difficulties involved in obtaining input data, and the current status of modeling efforts. Contaminant transport models which include the effects of dispersion have been applied to several field studies. Regional size models which limit the effects of dispersion have had limited success because of the scarcity and poor quality of field data. Another difficulty in the development of contaminant transport models is the current lack of knowledge regarding the quantification of chemical reaction terms.

Several examples of solute transport models which have, or could have, applicability to septic tank systems, can be cited. Khaleel and Redell (1977) developed a three-dimensional model describing two-phase (air-water) fluid flow equations in an integrated saturated-unsaturated porous medium. Also, a three-dimensional convective-dispersion equation describing the movement of a conservative, noninteracting tracer in a nonhomogeneous, anisotropic porous medium was developed. Finite difference forms of these two equations were solved using an implicit scheme to solve for water or air pressures, an explicit scheme to solve for water and air saturations, and the method of characteristics with a numerical tensor transformation to solve the convective-dispersion equations. The inclusion of air as a second fluid phase caused the infiltration rate to decrease rapidly to a value well below the saturated hydraulic conductivity when the air became compressed. This is in contrast to one-phase fluid flow problems in which the saturated hydraulic conductivity is considered to be the lower bound for the infiltration rate. A field-size problem describing the migration of septic tank wastes around the perimeter of a lake was also considered and solved using the total simulator.

Pickens and Lennox (1976) described the use of the finite element method based on a Galerkin technique to formulate the problem of simulating the two-dimensional transient movement of conservative or nonconservative wastes in a steady state saturated ground water flow system. The convection-dispersion equation was solved in two ways: in the conventional Cartesian coordinate system; and in a transformed coordinate system equivalent to the orthogonal curvilinear coordinate system of streamlines and normals to those lines. The two formulations produced identical results. Examples involving the movement of nonconservative contaminants described by distribution coefficients, and examples with variable input concentration are given. The model can be applied to environmental problems related to ground water contamination from waste disposal sites.

A final example of a solute transport model is the one developed by Konikow and Bredehoeft (1978). The model simulates solute transport in flowing ground water, and it was used in a field application described later in this chapter. The model is applicable to one- or two-dimensional problems having steady-state or transient flow. The model computes changes in concentrations over time caused by the processes of convective transport, hydrodynamic dispersion, and mixing (or dilution) from fluid sources. The model assumes that the solute is nonreactive and that gradients of fluid density, viscosity, and temperature do not affect the velocity distribution. However, the aquifer may be heterogeneous and (or) anisotropic. The model couples the ground water flow equation with the solute-transport equation. The digital computer program uses an alternating-direction implicit procedure to solve a finite-difference approximation to the ground water flow equation, and it uses the method of characteristics to solve the solute-transport equation. The model is based on a rectangular, block-centered, finite-difference grid. It allows the specification of any number of injection or withdrawal wells and of spatially varying diffuse recharge or discharge, saturated thickness, transmissivity, boundary conditions, and initial heads and concentrations. An analysis of several test problems indicated that the error in the mass balance will be generally less than 10 percent. The test problems demonstrated that the accuracy and precision of the numerical solution is sensitive to the initial number of particles placed in each cell and to the size of the time increment, as determined by the stability criteria. Mass balance errors are commonly the greatest during the first several time increments, but tend to decrease and stabilize with time.

Predictive Equations

In addition to general ground water flow and solute transport equations, specific predictive equations have been developed for virus removal in the subsurface environment beneath soil absorption systems. Sproul (1973) discussed methods of predicting the capacity of a septic tank-soil absorption system for removing viruses. Vilker (1978) conducted experiments and developed models for predicting the breakthrough of low levels of viruses from percolating columns under conditions of adsorption and elution. Breakthrough of viruses was illustrated by ion exchange/adsorption equations. Predictions were in qualitative agreement with observations from experiments that measured virus uptake by activated carbon or silty soil in columns.

SELECTION CRITERIA FOR TECHNICAL METHODOLOGIES

Septic tank systems may range from isolated systems to high densities (greater than one system per acre) serving single housing units. In addition, larger septic tank systems have been designed to serve up to several hundred housing units. Ground water modeling can be useful for evaluation of specific sites for systems, or even larger geographical areas that may be served by hundreds of systems. Modeling could be used to exclude septic tank system location on specific sites or in larger geographical areas. In addition, modeling can be useful in planning ground water monitoring programs for specific sites or geographical areas. As previously mentioned, available technical methodologies range from empirical assessment approaches to ground water flow and solute transport models. These methodologies differ in their input requirements, output characteristics, and general usability. Accordingly, certain selection criteria can be identified as basic to the selection of technical methodologies (TM) for usage in meeting particular needs. Examples of criteria statements include:

1. The TM should have been previously used for evaluation of septic tank systems.

2. The TM should be potentially usable, or adaptable for use, for evaluation of septic tank systems.

3. If the TM needs to be calibrated prior to use, the necessary data for calibration should be readily available.

4. The input data required for the TM should be readily available, thus the use of the TM could be easily implemented.

5. The resource requirements for use of the TM should be minimal (resource requirements refer to personnel needs and personnel qualifications, computer needs, and the time necessary for TM calibration and usage).

6. Usage of the TM for prediction of pollutant transport in the subsurface environment should have been previously documented.

7. The conceptual framework of the TM as well as its output should be understandable by non-ground water modeling specialists.

No single technical methodology (TM) which met all seven criteria was identified. Table 35 summarizes the criteria met by the technical methodologies described in this chapter. The Surface Impoundment Assessment and Waste-Soil-Site Interaction Matrix are empirical assessment methodologies. These two methodologies (1) provide indices of ground water pollution potential, (2) allow for direct comparison of different sites, (3) have their greatest utility in preliminary assessments, (4) are relatively easy to implement, (5) have low resource requirements, (6) are easily understood by nontechnical persons, and (7) can be easily adapted to septic tank systems due to their previous usage for projects with similar geometric configurations (area sources) such as waste stabilization ponds and chemical and sanitary landfills. The empirical assessment methodologies can be applied to septic tank systems serving single housing units, to larger systems serving multiple housing units, or to geographical areas characterized by high system densities, for example, greater than one system per square mile.

The Hantush analytical model listed in Table 35 (1) provides a quantitative prediction of the ground water flow increase, (2) allows for estimation of the qualitative changes in concentrations for conservative pollutants in ground water, (3) can be easily programmed for hand

Table 35: Comparison of Study Methodologies to Selection Criteria

Criteria	Surface Impoundment Assessment	Waste-Soil-Site Interaction Matrix	Hantush Analytical Model	Konikow and Bredehoeft Numerical Model
Previous Usage	0	0	0	0
Potential for Usage	3	3	3	2
Available Calibration Data	n.a.	n.a.	n.a.	1
Available Input Data	3	2	2	1
Minimal Resource Requirements	3	2	2	1
Documentation of Prediction Usage	n.a.	n.a.	3	3
Understandable	3	2	2	1

3 = high likelihood for satisfying criteria; 2 = moderate likelihood; 1 = low likelihood; 0 = criteria not satisfied.

calculators, (4) has relatively low resource requirements, (5) does not have extensive input data requirements, (6) is generally understandable by nontechnical personnel, and (7) can be adapted to septic tank systems due to its basic orientation to projects with similar geometric configuration. The Hantush analytical model can be applied to septic tank systems serving single housing units, or to larger systems serving multiple housing units. The Konikow and Bredehoeft numerical model listed in Table 35 (1) could provide the most accurate calculations for quantitative and qualitative changes to ground water resulting from septic tank systems, (2) requires extensive field data for model calibration and usage, (3) has extensive resource requirements, (4) is fully documented for prediction of pollutant transport, and (5) is difficult to understand by nontechnical persons. The Konikow and Bredehoeft model can be applied to septic tank systems serving single housing units, to larger systems serving multiple housing units, or to geographical areas with high system densities.

EMPIRICAL ASSESSMENT METHODOLOGIES

This section will provide a description of the two selected methodologies--surface impoundment assessment (U.S. Environmental Protection Agency, June 1978), and waste-soil-site interaction matrix (Phillips, Nathwani and Mooij, 1977). To serve as an illustration, both methods were applied to 13 septic tank system areas in central Oklahoma. Since neither methodology considers the total quantity of wastewater being discharged into the subsurface environment, a pollutant quantity adjustment factor was developed for the central Oklahoma areas. Finally, this section will summarize the results of a cursory field sampling program which was conducted in 4 of the 13 septic tank system areas in central Oklahoma.

Surface Impoundment Assessment

The surface impoundment assessment (SIA) method is based on work by LeGrand (1964). The method was developed for evaluating wastewater ponds (U.S. Environmental Protection Agency, June 1978) and it yields a sum index with numerical values ranging from 1 to 29. Due to the geometric similarity between pond leakage entering ground water and soil absorption system effluent entering ground water, the SIA method can be used for evaluating the ground water pollution potential of septic tank systems. The index is based on four factors--the unsaturated zone, the availability of ground water (saturated zone), ground water quality, and the hazard potential of the waste material (septic tank system effluent in this case). Numerical values for the unsaturated zone range from 0 to 9, for the availability of ground water from 0 to 6, for ground water quality from 0 to 5, and for hazard potential of waste from 1 to 9.

The unsaturated zone rating is based on considering earth material characteristics as well as zone thickness. Table 36 provides the basis for the evaluation, with the categories of earth materials based on permeability and secondarily upon sorption character. In rating a particular locality where hydrologically dissimilar layers exist, the septic tank system effluent

Table 36: Rating of the Unsaturated Zone in the SIA Method (U.S. Environmental Protection Agency, June 1978)

Earth Material Category	I	II	III	IV	V	VI
Unconsolidated Rock	Gravel, Medium to Coarse Sand	Fine to Very Fine Sand	Sand with < 15% clay, Silt	Sand with > 15% but ≤ 50% clay	Clay with < 50% sand	Clay
Consolidated Rock	Cavernous or Fractured Limestone, Evaporites, Basalt Lava Fault Zones	Fractured Igneous and Metamorphic (Except Lava) Sandstone (Poorly Cemented)	Sandstone (Moderately Cemented) Fractured Shale	Sandstone (Well Cemented)	Siltstone	Unfractured Shale, Igneous and Metamorphic Rocks
Representative Permeability in gpd/ft^2 -	>200	2 - 200	0.2 - 2	< 0.2	< 0.02	< 0.002
in cm/sec -	$>10^{-2}$	10^{-4} - 10^{-2}	10^{-5} - 10^{-4}	$<10^{-5}$	$<10^{-6}$	$<10^{-7}$
RATING MATRIX						
Thickness of the Unsaturated Zone (in Meters) >30	9A	6B	4C	2D	0E	0F
>10 ≤30	9B	7B	5C	3D	1E	0G
> 3 ≤10	9C	8B	6C	4D	2E	0H
> 1 ≤3	9D	9F	7C	5D	3E	1F
>0 ≤1	9E	9G	9H	9I	9J	9K

is more likely to move through the more permeable zones and avoid the impermeable zones. In such cases the earth material should be rated as the more permeable of the two or more layers which might exist. The availability of ground water factor considers the ability of the aquifer to transmit ground water, thus it is dependent upon aquifer permeability and saturated thickness. Table 37 provides information on the types of earth material and thicknesses for various ratings. The letters accompanying the rating matrices in Tables 36 and 37 are for the purpose of identifying the origin of the rating and documenting the process.

Table 37: Rating Ground Water Availability in the SIA Method (U.S. Environmental Protection Agency, June 1978)

Earth Material Category	I	II	III
Unconsolidated Rock	Gravel or sand	Sand with $\leq 50\%$ clay	Clay with $< 50\%$ sand
Consolidated Rock	Cavernous or Fractured Rock, Poorly Cemented Sandstone, Fault Zones	Moderately to Well Cemented Sandstone, Fractured Shale	Siltstone, Unfractured Shale and other Impervious Rock
Representative Permeability			
in gpd/ft^2	> 2	$0.02 - 2$	< 0.02
in cm/sec	$> 10^{-4}$	$10^{-6} - 10^{-4}$	$< 10^{-6}$
RATING MATRIX			
Thickness of Saturated Zone (Meters) ≥ 30	6A	4C	2E
3-30	5A	3C	1E
≤ 3	3A	1C	0E

The ground water quality factor is based upon criteria associated with the Underground Injection Control program of the U.S. Environmental Protection Agency. Table 38 contains information on the rating (U.S. Environmental Protection Agency, June 1978). If ground water has high total dissolved solids (TDS) the rating is lower since potential ground water uses which would be curtailed would be limited. If the ground water is serving as a drinking water supply the rating is 5 regardless of the TDS concentration. The waste hazard potential factor is associated with the potential for causing harm to human health. Examples of hazard potential ratings of waste materials classified by source are in Table 39. The ratings consider toxicity, mobility, persistence, volume, and concentration. Table 39 includes a range of ratings for several sources, with the concept being that in cases where there is considerable pretreatment, the rating may be lowered to the bottom of the range. The waste hazard potential rating based on wastes classified by type can also be used (U.S. Environmental

Protection Agency, June 1978). While no specific waste hazard rating was listed for septic tank system effluents, a rating of 5 can be used based on the fact that a rating of 4 to 8 was suggested for municipal sludges from conventional biological sewage treatment plants (U.S. Environmental Protection Agency, June 1978).

Table 38: Rating Ground Water Quality in the SIA Method (U.S. Environmental Protection Agency, June 1978)

Rating	Quality
5	$\leq$ 500 mg/l TDS or a current drinking water source
4	> 500 - $\leq$ 1000 mg/l TDS
3	> 1000 - $\leq$ 3000 mg/l TDS
2	> 3000 - $\leq$ 10,000 mg/l TDS
1	> 10,000 mg/l TDS
0	No ground water present

Table 39: Examples of Contaminant Hazard Potential Ratings of Waste Classified by Source in the SIA Method (U.S. Environmental Protection Agency, June 1978)

SIC Number	Description of Waste Source	Hazard Potential Initial Rating
02	AGRICULTURAL PRODUCTION - LIVESTOCK	
021	Livestock, except Dairy, Poultry and Animal Specialties	3 (5 for feedlots)
024	Dairy Farms	4
025	Poultry and Eggs	4
13	OIL AND GAS EXTRACTION	
131	Crude Petroleum and Natural Gas	7
132	Natural Gas Liquids	7
1381	Drilling Oil and Gas Wells	6

Table 39: (continued)

SIC Number		Description of Waste Source	Hazard Potential Initial Rating
20		FOOD AND KINDRED PRODUCTS	
	201	Meat Products	3
	202	Dairy Products	2
	203	Canned and Preserved Fruits and Vegetables	4
	204	Grain Mill Products	2
28		CHEMICALS AND ALLIED PRODUCTS	
	2812	Alkalies and Chlorine	7-9
	2816	Inorganic Pigments	3-8
	2819	Industrial Inorganic Chemicals, not elsewhere classified	3-9
29		PETROLEUM REFINING AND RELATED INDUSTRIES	
	291	Petroleum Refining	8
	295	Paving and Roofing Materials	7
	299	Miscellaneous Products of Petroleum and Coal	7

Summation of the ratings for each of the four factors in the SIA method yields an overall evaluation for the source. An additional consideration is the degree of confidence of the investigator as well as data availability for the specific site. An overall evaluation of the final rating is suggested, with the ratings being either A, B, or C. The rating of A denotes high confidence and is given when the data used has been site specific. Ratings of B and C denote moderate and low confidence, respectively, and are given when data has been obtained from a generalized source, or extrapolated from adjacent sites.

Central Oklahoma Study Area

The main aquifer in the study area where the two empirical assessment methodologies were applied was the Garber-Wellington aquifer in central Oklahoma. The surface area bounding the outcrop and underlying portions of the aquifer includes Oklahoma and Cleveland Counties as well as portions of Logan, Lincoln, Pottawatomie, McClain, Canadian, and Kingfisher Counties. This area is shown in Figure 20 (Canter, 1981). The Garber-Wellington aquifer contains over 50 million ac-ft of fresh water, with approximately two-thirds potentially available for development. The thickness of the fresh water zone ranges from about 50 to 275 m. Water well depths range from about 75 to 325 m, with the deeper wells located in the western half of the study area.

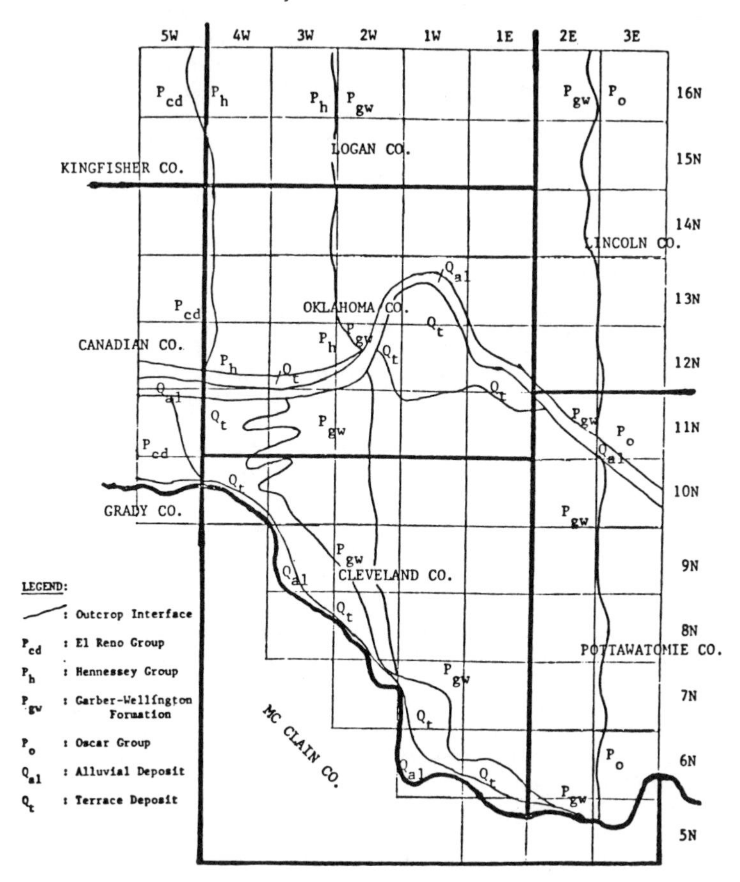

Figure 20: Surface Geology of Study Area

There are additional aquifers potentially influenced by septic tank system areas in central Oklahoma. For example, in Canadian County the Garber-Wellington aquifer is overlain by Permian-age rock formations such as the Hennessey shale and the El Reno group (Mogg, Schoff and Reed, 1969). Saturated zone thicknesses are generally greater than 35 m. In Logan, Oklahoma, and Cleveland Counties, septic tank system areas are surrounded and underlain by alluvial or terrace deposits. Where the alluvial or terrace deposits are underlain by the Hennessey Shale, as in Norman, Moore, and the western parts of Oklahoma City, the Garber-Wellington aquifer is confined (Bingham and Moore, 1975). The alluvial and terrace deposits in these areas are between 3 and 35 m thick. The eastern half of

Figure 20 represents the outcrop area for the Garber-Wellington aquifer (Bingham and Moore, 1975; Burton and Jacobsen, 1967).

There are 13 identifiable areas served by numerous individual septic tank systems in central Oklahoma. These areas were identified through discussions with Health Department personnel in Canadian, Logan, Oklahoma, and Cleveland Counties; and with personnel at the Oklahoma State Department of Health. Table 40 summarizes the populations served in the areas, and Figure 21 displays the general locations of the areas. The populations range from 150 (Sunvalley Acres) to over 12,000 (Midwest City). Assuming that an average of 4 persons is served by a septic tank system, the number of systems ranges from about 40 to 3000. In addition to population information there are five other characteristics common to all septic tank system areas that are required for usage in one or both of the empirical assessment methodologies. These characteristics are:

(1) Soil type in area and permeability measured in in/hr.

(2) The depth from the soil surface to the water table measured in ft.

(3) The land or water table gradient (slope) and the direction of flow.

(4) The distance from the septic tank area to the nearest public or private drinking water source (water well or lake) measured in ft.

(5) Thickness of the porous layer between the soil surface and bedrock measured in ft.

Data for these characteristics was collected for 13 septic tank areas in the central Oklahoma study area. Interpolation and engineering judgment had to be used in determining some of the characteristics for several of the septic tank areas. When information on specific characteristics was unavailable, a "worst case" condition was used.

Table 40: Populations Served by Septic Tank System Areas in Central Oklahoma Study Area

Location of System Area	Estimated Population Served and Year
Arcadia Oklahoma County	410 (1975)
Arrowhead Hills Oklahoma County	488 (1975

Table 40: (continued)

Location of System Area	Estimated Population Served and Year
Crutcho Oklahoma County	587 (1977)
Del City Oklahoma County	246 (1975)
Forest Park, Lake Hiwassee and Lake Alma Oklahoma County	1,200 (1975)
Green Pastures Oklahoma County	2,313 (1977)
Midwest City Oklahoma County	12,040 (1975)
Mustang Canadian County	3,550 (1975)
Nicoma Park Oklahoma County	3,000 (1975)
Norman (east of 24th Street) Cleveland County	8,000 (1980)
Seward Area Logan County	2,247 (1980)
Silver Lake Estates Oklahoma County	325 (1975)
Sunvalley Acres Canadian County	150 (1975)

Application of the surface impoundment assessment method to the general information about the 13 septic tank system areas yielded composite scores ranging from 12 to 24, with the specific results shown in Table 41 (Canter, 1981). Variations in the scores are primarily reflective of the geological features of the areas. Ten septic tank system areas are located on terrace deposits or in the Garber-Wellington outcrop area, and they received scores ranging from 22 to 24. The remaining three systems are located on outcrops of the Hennessey or El Reno groups, and they had scores between 12 and 15. An additional factor which could be utilized for evaluation of the pollution potential is the service area or estimated total

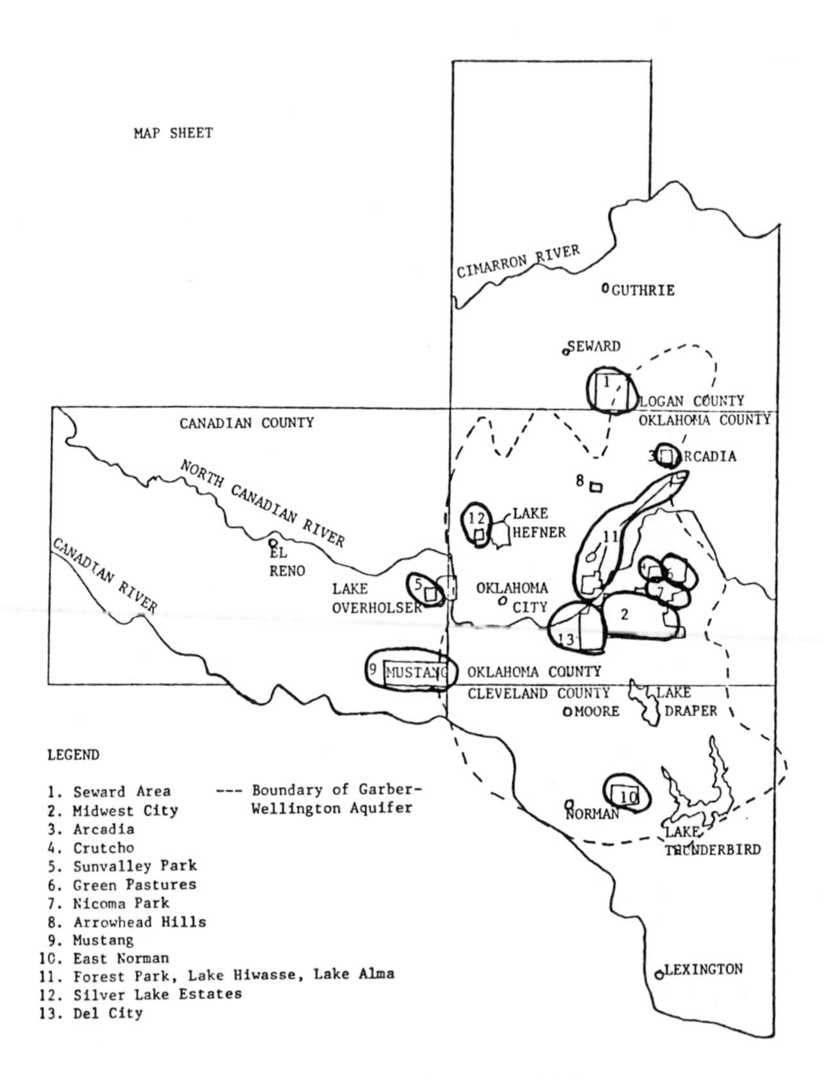

Figure 21: Septic Tank Areas in Central Oklahoma

flows into septic tank systems. Table 41 also contains estimates of the total wastewater flows for the respective service areas. The estimates were developed by multiplying the average wastewater flow per person by the number of persons served by septic tanks in the area. A flow of 52 gal/person/day was used in determining the total flow for a septic tank area (U.S. Environmental Protection Agency, September 1978). To serve as an illustration, the estimated population in the Arcadia area served by septic tank systems is 410 people. The flow expressed on an annual basis is: (410 people)(52 gal/person/day)(365 day/yr) = 7.8×10^6 gal/yr (use 8×10^6).

Table 41: Assessment of Septic Tank System Areas by Surface Impoundment Assessment Method (Canter, 1981)

Area (Underlying Aquifer)	Unsaturated Zone Rating	Ground Water Availability Rating	Ground Water Quality Rating	Waste Hazard Rating	Overall Ground Water Contamination Potential	Annual Wastewater Flow (10^6 gal/yr)
Maximum Value	9	6	5	9	29	
Minimum Value	0	0	0	1	1	
Confidence Level	B	C	B	B	--	
x Arcadia (G-W)*	8B	6A	5	5	24	8
x Seward (G-W)	8B	6A	5	5	24	175
Arrowhead Hills (G-W)	7B	6A	5	5	23	9
Crutcho (T, G-W)	7B	6A	5	5	23	11
Forest Park (G-W)	7B	6A	5	5	23	27
Green Pastures (T, G-W)	7B	6A	5	5	23	44
x Midwest City (G-W)	7B	6A	5	5	23	228
Nicoma Park (T, G-W)	7B	6A	5	5	23	57
East Norman (T, G-W)	6B	6A	5	5	22	152
Del City (G-W)	6B	6A	5	5	22	5
x Sunvalley Acres (ER)	5D	2E	3	5	15	3
Mustang (ER)	4D	2E	4	5	15	67
Silver Lake Estates (H)	2E	2E	3	5	12	6

x Denotes sampling conducted in area.

* G-W = Garber-Wellington, T = terrace deposits, ER = El Reno group, H = Hennessey group.

Based on considering the anticipated annual flows along with the ground water contamination potential rating, the following priority listing was obtained--Midwest City (highest potential), Seward, East Norman,

Nicoma Park, Green Pastures, Mustang, Forest Park, Crutcho, Arrowhead Hills, Arcadia, Del City, Silver Lake Estates, and Sunvalley Acres (lowest potential).

Waste-Soil-Site Interaction Matrix

The waste-soil-site interaction matrix was developed for assessing industrial solid or liquid waste disposal on land (Phillips, Nathwani and Mooij, 1977). Septic tank system effluent (liquid) is discharged to soil through the soil absorption system, hence this matrix is considered to be potentially applicable to the evaluation of septic tank system areas. The method involves summation of the products of various waste-soil-site considerations, with the resultant numerical values ranging from 45 to 4,830. The methodology includes ten factors related to the waste, and seven factors associated with the site of potential waste application. Table 42 contains a description of the waste factors and their numerical scores, and Table 43 lists the soil-site factors with their associated weights. Table 44 represents an example interation matrix resulting from this methodology, with the total summation of the products being 990.

Table 42: Waste Factors in Waste-Soil-Site Interaction Matrix (Phillips, Nathwani and Mooij, 1977)

Group	Factor
Effects	Human Toxicity (Ht)--ability of a substance to produce injury once it reaches a susceptible site in or on the body. Based on severity of effect all substances grouped into those with no toxicity, slight toxicity, moderate toxicity, and severe toxicity. The Ht values range from 0 (no toxicity) to 10 (maximum toxicity).
	Ground Water Toxicity (Gt)--related to minimum concentration of waste substance in ground water which would cause damage or injury to humans, animals, or plants. The Gt value is a function of the lowest concentration which would cause damage or injury to any portion of the ecosystem; the Gt values range from 0 (nontoxic) to 10 (very toxic).
	Disease Transmission Potential (Dp)--considers mode of disease contraction, pathogen life state, and ability of the pathogen to survive. Disease contraction includes direct contact, infection through open wounds, and infection by vectors (usually insects). Pathogen life state

Table 42: (continued)

Group	Factor
	includes pathogenic microorganisms with more than one life state (virus and fungi), one life state (vegetative pathogens), and those which cannot survive outside their host. The ability of the pathogen to survive includes survival in air, water, and soil environments. The Dp values range from 0 (no effect) to 10 (maximum effect).
Behavioral (Behavioral Performance)	Chemical Persistence (Cp)--related to the chemical stability of toxic components in the waste. Consideration is given to the concentration of toxic components after one-day and six-days contact with soil from potential disposal site. The Cp values range from 1 (very unstable toxic component) to 5 (very stable toxic component).
	Biological Persistence (Bp)--related to the biodegradability of the waste as determined by biochemical oxygen demand (BOD) and theoretical oxygen demand (TOD). The Bp values range from 1 (very biodegradable) to 4 (nonbiodegradable).
	Sorption (So)--related to the mobility of the waste in the soil environment. Consideration is given to initial concentration of toxic component(s) in waste as well as one-day following mixing with soil from potential disposal site. The So values range from 1 (very strong sorption) to 10 (no sorption).
Behavioral (Behavioral Properties)	Viscosity (Vi)--related to the flow of the waste toward the water table. Consideration is given to the waste viscosity measured at the average maximum temperature of the site during its proposed months of use. The Vi values range from 1 (very viscous) to 5 (viscosity of water).
	Solubility (Sy)--along with sorption, solubility relates to the mobility of the waste in the soil environment. Waste solubility is measured in pure water at 25°C and pH of 7. The Sy values range from 1 (low solubility) to 5 (very

Table 42: (continued)

Group	Factor
	soluble). In case the waste is miscible with water, Sy is equal to 5.
	Acidity/Basicity (Ab)--considers the influence of acidic or basic wastes on the solubility of various metals. Acidic wastes tend to solubilize metals whereas basic wastes tend to immobilize metals through precipitation. The Ab values range from 0 (no effect) to 5 (maximum effect).
Capacity Rate	Waste Application Rate (Ar)--related to the volumetric application rate of the waste at the site, the sorption characteristics of the site, and the concentration of toxic component(s) in the waste. The Ar values range from 1 (low volumetric application rate of a low concentration waste to a site having high sorptive properties) to 10 (high volumetric application rate of a high concentration waste to a site having low sorptive properties).

Table 43: Soil-Site Factors in Waste-Soil-Site Interaction Matrix (Phillips, Nathwani and Mooij, 1977)

Group	Factor
Soil	Permeability (NP)--relates to permeability of site materials. Clay is considered to have poor permeability, fine sand moderate permeability, and coarse sand and gravel good permeability. The NP values range from 2.5 (low permeability) to 10 (maximum permeability).
	Sorption (NS)--relates to sorption characteristics of site materials. The NS values range from 1 (high sorption) to 10 (low sorption).
Hydrology	Water Table (NWT)--considers the fluctuating boundary free water level and its depth. The zone of aeration occurs above the water table

Table 43: (continued)

Group	Factor
	and is important to oxidative degradation and sorption. The NWT values range from 1 (deep water table) to 10 (water table near surface).
	Gradient (NG)--relates to the effect of the hydraulic gradient on both the direction and rate of flow of ground water. The NG values range from 1 (gradient away from the disposal site in a desirable direction) to 10 (gradient toward point of water use).
	Infiltration (NI)--relates to the tendency of water to enter the surface of a waste disposal site. Involves consideration of the maximum rate at which a soil can absorb precipitation or water additions. A site with a large amount of infiltration will have greater ground water pollution potential. The NI values range from 1 (minimum infiltration) to 10 (maximum infiltration).
Site	Distance (ND)--relates to the distance from the disposal site to the nearest point of water use. The greater the distance the less chance of contamination because water dilution, sorption, and degradation increase with distance. The ND values range from 1 (long distance from disposal site to use site) to 10 (disposal site close to use site).
	Thickness of Porous Layer (NT)--refers to porous layer at the disposal site. The NT values range from 1 (about 100 ft or more of depth) to 10 (about 10 ft of depth).

Ten classes used for interpretation are as follows--Class 1 (45-100 points), Class 2 (100-200), Class 3 (200-300), Class 4 (300-400), Class 5 (400-500), Class 6 (500-750), Class 7 (750-1000), Class 8 (1000-1500), Class 9 (1500-2500), and Class 10 (greater than 2500). Classes 1-5 are considered acceptable, and classes 6-10 unacceptable. In the following detailed discussion the method for calculating each pertinent factor for the central Oklahoma study area is presented.

Table 44: Example of Waste-Soil-Size Interaction Matrix (Phillips, Nathwani and Mooij, 1977)

GROUP	WASTE \ SOIL	P*	SOIL GROUP: Permeability NP (2½-10)	SOIL GROUP: Sorption NS (1-10)	HYDROLOGY GROUP: Water Table WT (1-10)	HYDROLOGY GROUP: Gradient MG (1-10)	HYDROLOGY GROUP: Infiltration NI (1-10)	SITE GROUP: Distance ND (1-10)	SITE GROUP: Thickness of Porous Layer NT (1-10)	Total
	P*		5	4	5	2	6	7	1	30
EFFECTS GROUP	Human Toxicity Ht (0-10)	8	40	32	40	16	48	56	8	240
EFFECTS GROUP	Groundwater Toxicity Gt (0-10)	5	25	20	25	10	30	35	5	150
EFFECTS GROUP	Disease Transmission Potential DP (0-10)	0								
Behavioural Performance Subgroup	Chemical Persistence Cp (1-5)	3	15	12	15	6	18	21	3	90
Behavioural Performance Subgroup	Biological Persistence Bp (1-4)	4	20	16	20	8	24	28	4	120
Behavioural Performance Subgroup	Sorption So (1-10)	5	25	20	25	10	30	35	5	150

BEHAVIOURAL	Behavioural Properties Subgroup	Viscosity Vi (1-5)	2	10	8	10	4	12	14	2	60
		Solubility Sy (1-5)	1	5	4	5	2	6	7	1	30
		Acidity/ Basicity Ab (0-5)	1	5	4	5	2	6	7	1	30
	CAPACITY-RATE GROUP	Waste Application Rate Ar (1-10)	4	20	16	20	8	24	28	4	125
		Total	33	165	132	165	66	198	231	33	990

* P = point score

1. Effects Group

a. Human Toxicity (Ht): Human toxicity is based on classifying wastes or waste constituents into four categories regardless of the concentration of the waste. These categories are shown in Table 45, and the Ht value is determined as follows.

$$Ht = \alpha Sr$$

where Ht is the human toxicity rank, α is a constant, and Sr represents the toxicity rating. In the methodology α is considered to have a value of 10/3, and Sr can range from 0 to 3. Due to the potential for nitrates to cause methemoglobinemia in infants, an Sr of 3 was used for all 13 septic tank systems in the study area; therefore, the Ht value was 10 (maximum toxicity). Using an Sr of 3 represents a worst case approach.

Table 45: Toxicity Values for Waste-Soil-Site Interaction Matrix (Phillips, Nathwani and Mooij, 1977)

1. Sr=0, no toxicity

Applied for wastes that are categorized as follows:

(a) Materials that cause no harm under any conditions of use; or

(b) Materials which produce toxic effects on humans only under the most unusual conditions or overwhelming dosage.

2. Sr=1, slight toxicity

Materials produce changes in the human body which are readily reversible and which will disappear following termination of exposure, either with or without medical treatment.

3. Sr=2, moderate toxicity

Materials that produce irreversible as well as reversible effects in the human body but do not threaten life or cause serious permanent physical impairments.

4. Sr=3, severe toxicity

Materials that when absorbed into the body by inhalation, ingestion, or through the skin will cause injury or illness

Table 45: (continued)

of such severity to threaten life or to cause permanent physical impairment or disfigurement.

b. Ground Water Toxicity (Gt): Ground water toxicity is measured in terms of concentration of the waste or waste constituents. The concentration is the critical value which results in a detrimental effect on the ecosystem. Thus, a critical concentration is defined in terms of human toxicity, aquatic toxicity or plant toxicity, or the minimum concentration which would cause damage to humans, animals, or plants. The use of concentration in the toxicity term ensures that no overlap occurs with the human toxicity rank, which is based on severity of effect.

For toxicity to humans, the critical concentration can be chosen at the maximum allowable concentration in drinking water. For aquatic toxicity, the critical concentration can be taken as the lethal concentration (LC_{50}) value for fish in a standard bioassay. For plant toxicity, the critical concentration must be taken as the maximum concentration tolerated by the most sensitive plant in the area. Since any one of toxicity to humans, aquatic life, or plants may be limiting, the smallest critical concentration of the set is used to define the ground water toxicity, which is given by:

$$Gt = \frac{10}{7} (4 - \log_{10} C_c)$$

where Gt is the ground water toxicity rank, and C_c represents the smallest critical concentration in milligrams per liter (mg/l) for humans, aquatic life or plants. When C_c is larger than 10^4 mg/l, Gt will be equal to zero, and for C_c less than 10^{-3} mg/l, Gt will be equal to ten. Therefore, the range of Gt will be from 0 (nontoxic) to 10 (very toxic). In this study C_c was assumed to be 10 mg/l for nitrates in ground water; therefore, Gt was uniform at 4.3 for all 13 septic tank system areas.

c. Disease Transmission Potential (Dp): This factor is evaluated according to three specific disease transmission properties of the waste, denoted subgroup A, B and C. Subgroup A represents the mode of disease contraction, subgroup B represents the pathogen life state, and subgroup C represents the ability of the pathogen to survive. The final disease transmission potential factor is the sum of the contributions from the three groups. The estimation of each contribution from the three groups is as follows:

Subgroup A: mode of disease contraction; maximum value of factor is 4. Selection of one of the following three possible modes is made:

(a) direct contact: assigned a value of up to 4 on account of immediate threat;

(b) infection through open wounds: assigned a value of up to 3;

(c) infection by vector (usually insect): assigned a value of up to 1½ since site control can minimize this.

Subgroup B: pathogen life state; maximum value of factor is 3. Selection of one of the following three life state categories is made:

(a) pathogenic micro-organisms with more than one life state (virus, fungi): assigned a value of up to 3;

(b) pathogenic micro-organisms with only one life state (vegetative pathogens): assigned a value of up to 2;

(c) pathogenic micro-organisms which cannot survive outside their host: assigned zero value.

Subgroup C: ability of pathogen to survive in various environments; maximum value of factor is 2½. Selection is made as follows:

(a) able to survive in air: assign value of 1.5;

(b) able to survive in water: assign value of 1;

(c) able to survive in soil: assign value of 1/2.

A uniform Dp value of 8.5 was used for each of the 13 septic tank system areas in the central Oklahoma study area. The value of 8.5 was derived by considering that bacterial and viral contamination of nearby wells can occur from septic tank systems. The 8.5 points resulted from 4 points from direct contact as the mode of disease contraction, 3 points from more than one pathogen life states, and 1.5 points due to pathogen ability to survive in water and soil. Again, this represents a worst case approach.

2. Behavioral Group (Behavioral Performance Subgroup)

a. Chemical Persistence (Cp): This factor relates to the persistence over time of chemical constituents in the waste. The factor is expressed by assuming that the decay of the toxic component(s) of a waste can be specified by a single parameter. This is chosen to be a pseudo-first order rate constant k. Then the chemical persistence factor Cp will be given by:

$$Cp = 5 \exp(-kt), \text{ but if } Cp < 1 \text{ then } Cp = 1$$

where t is the time and k is determined from the following equation:

$$C_6/C_1 = \exp(-kt)$$

where C_1 is the concentration of toxic component(s) at one day, and C_6 is the concentration of toxic component(s) at 6 days. A mixture of waste and soil in question is prepared to make a 50 percent by weight soil mixture used to determine C_1 and C_6. The contact must occur at the average minimum temperature for the site, and the mixture must be in the open. The chemical persistence factor Cp will then range from 1 (for very unstable toxic component) to 5 (for a very stable toxic component). No laboratory studies were made to determine Cp in this case study related to central Oklahoma. Instead, a worst case approach was used in that a Cp value of 5 was used for all 13 septic tank system areas.

b. Biological Persistence (Bp): This factor relates to the biodegradability of waste components over time. Biological degradability is measured in terms of biochemical oxygen demand, BOD, usually measured over 5 days. For highly biodegradable waste, the BOD is approximately equal to the theoretical oxygen demand, TOD, measured by chemical oxidation methods. The ratio of BOD:TOD is then a measure of degree of biodegradability. The following equation expresses the biological persistence in quantitative values:

$$Bp = 4 (1 - BOD/TOD)$$

but if Bp is less than 1, then use Bp = 1. The range of values of this factor is from 1 (very biodegradable) to 4 (unbiodegradable). No laboratory studies were conducted to determine Bp in this case study related to central Oklahoma. However, several other published studies have indicated that the BOD of septic tank effluent is about 60 percent of the TOD; therefore, a Bp value of 1.6 was used for all 13 septic tank system areas.

c. Sorption (So): This factor reflects the adsorption properties of both the waste and the soil receiving the waste. It is measured in the same manner as the chemical persistence factor with a 50 percent by weight mixture of waste and soil. The only difference between the two factors is the length of time between measurements of the waste concentration. The function for determination of the sorption parameter is:

$$So = 11 - Co/C_1$$

where Co = concentration of toxic components in waste initially, and C_1 = concentration of toxic components in waste after 1 day. This short length of time effectively eliminates the effect of rate processes. If Co/C_1 is larger than 10, then So is set equal to 1. The range of this factor is, therefore, from 1 (very strong adsorption) to 10 (no adsorption). The sorption parameter receives a high point value because it is an important determinant of the capacity of a site for neutralizing wastes. No laboratory studies were conducted to determine So in this case study related to central Oklahoma. In order to present the worst case approach, an So value of 10 was used for all 13 septic tank system areas. No adsorption was assumed based on the general mobility and low adsorption of nitrates in soils and ground water. A different So value would need to be used if organic pollution from septic tank systems was of concern.

3. Behavioral Group (Behavioral Properties Subgroup)

a. Viscosity (Vi): This factor is a measure of the ability of the waste for rapid or slow movement through the soil to the water table. The flow of the waste towards the water table is a function of the waste viscosity. Wastes with low viscosities will more rapidly contaminate ground water. This factor is defined as follows:

$$Vi = 5 - \log_{10} \mu$$

where μ is the viscosity of the waste in centipoises, but if μ is larger than 10, then Vi should equal to one (Vi = 1) and for μ less than one, Vi is equal to 5 (Vi = 5). The viscosity is measured at the average maximum temperature of the site during its proposed months of use. The range of Vi is from 1 (very viscous) to 5 (viscosity of water). No laboratory studies were conducted to determine Vi in this case study related to central Oklahoma. Since water is the primary medium in septic tank effluents, a Vi value of 5 was used for all 13 septic tank system areas. Use of a Vi of 5 represents a worst case approach.

b. Solubility (Sy): This factor reflects the solubility of the waste in water and is a measure of waste mobility with the water phase in the subsurface environment. The solubility factor is defined as follows:

$$Sy = 3 + 0.5 \log_{10} S$$

where S is the solubility of waste in pure water at 25°C and pH of 7, S is measured in milligrams per liter. If S is less than 10^{-4} mg/l, then Sy is equal to one, and if S is larger than 10^4 mg/l, then Sy = 5. Therefore, the solubility term has a range of 1 (low solubility) to 5 (very soluble). In case the waste is miscible with water, Sy is equal to 5, the maximum value. The term is equally applicable to dissolved solids and dissolved liquids. No laboratory studies were conducted to determine Sy in this case study related to central Oklahoma. Since nitrates are highly soluble in water, an Sy value of 5 was used for all 13 septic tank system areas. Use of an Sy value of 5 represents a worst case approach.

c. Acidity/basicity (Ab): Highly acidic or basic wastes are undesirable in the environment. Highly acid wastes will solubilize heavy metal precipitates and allow them to migrate and enter the ground water, while highly basic wastes will precipitate the metals and thus immobilize them. The acidity/basicity factor, Ab, can be determined as follows:

pH of waste	≤ 0	1	2	3	4	5	6	7	8	9	10	11	12	13	≥ 14
Ab value	5	5	5	4	3	2	1	0	0	1	1	2	2	3	3

The range of the acidity/basicity factor is then from 0 (no effect) to 5 (maximum effect). In case the waste is a solid, the pH of a 50 percent by weight mixture of the waste in water is measured and used to deduce the factor Ab. No laboratory studies were conducted to determine Ab in this case study related to central Oklahoma. However, since the pH of septic tank effluents is in the range of 7 to 8, an Ab value of 0 was used for all 13 septic tank system areas.

4. Capacity Rate Group

a. Waste application rate (Ar): This factor measures the attenuation effect of the soil based on the waste load that is being applied. If the waste loading rate is too high, then the attenuating ability of the soil will be exceeded, causing the excess waste to penetrate deeper into the soil and evantually to the ground water. The quantity of contaminant per unit volume of waste multiplied by the volumetric rate of application of the waste per unit area is equal to the application rate of the contaminant (quantity applied per unit area per unit time). The waste application rate factor is defined as follows:

$$Ar = (9/2) \log_{10} ((Rf \cdot Co)^{1/2} \cdot NS) + 1$$

where Ar is the waste application rate, NS is the sorption parameter for site (defined later), Rf is the volumetric rate factor determined from the following table:

Rf	1	2	3	4	5	6	7	8	9	10
Volumetric application rate gal/ft^2, day	<0.1	0.1–0.5	0.5–1.0	1–2	2–3	3–4	4–5	5–6	6–7	>7

and Co is the function of concentration of toxic component in waste, determined from:

$$Co = 5 + 1.25 \log_{10} C$$

where C is the concentration of waste in mg/l. But if C is less than 10^{-4} mg/l, then Co is equal to 1, and if C is larger than 10^4 mg/l, then Co is equal to 10. The concentration term Co has a range of from 1 (low solubility) to 10 (high solubility), and the waste application rate factor (Ar) ranges from 1 (low volumetric application rate of a low concentration waste to a site having high sorptive properties) to 10 (high volumetric application rate of a high concentration waste to a site having low sorptive properties). If one contaminant in a mixture of wastes has a dominant effects factor, then the waste behavior may be based on that component, otherwise each component should be considered separately. For this study a volumetric application rate of 4 gal/ft^2/day was used, thus Rf is 6. Since nitrates are of concern, a value of C of 34.6 mg/l was used since this represents an average tank effluent concentration (U.S. Environmental Protection Agency, September 1978). Therefore, the Co value is 6.9. The value for NS will be subsequently discussed. Substitution of the Rf and Co values into the above equation yields the following:

$$Ar = \frac{9}{2} \log ((6 \times 6.9)^{1/2} \times NS) + 1$$

$$Ar = \frac{9}{2} \log (6.4 \times NS) + 1$$

5. Soil Group

   a. Permeability (NP): In this method the ranges in permeability are broadly estimated. Clay is assigned a poor permeability, fine sand a moderate permeability, and coarse sand and gravel good permeability. It should be noted that good permeability denotes poor conditions in terms of ground water quality protection. The sites considered in this study fall into two categories: (1) the one-medium site with disposal in loose granular earth materials extending to about 100 ft below ground surface; and (2) the two-media site with disposal in unconsolidated granular materials at the ground surface underlain at shallow depths by dense rocks and linear openings. The normalized permeability factor is defined as:

$$NP = \frac{10}{Pmax + 1} \; (Pmax + 1 - P)$$

where NP is the normalized permeability factor, P is the permeability point score from the below charts, and Pmax is the maximum value of P from LeGrand. For loose granular single and two media sites, Pmax is equal to 3.

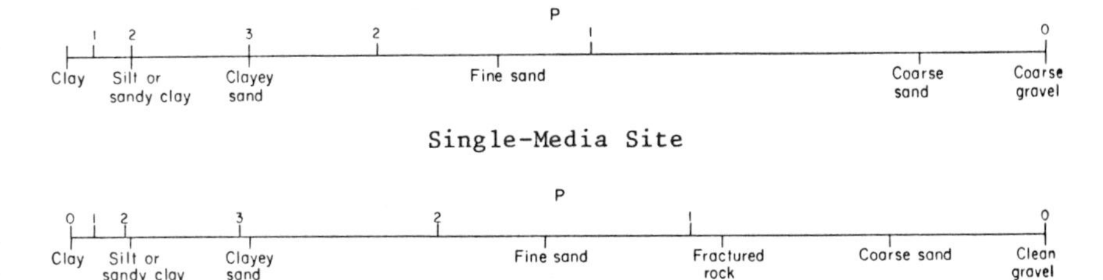

Double Media Sites

Therefore, NP ranges from 2.5 (low permeability) to 10 (maximum permeability). In this study, 11 of the 13 septic tank system areas were underlain by fine sand, and 2 areas were underlain by clayey sand. The NP values for the fine sand areas were 5.9, while for the areas with soils nearer clayey sand the NP values ranged from 2.5 to 3.1.

b. Sorption (NS): The normalized sorption value is determined as follows:

$$NS = \frac{10}{Smax + 1} \; (Smax + 1 - S)$$

where NS is the normalized sorption factor, S is the sorption point score (LeGrand) determined from the below charts, and Smax is the maximum value of S from LeGrand. For loose granular site or for two-media site Smax is equal to 6.

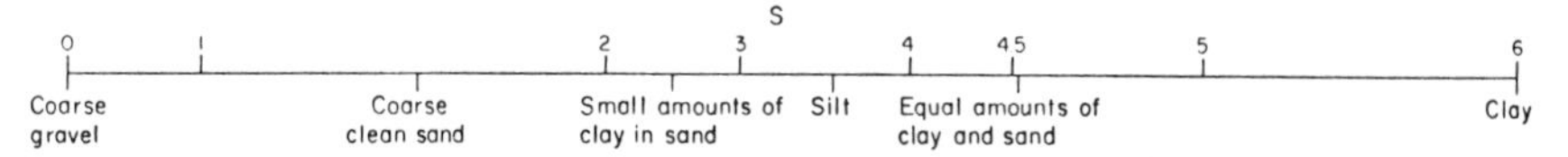

Loose Granular Sites

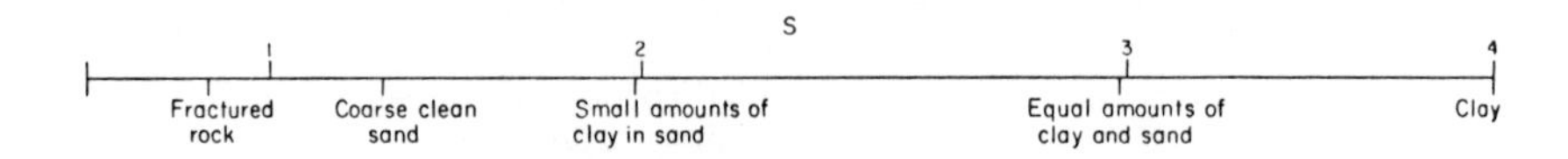

Two-Media Sites

The range of NS is therefore from 1 (high sorption) to 10 (low sorption). In this study, 10 of the 13 septic tank system areas were characterized by small amounts of clay in sand, thus the NS values equalled 7.1. Two of the 13 areas had equal amounts of clay and sand, thus the NS values equalled 5.7. Finally, one area with silt had an NS value of 5.0. The NS values are also used in the waste application rate factor (Ar). Based on the above discussion of Ar, the Ar values were 8.5 when NS was 7.1, 8.0 when NS was 5.7, and 7.8 when NS was 5.0.

6. Hydrology Group

a. Water Table (NWT): The water table is the fluctuating boundary free water level and its depth is determined by observing the free water in a well. The zone of aeration generally occurs above the water table and is important to oxidative degradation and sorption. The normalized water table value used in this method is given by:

$$NWT = \frac{10}{WTmax + 1} (WTmax + 1 - WT)$$

where NWT is the normalized water table factor, WT is the water table point score (LeGrand) from the below chart, and WTmax is the maximum value of WT from LeGrand. For loose granular and two-media sites, WTmax is equal to 10.

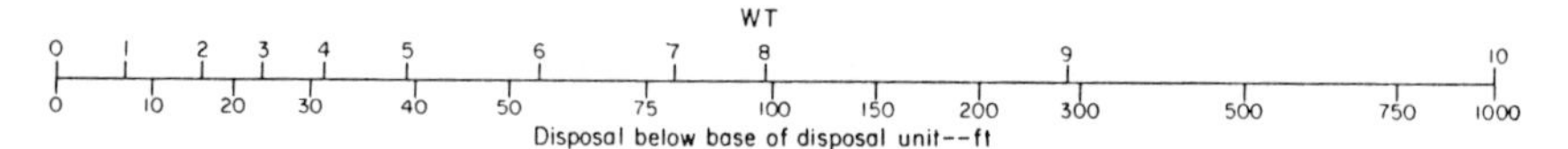

Loose Granular Materials

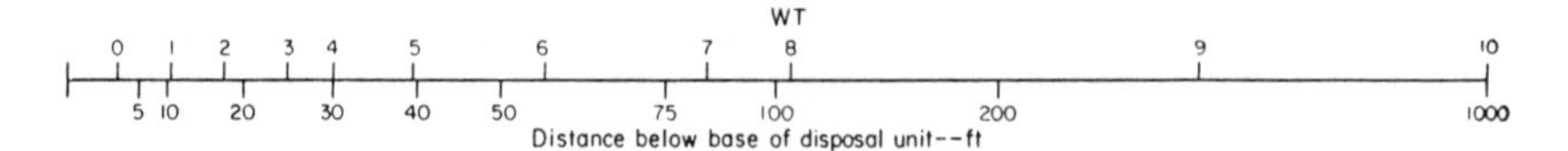

Two-Media Sites

The range of NWT is therefore from 1 (best case: deep water table) to 10 (worst case: water table near surface). In this study the WTmax is equal to 10, hence

$$NWT = \frac{10}{11} (11 - WT)$$

In this study the NWT values range from 2.4 to 9.5. For an NWT of 2.4, the WT is about 8.3, and the distance below the septic tank area to the water table is about 150 ft. For an NWT of 9.5, the WT is about 0.5, and the distance to the water table is about 7 ft. Data on the depth to the water table for each of the 13 septic tank system areas is in Appendix B (Carriere, 1980).

b. Gradient (NG): The gradient has an effect on both the direction and the flow rate of ground water. Movement of water away from the septic tank system area is much more desirable than movement towards it. A water table may be lowered by pumping from a well, thus increasing the gradient and flow rate. The gradient for the matrix is given by:

$$NG = \frac{10}{Gmax + 1} (Gmax + 1 - G)$$

where NG is the normalized gradient factor, G is the gradient point score (LeGrand) from the chart below, and Gmax is the maximum value of G from LeGrand. Gmax is equal to 7 for loose granular and two media sites.

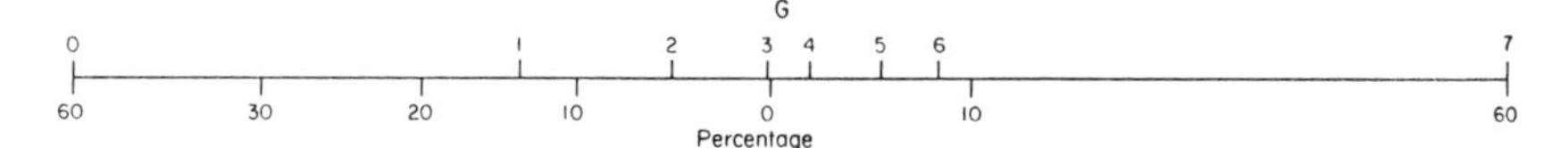

Loose Granular Materials

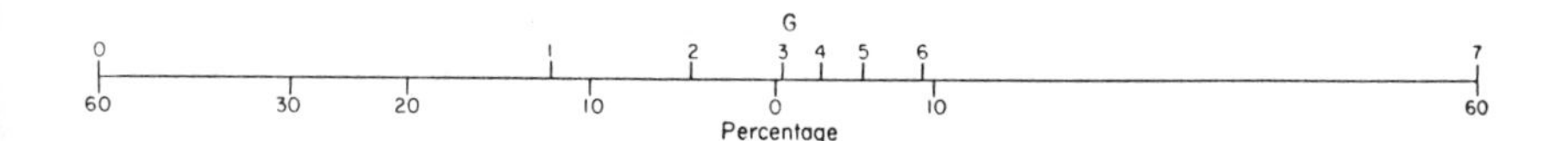

Two-Media Sites

Therefore, NG ranges from 1 (gradient away from the disposal site in a desirable direction) to 10 (gradient towards point of water use). It should be noted that in large septic tank system areas there could be some locations with gradients away from the site, and other locations with gradients toward points of water use. In this study the Gmax is equal to 7, hence

$$NG = \frac{10}{8} (8 - G)$$

In this study the NG values range from 3.0 to 5.6. For an NG of 3.0, the G value is 5.6, and the percentage of gradient slope is about 7 percent. For an NG value of 5.6, the G value is 3.5, and the percentage of gradient slope is about 2 percent. Data on the land and water table gradient slope is in Appendix B for the 13 septic tank system areas (Carriere, 1980).

c. Infiltration (NI): This factor describes the tendency of moisture from precipitation to enter the surface of a disposal site. The application of this factor to septic tank systems is analogous to the percolation test to determine the rate at which water will percolate or infiltrate the soil in inches per hour. A septic tank site with a high percolation rate or infiltration rate (in/hr) is more likely to contaminate ground water than a site with a low infiltration rate. The infiltration factor as included in this method represents the tendency of water to enter the surface of a waste disposal site. The infiltration (i) is the maximum rate at which a soil can absorb precipitation or water additions. In the case of seepage beds or fills, it would be considered as the maximum rate that liquid or fluid enters the soil at the bed interface. The normalized infiltration factor used in this method is determined as follows:

i, inches	<2	2–4	4–6	6–8	8–10	10–12	12–14	14–16	16–18	18–20
NI	1	1	2	2	3	3	4	4	5	5

i, inches	20–22	22–24	24–26	26–28	28–30	30–32	32–34	34–36	36–38	38–40	>40
NI	6	6	7	7	8	8	9	9	10	10	10

Thus the range of the infiltration factor is from 1 (minimum infiltration) to 10 (maximum infiltration). The infiltration (i) for the central Oklahoma study area is estimated at 2 inches per year due to the moderate precipitation and high evaporation on an annual basis. Therefore, an NI value of 1.0 was used for all 13 septic tank system areas in this study.

7. Site Group

a. Distance (ND): This factor is a measure of the distance from a disposal site to any point of water use, e.g. lake, city water well, or private water well. The greater the distance from the disposal site to the point of use the less will be the chance of contamination. This is because dilution occurs with distance traveled, sorption becomes more complete, time of travel increases with distance, and thus decay or degradation is more complete, and the water table gradient tends to decrease so that the velocity of flow decreases. The normalized distance factor is given by:

$$ND = \frac{10}{Dmax + 1}\ (Dmax + 1 - D)$$

where ND is the normalized distance factor, D is the distance point score (LeGrand) determined from the chart below, and Dmax is the maximum value of D from LeGrand. Dmax is equal to 11 for loose granular single media sites and two-media sites.

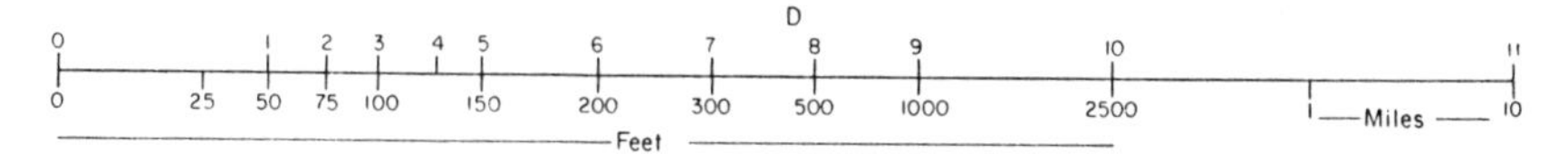

Loose Granular Materials

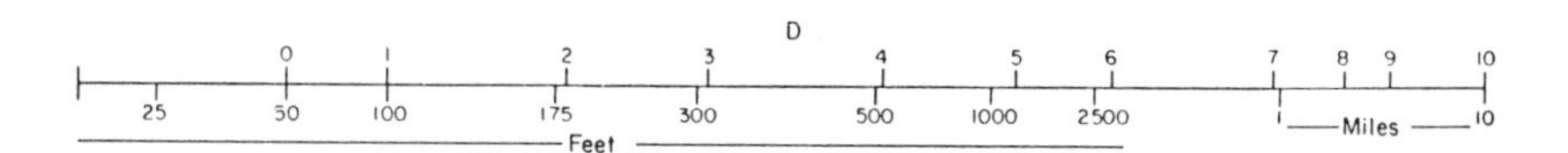

Two-Media Sites

The range of ND is from 1 (long distance from disposal to effect site) to 10 (disposal site close to effect site). In this study the Dmax is equal to 11, hence

$$ND = \frac{10}{12}\ (12 - D)$$

In this study the ND values range from 2.5 to 9.2. For an ND value of 2.5, the D value is 9, and the distance from the site to the nearest water use is about 5 miles for a two-media site. For an ND value of 9.2, the D value is about 1, and the distance from the site to the nearest water use is about 50 ft for loose granular material, and 100 ft for two-media sites. Data on distances to public/private water wells is included in Appendix B for each of the 13 septic tank system areas (Carriere, 1980).

b. Thickness of porous layer (NT): This factor is a measure of the unsaturated zone above the bedrock at each site. The porous layer is defined as being greater than 100 ft. In case the layer is less than 100 ft in thickness, then the site is classified as a two-media site, the underlying media being considered relatively impermeable. In the second case, an additional rating factor is needed, defined as follows:

$$NT = \frac{10}{Tmax + 1}\ (Tmax + 1 - T)$$

where NT is the thickness of porous layer factor (less than 100 ft thick) for two-media sites, T is the thickness point count (LeGrand) determined from the chart below, and

Tmax is the maximum value of T from LeGrand. Tmax is equal to 6.

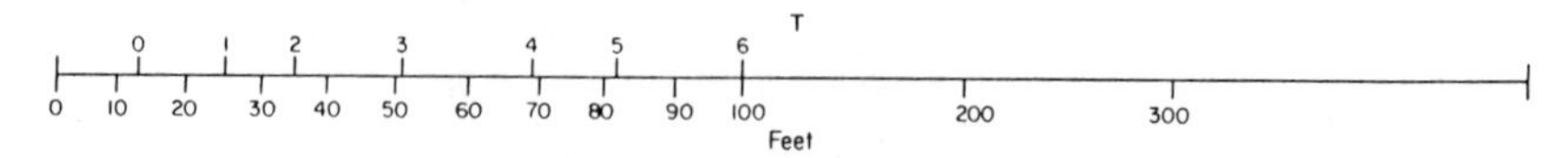

Two-Media Sites

NT ranges from 1 (about 100 ft of depth porous layer) to 10 (about 10 ft of depth of porous layer). In this study the Tmax is equal to 6, hence

$$NT = \frac{10}{7}(7 - T)$$

In this study an NT value of 10 was used for all 13 septic tank system areas since the thickness of the porous layer to bedrock was typically less than 10 ft.

Application of the waste-soil-site interaction matrix to the 13 septic tank system areas in central Oklahoma yielded composite scores ranging from 2005 to 2641. Table 46 displays the matrix results for the Arcadia area as an example, and Table 47 lists the assessment scores for all 13 areas. The interaction matrices for the other 12 areas are in Appendix C (Carriere, 1980). Ten of the 13 areas are rated in Class 9 (1500-2500) for interpretation, and three are in Class 10 (greater than 2500). The lowest score represents the area least likely to contaminate the ground water, and the highest score represents the area most likely to contaminate the ground water. Based on the normal usage of the waste-soil-site interaction matrix, both Classes 9 and 10 would be unacceptable as waste sites. However, since the areas are already being used for septic tank systems, the assessment scores can be viewed as indicating ground water pollution potential, with the areas with lower scores having lower potential. Based on considering the assessment scores along with the anticipated annual wastewater flows into the septic tank systems, the following priority listing was obtained: Midwest City (highest ground water pollution potential), Seward, East Norman, Mustang, Nicoma Park, Green Pastures, Forest Park, Crutcho, Arrowhead Hills, Arcadia, Silver Lake Estates, Del City, and Sunvalley Acres (lowest potential).

Comparison of Empirical Assessment Methodologies

Table 48 provides a comparative display of the rank order ground water pollution potential of the 13 septic tank system areas as determined by the two selected empirical assessment methodologies adjusted by considering the annual wastewater flows in the areas. The two adjusted methodologies provided similar rank orderings of the 13 septic tank system areas. Midwest City, Seward, and East Norman were ranked as having the highest ground water pollution potential, while Sunvalley Acres was ranked as the lowest. Following are some summary comments relative to these two methodologies:

Table 46: Waste-Soil-Site Interaction Matrix Assessment for Arcadia, Oklahoma County, Oklahoma

WASTE \ Soil	P \ P	NP	NS	WT	G	I	D	T	TOTAL
		5.9	7.1	7.3	3	1	9.2	10	---
Ht	10	59	71	73	30	10	92	100	435
Gt	4.3	25	31	31	13	4	40	43	187
Dp	8.5	50	60	62	26	9	78	85	370
Cp	5	30	36	37	15	5	46	50	219
Bp	1.6	9	11	12	5	2	15	16	70
So	10	59	71	73	30	10	92	100	435
Vi	5	30	36	37	15	5	46	50	219
Sy	5	30	36	37	15	5	46	50	219
Ab	0	0	0	0	0	0	0	0	0
Ar	8.5	50	60	62	26	9	78	85	370
TOTAL	---	342	412	424	175	59	533	579	2524

P = normalized score

Table 47: Assessment of Septic Tank System Areas by Waste-Soil-Site Interaction Matrix Methodology

Area (Underlying Aquifer	Assessment Score	Annual Wastewater Flow (10^6 gal/yr)
[x]Sunvalley Acres (ER)	2641	3
Crutcho (T, G-W)	2556	11
[x]Arcadia (G-W)*	2524	8
[x]Seward (G-W)	2479	175
Arrowhead Hills (G-W)	2379	9
Green Pastures (T, G-W)	2366	44
[x]Midwest City (G-W)	2363	228
Nicoma Park (T, G-W)	2319	57
Forest Park (G-W)	2310	27
Mustang (ER)	2288	67
Silver Lake Estates (H)	2203	6
Del City (G-W)	2030	5
East Norman (T, G-W)	2005	152

[x]Denotes sampling conducted in area.

*G-W = Garber-Wellington, T = terrace deposits, ER = El Reno group, H = Hennessey group.

Table 48: Comparison of Rank Order of Septic Tank System Areas

Adjusted Surface Impoundment Assessment Methodology	Adjusted Waste-Soil-Site Interaction Matrix Methodology
[x]Midwest City (1)	[x]Midwest City (1)

Table 48: (continued)

Adjusted Surface Impoundment Assessment Methodology	Adjusted Waste-Soil-Site Interaction Matrix Methodology
[x]Seward	[x]Seward
East Norman	East Norman
Nicoma Park	Mustang
Green Pastures	Nicoma Park
Mustang	Green Pastures
Forest Park	Forest Park
Crutcho	Crutcho
Arrowhead Hills	Arrowhead Hills
[x]Arcadia	[x]Arcadia
Del City	Silver Lake Estates
Silver Lake Estates	Del City
[x]Sunvalley Acres (2)	[x]Sunvalley Acres (2)

[x]Denotes ground water sampling conducted.

(1) Highest ground water pollution potential.

(2) Lowest ground water pollution potential.

(1) The final ranking of the 13 septic tank system areas is largely dependent upon the annual wastewater flow in the area, and this is directly related to the number of persons and septic tank systems in the area.

(2) Both the surface impoundment assessment method and the waste-soil-site interaction matrix can be used to develop a priority ranking of existing or planned septic tank system areas. Since the surface impoundment assessment method has 6 items of needed information versus 17 items in the interaction matrix, the SIA method is easier to use. However, it should be noted that neither methodology accounts for wastewater flow, and this

is an important factor which should be given consideration in the use of either method for septic tank system areas.

(3) A methodology specifically developed for septic tank system areas would be useful. The methodology could use some factors from both the SIA method and the interaction matrix, and should include some additional factors such as wastewater flow, percolation rate, septic tank density, and average life of septic tank systems.

As part of the case study reported herein, a modest field sampling program was conducted to evaluate the pollution potential predictions for 4 of the 13 septic tank systems areas in the study area. The program consisted of locating 11 existing wells in the 4 areas, pumping the wells for several minutes, and then collecting one-liter samples. Field measurements included pH, salinity, and conductivity. Subsequent laboratory analyses were performed for orthophosphates, total phosphorus, Kjeldahl (organic) nitrogen, nitrate-nitrogen, alkalinity, hardness, and TDS. Specific areas monitored during the program are identified in Table 48. Two areas had high ground water pollution potential (Midwest City and Seward), and two had lower potential (Arcadia and Sunvalley Acres). The following criteria were used in interpreting the key analytical results (U.S. Environmental Protection Agency, July 1976):

pH--value should be between 6.5 - 8.5

Orthophosphate--4 mg/l represents weak domestic sewage

Total phosphorus--6 mg/l represents weak domestic sewage

Nitrate nitrogen--10 mg/l is drinking water standard

TDS--500 mg/l is drinking water standard

Eleven wells were sampled in the four septic tank system areas, and the results are in Table 49. All wells were within the septic tank system areas, and none can be considered as background wells. Seven of the 11 wells exceeded the Oklahoma nitrate-nitrogen standard; four wells exceeded the USPHS TDS standard, and nine wells had organic phosphorus concentrations of greater than 1 mg/l (weak domestic sewage has 2 mg/l). Therefore, ground water contamination appears to be occurring in each of the four septic tank system areas. In terms of rank ordering of the areas and considering nitrates only, the average concentrations for the wells sampled were as follows:

Midwest City	25 mg/l (3 wells)
Seward	39 mg/l (3 wells)
Arcadia	13 mg/l (3 wells)
Sunvalley Acres	9 mg/l (2 wells)

Table 49: Well Samples and Analysis for Septic Tank System Areas

	LOCATION										
	Arcadia			Seward			Midwest City			Sunvalley Acres	
Parameter \ Well Number	1	2	3	9	10	11	25	26	27	18	19
pH	7.45	7.75	8.00	7.60	7.70	7.55	7.30	7.55	7.35	7.60	7.65
Salinity (ppt)	<1	<1	<1	0	<1	0	0	0	0	<1	<1
Conductivity hos/cm)	720	705	750	500	800	550	375	500	450	975	925
Orthophosphate (mg/ℓ)	5.2	1.0	1.0	2.1	0.2	2.1	2.55	1.60	2.55	1.2	0.2
Total Phosphorus (mg/ℓ)	5.4	1.8	3.0	5.0	2.0	4.0	8.00	6.00	6.50	3.4	3.4
Total Kjeldahl Nitrogen (mg/ℓ)	1.4	<1	<1	1.12	0	2.8	1.12	0.56	0	0.28	0.15
Nitrate-N (mg/ℓ)	30.1	3.3	5.2	17.6	83.7	16.8	18.0	37.6	20.5	9.7	7.9
Alkalinity (mg/ℓ as $CaCO_3$)	349	343	376	240	285	200	149	201	217	350	345
Hardness (mg/ℓ as $CaCO_3$)	316	322	362	256	330	240	150	212	220	322	374
Total Dissolved Solids (mg/ℓ)	565	478	556	382	565	273	306	376	187	422	529

Depth (meters)	<30	<30	65	25	35	20	27	45	45	31	30
Distance (meters) and Direction from Source	8 -	12 -	25 -	25 W	20 W	27 W	25 -	25 -	25 -	30 N	25 S

As noted earlier, Midwest City and Seward were considered to have higher ground water pollution potential than Arcadia and Sunvalley Acres. Both Midwest City and Seward exhibited higher nitrates in ground water than did Arcadia and Sunvalley Acres. However, it is stressed that this was a cursory sampling program, and a more extensive and systematic field sampling program would need to be conducted to confirm the general assumptions of the utilized empirical assessment methodologies.

HANTUSH ANALYTICAL MODEL

The Hantush analytical model was developed to determine the rise and fall of the water table under circular, rectangular, or square recharge areas; it does not address ground water quality (Hantush, 1967). A septic tank system serving an individual home can be considered as a rectangular recharge area since it introduces septic tank effluent into the soil through a subsurface drain system. A larger septic tank system serving up to several hundred homes can also be considered as a rectangular recharge area due to its subsurface drain system. The assumptions basic to the Hantush model are:

(1) the aquifer is homogeneous, isotropic, and resting on a horizontal impermeable base;

(2) the formation coefficients are constant in time and space; and

(3) the constant rate of deep percolation relative to the hydraulic conductivity is so small that the vertically downward percolation is almost completely refracted in the direction of the tilt of the water table.

The rise of the water table in response to a vertically downward uniform rate of recharge that is supplied from a rectangular area can be estimated with the following equation (Hantush, 1967):

$$h_m^2 - h_i^2 = \frac{W_m \bar{m} t}{15 S_y} \left\{ W^*\left[1.37 (b_m + x)\sqrt{\frac{S_y}{Tt}},\ 1.37(a_m + y)\sqrt{\frac{S_y}{Tt}}\right] \right.$$

$$+ W^*\left[1.37 (b_m + x)\sqrt{\frac{S_y}{Tt}},\ 1.37 (a_m - y)\sqrt{\frac{S_y}{Tt}}\right]$$

$$+ W^*\left[1.37 (b_m - x)\sqrt{\frac{S_y}{Tt}},\ 1.37 (a_m + y)\sqrt{\frac{S_y}{Tt}}\right]$$

$$\left. + W^*\left[1.37 (b_m - x)\sqrt{\frac{S_y}{Tt}},\ 1.37 (a_m - y)\sqrt{\frac{S_y}{Tt}}\right] \right\}$$

where

h_i = initial height of water table above aquiclude, in feet

h_m = height of water table above aquiclude with recharge, in feet

W_m = recharge rate, in gpd per unit area

m = $0.5 (h_i + h_m)$, in feet

t = time after recharge starts, in days

S_y = specific yield of aquifer, fraction

b_m = one-half width of recharge area, in feet

x,y = coordinates of observation point in relation to center of recharge area, in feet

T = coefficient of transmissibility, in gpd/ft

a_m = one-half length recharge area, in feet

The function of W* is defined by

$$W^* (\alpha,\beta) = \int^{1} \text{erf} \frac{\alpha}{\sqrt{T}} \text{ erf } \left(\frac{\beta}{\sqrt{T}}\right) dT$$

where $\alpha = 1.37 (b_m \pm x) \sqrt{\frac{S_y}{Tt}}$

$$\beta = 1.37 (a_m \pm y) \sqrt{\frac{S_y}{Tt}}$$

The Hantush analytical model was developed based on the assumption of a uniform recharge rate. The effluent from a septic tank system is not uniform; however, the results from applying the model to septic tank systems can be considered conservative, i.e., the actual rise of the water table will be less than or equal to that predicted by the Hantush model.

To illustrate the application of the Hantush analytical model an example will be presented for an individual, mound-type septic tank system. This example was chosen since mound-type systems are used in areas with high water tables, thus the water table rise would be of particular concern. The data used in the following example are hypothetical, however, they are typical for the Wisconsin area where mound-type septic tank systems are used (Harkin et al., 1979).

Problem: A 3-bedroom home with a daily wastewater load of 450 gallons is to use a mound-type septic tank system. Using the recommended loading rate of 1.2 gal/day/ft^2, the derived bottom area of the mound is calculated to be 375 ft^2. Using a 20 ft by 20 ft square mound system, calculate the water table rise under the system. The mound system will be located on low permeable (P = 100 gal/day/ft^2) sand, 2 ft above the water table that extends down 5 ft. The specific yield of the aquifer is assumed to be 0.1; and the coefficient of transmissibility is assumed to be 500 gal/day/ft. Calculate the rise in the water table after 3 days at distances up to 20 ft downgradient from the system recharge area. The key data for solving the problem is summarized in Table 50.

Solution: A seven step procedure for using the Hantush analytical model has been developed and is listed in Table 51 (Kincannon, 1981). Applying this procedure to the data for the example problem shown in Table 50 yields the calculated water table rises as shown in Table 52.

Table 50: Data for Example Problem Using Hantush Analytical Model

Parameter		Value
Recharge Rate (W_m)	--	1.2 gal/day/ft^2
Time after recharge (t)	--	3 days
Specific Yield (S_y)	--	0.1
1/2 width of recharge area (b_m)	--	10 ft
1/2 length of recharge area (a_m)	--	10 ft
Transmissibility (T)	--	500 gal/day/ft
Y coordinate of observation well	--	0 ft
X coordinate of observation well	--	from 0 to 20 ft

Table 51: Calculation Procedure for Hantush Analytical Model (Kincannon, 1981)

Step 1. Collect the following data:

(A) Recharge rate (W_m)

(B) Net time after recharge starts (t)

(C) Specific yield of aquifer (S_y)

(D) One-half width of recharge area (b_m)

Table 51: (continued)

(E) One-half length of recharge area (a_m)

(F) Coefficient of transmissibility (T)

(G) Coordinates of observation point in relation to the center of recharge area (x,y)

Step 2. Calculate α, and β.

(A) $\alpha_1 = 1.37\ (b_m + x)\sqrt{\frac{S_y}{Tt}}$

$\alpha_2 = 1.37\ (b_m - x)\sqrt{\frac{S_y}{Tt}}$

(B) $\beta_1 = 1.37\ (a_m + y)\sqrt{\frac{S_y}{Tt}}$

$\beta_2 = 1.37\ (a_m - y)\sqrt{\frac{S_y}{Tt}}$

Step 3. Obtain $W^*\ (\alpha_1, \beta_1)$ from the Tables in Appendix D.

Step 4. Repeat steps 2 and 3 for

$W^*\ (\alpha_1, \beta_2)$, $W^*\ (\alpha_2, \beta_1)$ and $W^*\ (\alpha_2, \beta_2)$.

Step 5. Calculate the rise in water table

$(H_m - h_i)$ by multiplying the total of Steps 3 and 4 by $\frac{W_m\ t}{30\ S_y}$.

Step 6. Repeat the above steps for varying values of x and y until a rise in the water table of 0.5 ft is achieved.

Step 7. Plot the rise in the water table versus the distance to the observation well.

Table 52: Water Table Rise Under Mound-type Septic Tank System

Observation Well Coordinates (ft) x	y	Water Table Rise* (ft) $(h_m - h_i)$
20	0	.19
16	0	.23
12	0	.26
8	0	.31
4	0	.34
2	0	.35
0	0	.375 (.552[a]) (.656[b])

*after 3 days unless otherwise noted

a = after 50 days

b = after 100 days

As seen from Table 52, the maximum water table rise after 3 days will be 0.375 ft (4.5 in). An increase in the time after recharge of more than 30-fold (to 100 days) does not even double the rise in the water table. This suggests that the water table rise will approach some equilibrium value somewhere around 3 or 4 months after initiation of discharge. Although the water table rise in any case only approaches a maximum of 8 in, this becomes a significant rise in view of the fact that mound systems are used in areas of high water tables. Actual loadings by septic tanks will be intermittent which will decrease the actual rise of the water table, however increases in loading rates (either by malfunctioning or overloaded systems) could increase the water table rise.

KONIKOW-BREDEHOEFT NUMERICAL MODEL

The Konikow-Bredehoeft (K-B) numerical model was applied to a septic tank system study area near Edmond, Oklahoma, to determine its usefulness in predicting nitrate concentrations in ground water from this source type. The K-B model is a two-dimensional solute transport model which has been used in the analysis of ground water pollution from a variety of source types. The K-B model, which exists as a packaged program available for the user, solves both the flow equation and the solute transport equation. A general discussion of these two equations is as follows.

The equation describing the transient ground water flow in two dimensions (areal flow) for an inhomogeneous anisotropic confined aquifer may be written as follows (Bredehoeft and Pinder, 1971):

$$\frac{\partial}{\partial x}\left(T_{xx}\frac{\partial h}{\partial x}\right) + \frac{\partial}{\partial y}\left(T_{yy}\frac{\partial h}{\partial y}\right) = S\frac{\partial h}{\partial t} + W(x,y,t) \qquad (1)$$

where

x and y are the coordinate directions,

T_{xx} and T_{yy} are the transmissivity in x and y direction, respectively (L^2T),

h = h(x,y,t) and is the hydraulic head (L),

S is the storage coefficient of the aquifer (dimensionless),

t is the time (T), and

W(x,y,t) is the volumetric flux of recharge or withdrawal per unit surface area of the aquifer (L/T).

Expanding the second term on the right side to reflect possible withdrawal or recharge and steady state leakage, the flow equation can now be expressed as:

$$\frac{\partial}{\partial x}\left(T\frac{\partial h}{\partial x}\right) + \frac{\partial}{\partial y}\left(T\frac{\partial h}{\partial y}\right) = S\frac{\partial h}{\partial t} + Q(x,y,t) - \frac{K_s}{m}(H_s-h) \qquad (2)$$

where the second term on the right, X(x,y,t) (LT^{-1}), is the direct withdrawal or recharge, such as pumpage from a well, well injection, precipitation, or evapotranspiration. The third term on the same side, $\frac{K_s}{m}(H_s-h)$, shows a steady state leakage bed, in which K_s is the vertical hydraulic conductivity of the confining layer, stream bed, or lake bed (LT^{-1}), m(L) is the thickness of the confining layer, stream bed, or lake bed, and H_s is the hydraulic head in the source bed, or lake. T = T(x,y) is the transmissivity (L^2/T).

The dispersion of a tracer in fluid flow through saturated homogeneous porous media (solute transport) may be described by the differential equation as (Khaleel and Reddell, 1977):

$$\frac{\partial C}{\partial t} - \frac{\partial}{\partial x_i}(V_iC) = \frac{\partial}{\partial x_i}\left(D_{ij}\frac{\partial C}{\partial x_j}\right) \qquad i,j = 1,2,3 \qquad (3)$$

where

C is the tracer concentration (M/L^3),

D_{ij} is the coefficient of hydrodynamic dispersion (L^2/T), (a second-order tensor),

V_i is the component of velocity vector (L/T),

i and j subscripts are used to denote tensor, where i and $j = 1,2,3$, and

x_i, x_j are the Cartesian coordinates (L).

The differential equation used to provide a model for studying ground water pollution patterns in a given aquifer system is a combination of two equations: (1) ground water flow (2), and (2) convection-dispersion (3). As mentioned previously, these two equations must be solved simultaneously because both convective transport and hydrodynamic dispersion are functions of the ground water flow velocity. After the ground water flow velocity is obtained from the head distribution, it is used as an input parameter in the solute-transport model. The following is the combination of two equations, (1) and (3), which describe the two-dimensional solute-transport in a transient ground water flow (Konikow and Grove, 1977):

$$\frac{\partial C}{\partial t} = \frac{1}{b}\frac{\partial}{\partial x_i}\left(bD_{ij}\frac{\partial C}{\partial x_j}\right) - V_i\frac{\partial C}{\partial x_i} + \frac{C\left[S\frac{\partial h}{\partial t} + W - n\frac{\partial b}{\partial t}\right] - C'W}{nb} \qquad i,j=1,2 \tag{4}$$

where

C is the concentration of the dissolved chemical species in the aquifer (M/L^3),

t is the time (T),

b is the saturated thickness of the aquifer (L),

D_{ij} is the coefficient of hydrodynamic dispersion (a second order tensor) (L^2/T),

V_i is the velocity component in the x and y direction (L/T),

h is the hydraulic head (L),

W is the sink/source term (L/T),

n is the effective porosity (dimensionless),

C' is the concentration of the dissolved chemical in a source (when it gets into ground water) or sink fluid (M/L^3), and

S is the storage coefficient (dimensionless).

The key to all the available ground water models is to represent Equation 4 in a finite difference form, that is, to approximate the partial derivatives with finite differences between two points. If this operation is applied to a collection of points (represented by a grid) in the area of interest, a set of simultaneous equations results. The various models usually differ in their approach to solving this set of equations. The terms W and C^1 are characteristics of the septic tanks that will have to be determined and inserted into the model as input. As noted previously, these two terms are not always readily available and may represent a significant detriment to the use of numerical models for septic tank systems.

As noted earlier, the K-B model exists as a packaged computer program available in Fortran IV from the U.S. Geological Survey (Konikow and Bredehoeft, 1978). A complete listing of the program and definitions of selected program variables is in Appendix E. The K-B model was used to determine the feasibility of modeling the effects of septic tank systems on ground water quality by direct application to an existing situation. The scope of this analysis involved three phases. First, all available information concerning a selected area of intense septic tank use and its underlying aquifer had to be gathered. Second, any information gaps had to be identified and filled as accurately as possible through the use of estimates or assumptions. Third, the information and data gathered on the study area was used as input to a model for predicting the long term effects of the septic tank systems on the ground water quality (Sohrabi, 1980).

Study Area Near Edmond, Oklahoma

The area selected for this study is located from Latitude 35°36' to 35°42'15", and from Longitude 97°23'15" to 97°27'30", in T14N, T13N, R2W, Edmond, Oklahoma County (see Figure 22). The east and west boundaries of the study area are the Edmond City limits both east and west of I-35 (about two miles west and east); the north boundary is one and one-half miles from the Logan County line; and the south boundary is one-half mile south of Memorial Road. The study area includes about 28 square miles. Homes are located on one-half to one-acre lots; approximately 17,000 people live in the selected area, with 73 percent served by septic tank systems and individual wells.

This area was chosen because: (1) it has been classified as having concentrated areas of septic tanks which are potential sources of nitrate contaminants into the Garber-Wellington aquifer; (2) the hydrogeology of the area is fairly well understood; and (3) the area is in the outcrop region of the Garber-Wellington aquifer. Being located in the outcrop region places the study area in a recharge zone for the Garber-Wellington aquifer, thus the septic tank systems represent a potential threat to the ground water quality. A significant portion of Edmond, especially the east side of the city, is served by septic tank systems. The use of the septic tank systems raises some significant issues which must be addressed in water quality management planning because of the importance of ground water in future

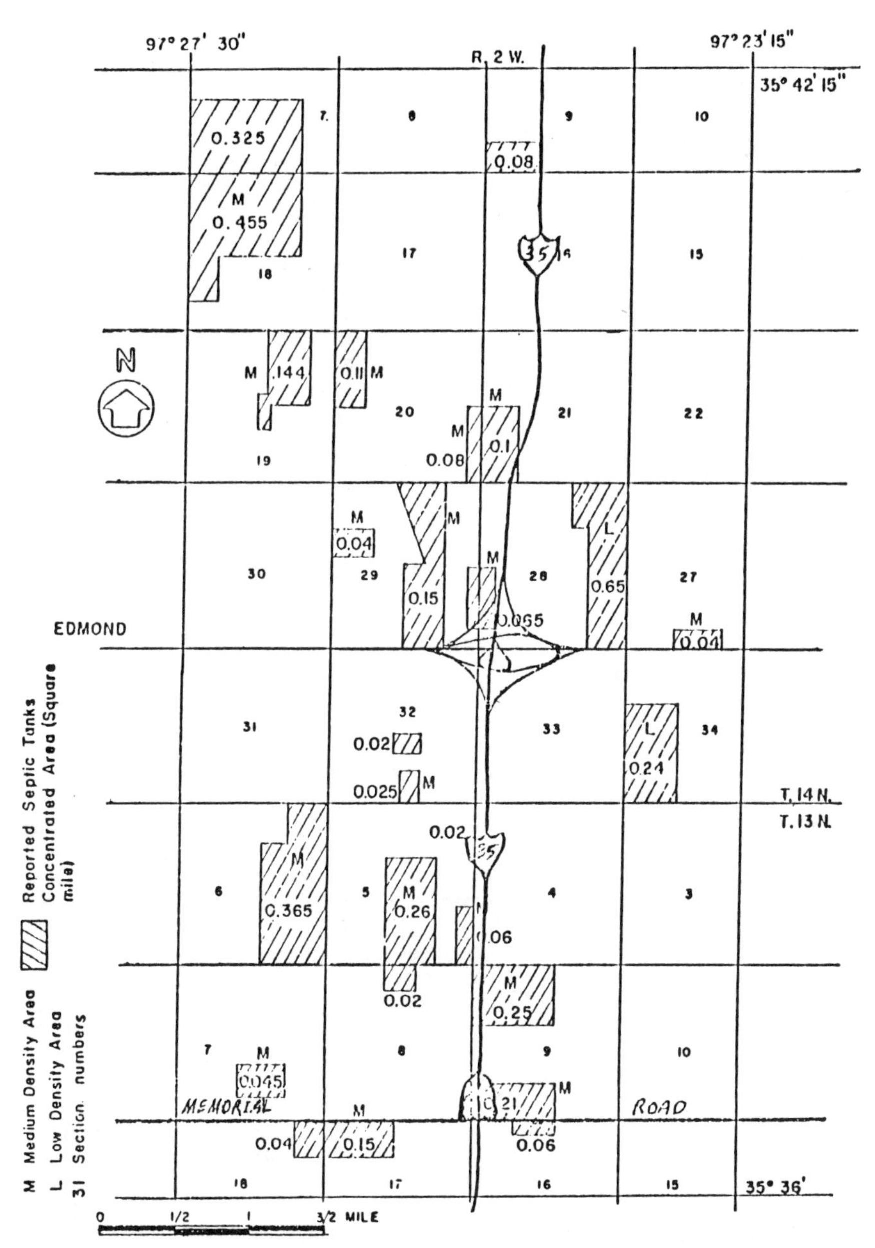

Figure 22: Map Shows Residential Areas Which Are Served By Septic Tank Systems in Modeled Area (areas showing reported septic tanks have not been drawn to scale)

development of the region. In Edmond, 100 percent of the needed water is supplied by well water. The main sources of nitrate contamination in the study area are from septic tank systems.

The nitrate levels of the deeper artesian (confined) aquifer in the Edmond area range from about 1.0 to 14.3 mg/l (measured in 1971). High nitrate concentrations (18 mg/l NO_3) have been reported near the Arrowhead Hills development in the study area. The Arrowhead Hills development has a number of septic tank systems. While 18 mg/l nitrates is not a dangerous concentration, it could be an indication of the beginning of nitrate contamination of the water table aquifer by septic tank systems. If the trend toward housing additions with septic tanks is continued, it is possible that health effects from such contamination will be experienced.

At present, a large percentage of the public water supply in Edmond is obtained from the confined portion of the Garber-Wellington aquifer. However, future growth will necessitate greater use of the unconfined portion as the confined portion is pumped beyond safe and economical limits. Edmond is now using almost 20 percent of the water from the unconfined portion, and this is going to increase as a result of city development toward the east side. Areas of potentially high nitrate concentration due to present or proposed densities using septic tank systems are shown in Figure 22.

Hydrogeology of Study Area

As shown in Figure 23, the study area is underlain everywhere by the Garber Sandstone and Wellington Formation, which have a combined maximum thickness of about 340 ft as determined by geophysical logs. The Garber and Wellington constitute a single aquifer, or water-bearing unit. The regional dip is 30-35 ft/mi westward and southward toward the Anadarko Basin (Wood and Burton, 1968). These two units, Permian in age, were deposited under similar conditions, and both consist of lenticular beds of sandstone, siltstone and shale that may vary greatly in thickness within short lateral distances (see Figure 24). The two units have similar hydrologic properties and are hydrologically interconnected (Carr, 1977). The sandstone layers are fine to very fine-ground and loosely cemented and crumble easily. None of the sand in the Garber and Wellington is coarser than 0.35 mm (millimeter), and the average diameter of the grains is 0.155 mm. The sandstone is composed almost entirely of subangular to subrounded fragments of fine-grained quartz (Wood and Burton, 1968).

The study area overlies an aquifer outcrop characterized by rolling, steep-sided hills that are forested with scrub oak and other small, slow-growing deciduous trees (Wood and Burton, 1968). The Hennessey Group, which overlies the Garber-Wellington aquifer in the western part of the study area as shown in Figure 24, consists of shale, siltstone, and thin beds of very fine grained sandstone. In general, the unit thickens to the west and south. The thickness about 4 miles west of the study area is about 90 ft, but it increases to about 400 ft at the Canadian County line (Carr and Marcher, 1977).

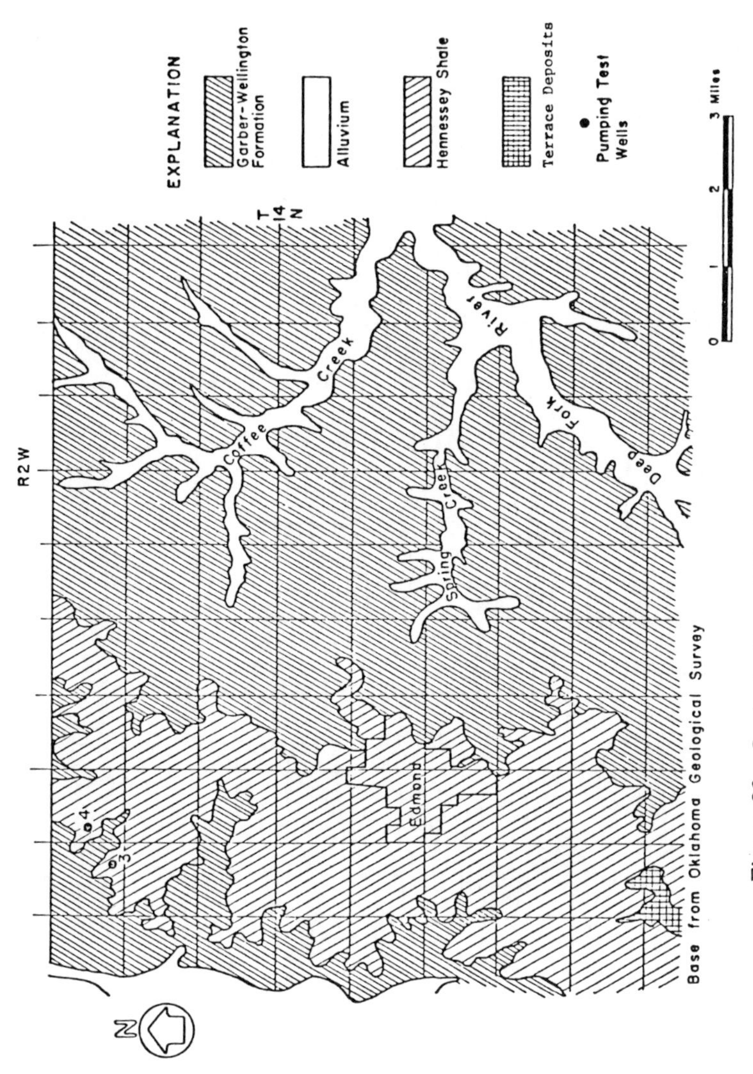

Figure 23: Geologic Map of Modeled Area

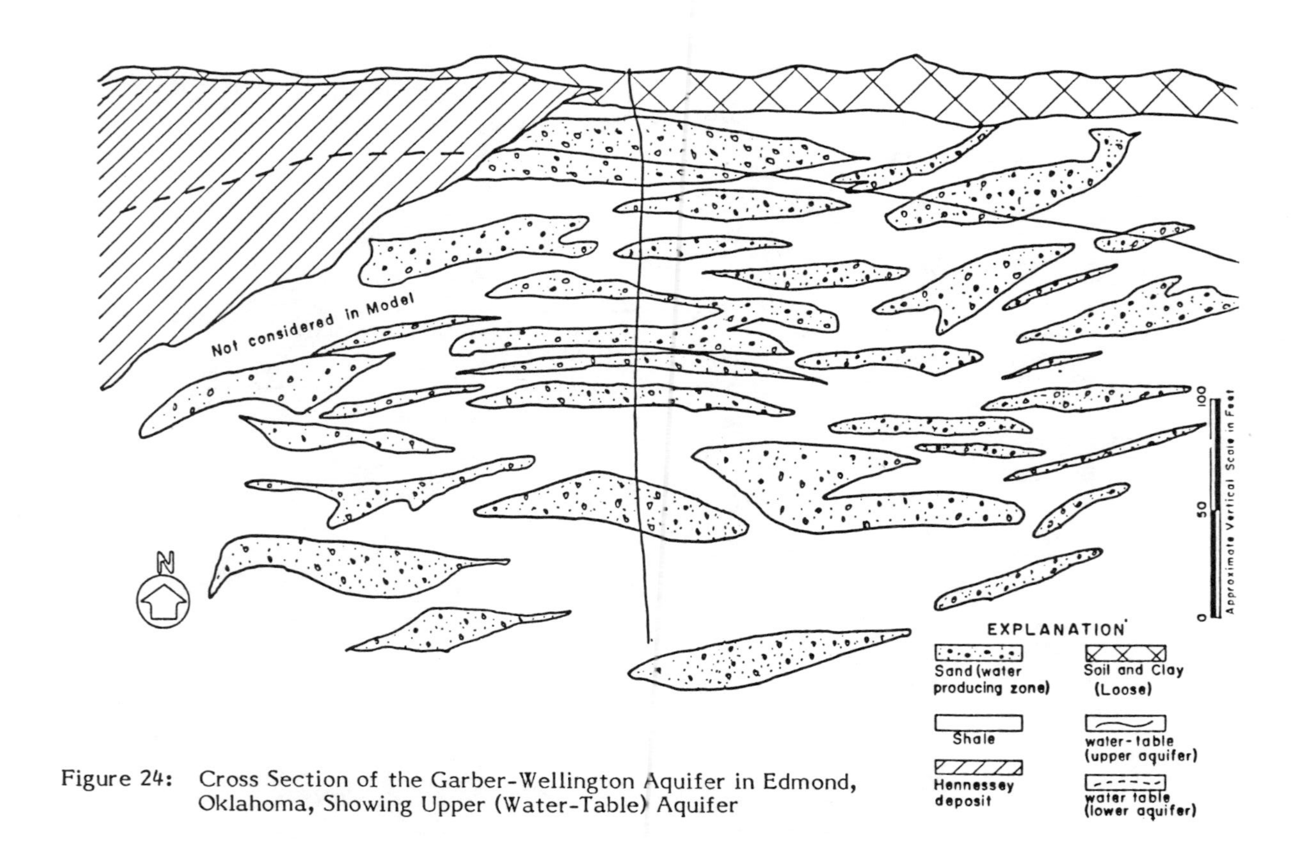

Figure 24: Cross Section of the Garber-Wellington Aquifer in Edmond, Oklahoma, Showing Upper (Water-Table) Aquifer

The Hennessey-Garber contact is a plane between the two formations which separates them from each other in the region. This contact can be a determining line for classifying aquifers. The available data from well logs suggest that the upper part of the aquifer is not saturated in a belt of about 400 miles west of and parallel to the Hennessey-Garber contact (Carr and Marcher, 1977). As a result, water in this belt is under water table or semi-artesian conditions as shown in Figure 23. West of this belt, the aquifer is fully saturated and the Hennessey serves as a confining layer so that artesian conditions prevail. But from the contact toward the east, unconfined conditions prevail at depths of less than 250 ft where the aquifer is exposed at the surface (Carr and Marcher, 1977). In addition, wells drilled east of the contact encounter water under water-table conditions. Most wells in the Edmond study area, where the aquifer crops out at the surface, are drilled to depths ranging from 600 to 750 ft. The thickness of the saturated portion of the aquifer in the study area, which was determined by examining the geophysical logs of water wells, ranges from about 50 to 215 ft in the area as shown in Figure 25. The hydrology of the Garber-Wellington aquifer is not that of a continuously uniform saturated body of rock, each with its own capacity to store and transmit water to wells. Almost one-third of the thickness has a significant role in transmitting water (Engineering Enterprises, Inc., data on file, 1980). This thickness is from the first effective shale layer, which is assumed to be an impermeable layer for the selected upper water-table aquifer, to the phreatic surface.

Water in the upper part of the aquifer has two components of movement: (1) lateral movement from areas of recharge to points of discharge, which is the principal component; and (2) vertical movement downward due to differences between the piezometric and potentiometric heads of the confined and unconfined zones, respectively. The rate of downward movement is probably very slow under natural conditions in most places because the upper and lower parts of the aquifer are interconnected by a shale bed of low hydraulic conductivity (Carr and Marcher, 1977).

About one mile of the Deep Fork River in the southwest corner of the Edmond study area and its two tributaries, Coffee and Spring Creeks, drain the surface waters in the area. Coffee and Spring Creeks are seasonal streams and most of the year are dry. The Deep Fork River had an average of 30.9 cfs discharge at the Arcadia gaging station in the spring of 1977. Almost all of the water discharged in it is from industrial manufacturers and wastewater treatment plants.

Input Data for Model

The structure of the Konikow-Bredehoeft (K-B) model is such that the flow equation is solved by employing a finite-difference approximation to the partial differential equation and an alternating direction implicit procedure for solving the resulting simultaneous equations. The mass transport equation is solved in two parts: (1) first, the effects of convective transport are evaluated using the method of characteristics; and (2) the effects of hydrodynamic dispersion are evaluated using a finite-difference scheme. The structure of the K-B model is such that the outermost nodes of

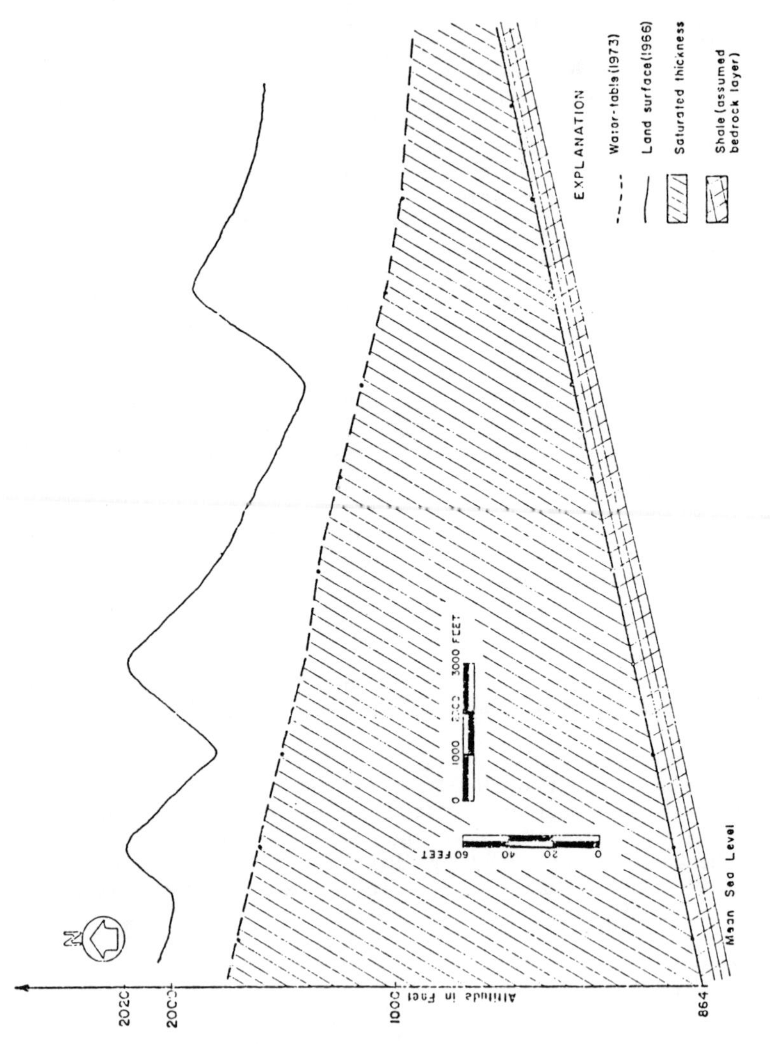

Figure 25: Diagrammatic West-East Cross-section of Modeled Area Showing Land Surface and Saturated Thickness of Upper Part of Garber-Wellington Aquifer Above Assumed Layer for this Study

the grid approximating the aquifer are designated as "constant head" or "no-flow". Sources of constant recharge and constant solute concentration can also be designated. The model also allows two direction hydraulic conductivities (K_x and K_y) and dispersivities (D_L and D_T) to be specified. The input information needed for the K-B model, and the approaches or assumptions associated with obtaining the information, is delineated in the following subsections.

Hydraulic Conductivity

The hydraulic conductivity of the aquifer was assumed to be constant and uniform over time and space at a value of 15 gallons per day per square foot. This constant value was determined from pump tests performed in the area in 1979.

Transmissivity, Aquifer Thickness and Water Table Elevation

The transmissivity values of the aquifer were calculated within the model by multiplying the hydraulic conductivity by the saturated thickness of the aquifer. The saturated thickness of the aquifer was determined from analysis of seven available drillers' logs. The thickness of the aquifer was computed as about 214 ft on the west side, and was assumed to decrease at a uniform rate to about 70 ft on the east side of the study area. Water table elevations were obtained from an interpolated potentiometric surface plotted by Carr and Marcher (1977).

Specific Yield

Little information is available on the specific yield of the Garber-Wellington aquifer; therefore, it was estimated from the Upper Permian Rush Springs Sandstone which is similar to the study aquifer. The conservatively estimated value of specific yield for the Garber-Wellington aquifer is 0.20 (Carr and Marcher, 1977). The specific yield determined from 32 analyses in the similar aquifer has an average value of 0.22.

Effective Porosity

There was no data available to describe this parameter in the study area. Therefore, the effective porosity was assumed to be 0.35 based on estimates provided by the U.S. Geological Survey.

Dispersivity

The value of the longitudinal dispersivity (100 meters) and the ratio of transverse to longitudinal dispersivities (0.33) were estimated on the basis of review of available literature. No determinations of this parameter (by either field or laboratory analysis) have been done in the Edmond study area.

Recharge and Discharge

The most important source of recharge in the study area is precipitation. The estimated percolation into the ground water basin is about 10 percent of the annual precipitation, or 3.6 in/yr (Carr and Marcher, 1977). This value is assumed to be uniform over the study area despite the fact that parts of the area (streets, highways, impervious top soil, etc.) do not contribute to the natural replenishment of the aquifer. According to the Association of Central Oklahoma Governments and the Edmond Water Department, there are no intensive agricultural activities in the study area; therefore, no return flow from irrigation sources was considered.

The other source of recharge is return flow from septic tanks; however, this is minimal from a quantitative viewpoint, in comparison to precipitation. Evapotranspiration losses (annually 87 in from a Class A Pan) from soil absorption field systems which lie under the ground surface have a significant role in decreasing effluent percolation through the soil column. Septic tank system effluents are subject to two processes in the subsurface environment: (1) part of the applied septic tank effluent is consumed by crops or grasses and by evapotranspiration, and (2) part percolates below the root zone. A large part of the cumulative percolate may be stored in the unsaturated zone, and the recharge of the zone of saturation by this effluent is very small (this value is estimated as 1 percent based on information obtained from the State Health Department and published literature). As mentioned previously, this amount exerts only a small effect from a quantitative viewpoint; however, it does have significant effect from a qualitative viewpoint.

Presently, the principal means of discharge from the aquifer in the study area is believed to be pumping from seven domestic water supply wells; the rates have ranged from approximately 150 to 350 gpm (ACOG and Oklahoma Water Resources Board, data on file, 1980). The other means of discharge from the aquifer is believed to be a segment of the Deep Fork River which serves as a gaining reach. Specific discharge (the volume of water flowing per unit time through a unit cross-sectional area) through this natural boundary was estimated as 1.052×10^{-9} ft/s by use of Darcy's Law, $q = K \frac{\partial h}{\partial l}$, where q is the specific discharge, K is the hydraulic conductivity, and $\frac{\partial h}{\partial l}$ is the hydraulic-head gradient. The area of the stream bed was less than the area of the stream bed nodes by a ratio of 1 to 100, and it was considered in the calculation of the discharge. The Edmond study has some houses with individual wells having yields ranging from about 30 to 50 gpd. The reported annual pumpage, assuming pumps are in operation 20 hours per day, was used in the K-B model for domestic water supply wells instead of using the reported yield (Edmond Water Department, data on file, 1980).

Another means by which ground water can leave the aquifer is by evaporation, but this process has an insignificant effect because of the relatively low water table in the investigated area.

Concentration of Sinks and Sources

No nitrate data for septic tank effluents existed for the Edmond study area. Therefore, published literature was used to approximate the concentration of nitrates in septic tank system effluents when it reaches the ground water. It is assumed that only nitrates leaching from dense numbers of septic tank systems may have a significant future impact on the ground water in the Edmond study area. Nitrogen in the effluent from septic tanks is primarily in the form of organic nitrogen (25 percent) and ammonium ions (NH_4^+) (75 percent) (Peavy, 1978). Several studies have been conducted which show that the range of ammonia-nitrogen and nitrate-nitrogen in septic tank effluents is between 77-111 mg/l and 0.00-0.10 mg/l, respectively (the mean values were 97 mg/l and 0.026 mg/l) (Viraraghavan, 1976).

The major part of the nitrogen reaching the water table from septic tanks is in the form of negatively charged nitrate ions (NO_3^-). Figure 26 displays the sources, transformations and pathways of nitrogen in the subsurface environment. Ammonia-nitrogen is converted to nitrate when aerobic conditions exist. Since nitrates are quite soluble they remain dissolved in the water as they percolate through the soil.

The transit time of pollutants from the land surface to the ground water table was developed for the percolation of irrigation water in the sandstone coastal aquifer of Israel (Mercado, 1976). This aquifer has a total thickness of 51 ft and consists of sand and calcareous sandstone layers of Plio-Pleistocene age, intersected with clay and loam layers. The transit time formula for irrigation water pollutants may be written as (Mercado, 1976):

$$T \simeq \frac{L\sigma}{R + \delta q_{ir}} \tag{5}$$

where

T is the retention time of pollutants in the unsaturated soil column (T),

L is the average depth of the water table below the land surface (L),

σ is the relative volumetric moisture content (dimensionless),

R is the precipitation rate (L/T),

q_{ir} is the irrigation rate (L/T), and

δ is the return flow ratio (dimensionless).

To approximate the retention time for septic tank effluent percolation into ground water, equation (5) is modified as:

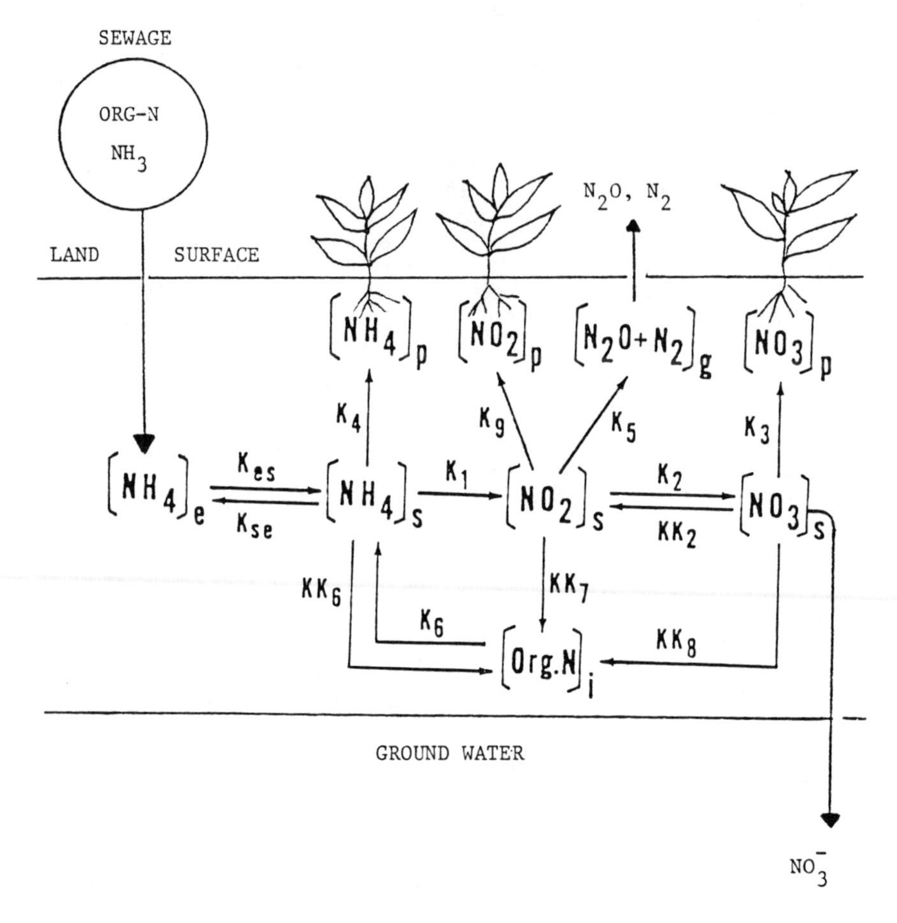

Note: Subscripts K and KK denote rate constants, while subscripts e, s, p, i and g respectively refer to exchangeable, solution, plant, immobilized, and gaseous phase.

Figure 26: Possible Transformations and Pathways of Nitrogen from Septic Tank Systems (Tanji and Gupta, 1978; and Freeze and Cherry, 1979)

$$T \simeq \frac{L\sigma}{R + q_s\varepsilon} \tag{6}$$

where

q_s is the septic tank effluent (L/T), and

ε is the evapotranspiration coefficient.

The transit time for septic tank effluent is calculated on the basis of the following information from the Edmond study area:

L = 42 ft for the node at i = 6, j = 16

σ = 12 percent (Fisher et al., 1969)

R = 36.13 in/yr

q_s = 6.09 in/yr (based on calculation)

ε = 40 percent (estimated value)

According to equation (6) and the above information, the septic tank effluent as influenced by precipitation takes about 1.6 years to reach the ground water. In Mercado's (1976) study concerning nitrate pollution of aquifers, he assumed the relationship between nitrogen quantities released in the surface and nitrogen quantities reaching the water table to be linear. The assumed linear relationship between potential nitrogen on the surface and actual contributions to the aquifer is expressed by

$$\text{sewage contribution} = \beta\,(SWG) \tag{7}$$

where

is the linear proportion coefficient for sewage contribution ($\beta \leq 1$) (dimensionless), and

SWG is the sewage disposal concentration (mg/l).

The following estimation of the concentration of the septic tank effluent when it reaches the ground water is based on the Mercado studies on nitrate pollution in aquifers in Israel in 1976. The mean value of ammonia-nitrogen in the septic effluent is 97 mg/l, and the β value is assumed to be 70 percent for the study area; therefore, according to equation (7), the nitrogen contribution to the ground water will be 67.9 mg/l (NH_4^+-N). In order for ammonia-nitrogen to be converted to nitrate (NO_3^-), the following chemical reactions will occur (assuming all the contributed ammonia-nitrogen will be converted to nitrates):

$$NH_4^+ \rightarrow NO_2^- \rightarrow NO_3^- \text{ (nitrification process)}$$

The resulting concentration was 300 mg/l (NO_3^-). Once again, according to the Mercado studies, nitrate removal in the unsaturated soil column will consist of a 50 percent loss by uptake in surface vegetation, and a 33 percent loss due to denitrification, adsorption, and fixation. Only 16 percent of the total nitrates will reach the aquifer (50 mg/l).

The locations and number of septic tanks in each grid was determined by using an existing land-use map and communication with the city engineer in Edmond. The following assumptions were made in the use of the existing land-use map:

(1) in low-density areas there is one dwelling unit per acre with 3 persons per house;

(2) in medium-density areas there are two dwelling units per acre with 3 persons per house;

(3) in high-density areas there are five dwelling units per acre with 3 persons per house; and

(4) a minimum allowable lot size of one acre is needed to provide an area for dilution of septic tank effluents. This was assumed based on a percolation rate of 2 to 6.30 in/hr (sandstone bedrock is about 1 ft below the land surface, therefore, there will be less percolation through this bedrock) (Fisher et al., 1969).

Boundary and Initial Conditions

In order to obtain a specific solution for the K-B model, it is necessary to define the boundary and initial conditions for the aquifer. Two types of boundary conditions were used in the K-B model: (1) no-flow (a specific case of constant-flux boundary conditions); and (2) a constant-head. The K-B model requires that the project area be surrounded by a no-flow boundary because of the applied numerical procedure (Konikow and Bredehoeft, 1978). A constant-head boundary was specified for a portion of the Deep Fork River which passes through the southwestern corner of the modeled area. It is a physical boundary and a gaining reach.

Constant-head and no-flow boundaries used in the modeled area and their locations are shown in Figure 27. A zero flux boundary was created by assigning a value of zero transmissivity to nodes surrounding the Edmond study area, and the head values used for the constant-head boundary were taken from the 1973-1974 potentiometric surface map. After existing water-level contours (1973-1974) were interpolated from 50 ft intervals to 10 ft intervals, the head values were determined for each node on the basis of this modification. They were used as initial heads in the model for 1973. According to verbal communication with the U.S. Geological Survey in Oklahoma City, the water level data between 1973 and 1974 showed minimal differences.

Initial nitrate concentrations were obtained from the Edmond Water Department. This data was taken from wells number 16 and 17 in 1971 (the nitrate concentration was 1 mg/l in both wells). Because of lack of data

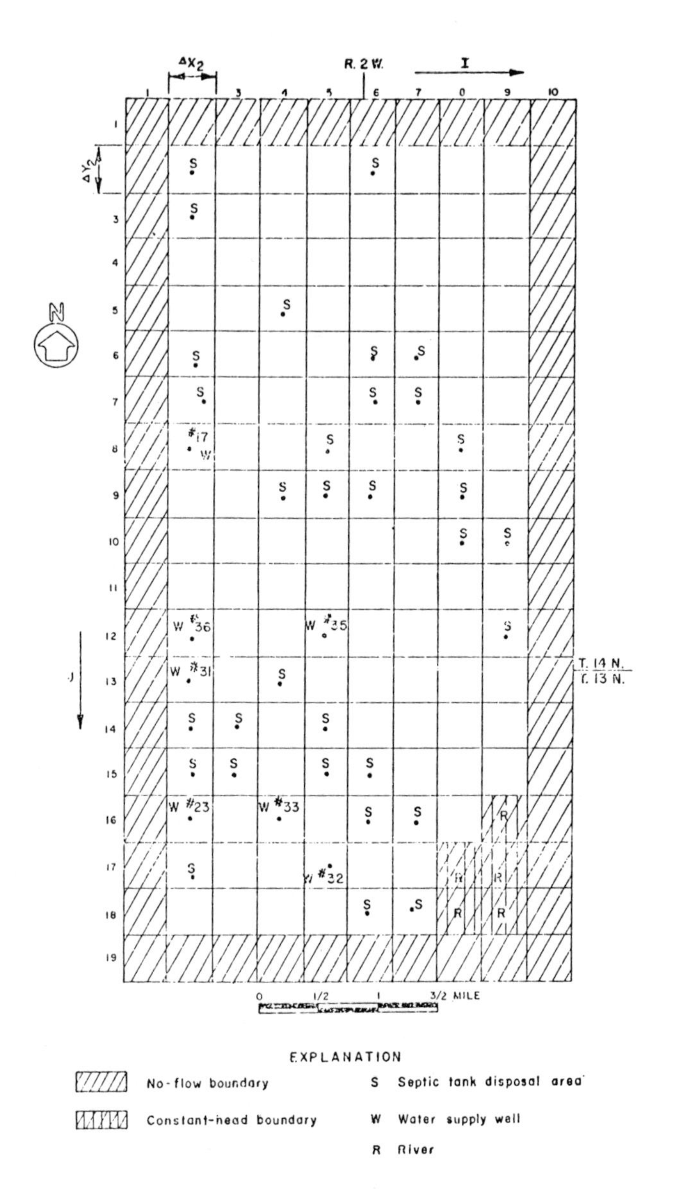

Figure 27: Finite-difference Grid Used to Model the Study Area

from the other wells, it is assumed that the aquifer had a uniform concentration during this year.

Results and Discussion

The results of analysis by the K-B model of the Edmond study area must be classified as disappointing and frustrating. Disappointment stems from the fact that the model was unable to be calibrated, even for ground water flow (water levels). Frustration stems from the fact that the difficulties encountered with the model were due solely to the lack of and questionable validity of input data. These points are examined in detail below.

The first task in the K-B model usage was to calibrate the model, that is, adjust the input data so that the results produced by the model parallel those actually measured. As stated previously, the model was unable to be calibrated for ground water flow, hence, attempts to calibrate it for solute transport were not made. The model actually reproduced some moderately accurate head values for the southwestern part of the study area, but this positive note was more than offset by the fact that ground water levels were predicted to be above the ground surface in the southeastern part of the region.

Difficulties in calibrating the model can be attributed to two main sources: (1) aquifer characterization information; and (2) input data. The ability to transform the hydrologic behavior of the aquifer into a numerical description for the computer model is critical. Placement of no-flow and constant-head boundaries such that they accurately reflect the aquifer and/or they do not adversely affect the results of the numerical analysis is important. The initial status of the aquifer (in terms of heads, concentrations, etc.) also needs to be accurately identified. Both of these aspects of characterization are dependent upon not only a firm grasp of the study area hydrogeology, but also a firm understanding of the relationships and complexities within the numerical model.

The K-B model, like most ground water models, also requires the input of a number of specific aquifer parameters. These parameters should accurately reflect the aquifer. In other words, a parameter determined from a particular test in a particular spot may be accurate but may not reflect the gross properties of the aquifer. The input parameters are not constant and, in fact, nor are the values adjusted in order to calibrate the model. However, if the calibration procedure requires that certain values be adjusted out of the range of real life values, it must be concluded that errors exist somewhere. Such is the case in this analysis. As discussed previously, most of the input data was either estimated, assumed or obtained from minimal information. The total uncertainty of these values defeated the calibration process. In summary, the analysis of this study area was precluded by a combination of the detailed input data requirements of the model and the lack of accurate information concerning this study area.

The only conclusion to be drawn concerning the applicability of sophisticated ground water models to the problem of septic tank systems is that the utility of the models may be outweighed by their significant data requirements. In other words, before an analysis of the septic tank systems can ever begin, the aquifer must be understood in great detail.

Even though water quality considerations were foregone in this analysis, a few comments can be made concerning the knowledge of the influence of septic tank systems on ground water quality. As seen in the development of the input data for the septic tank areas, the actual functioning of a septic tank system is understood theoretically, but quantitative information on the various subsurface processes is almost non-existent. Only two parameters concerning septic tank system behavior are needed for the K-B model--the amount and concentration of recharge reaching the ground water. Both of these parameters had to be calculated using estimates for input data. A better understanding of the subsurface behavior of septic tank system effluents would be needed for any detailed ground water quality modeling study.

HIERARCHICAL STRUCTURE FOR MODEL USAGE

Three types of models for evaluating the potential effects of septic tank systems on ground water have been described in this chapter. The three types of models are:

(1) An empirical assessment model for developing a ground water pollution potential index based on site hydrogeological information, wastewater characteristics and flows. Two existing models were described (surface impoundment assessment, and waste-soil-site interaction matrix), and it was suggested that both be modified by including consideration of the annual wastewater flow in prioritizing the ground water pollution potential of septic tank system areas. Further, it was determined that the SIA model was perhaps a better choice since it required less site and wastewater information, and it yielded essentially the same priority ranking of 13 septic tank system areas in a central Oklahoma study area as did the interaction matrix model. The site and wastewater information needed for the SIA model should be fairly readily available, or it can be based on defensible assumptions. However, it was noted that the SIA model was developed for surface impoundments containing liquids, and although similar in physical arrangements to a septic tank system (liquids introduced to the subsurface from a surface or near-surface source), the model does not focus on some key issues relevant to septic tank systems. Accordingly, an empirical assessment model specific for septic tank systems and system areas is needed. This model should include such factors as percolation rate, septic tank density, septic tank age, wastewater flow, depth to ground water, distance to nearest water well and gradient relationship to well.

(2) An analytical model developed by Hantush for determining the rise in the water table underneath a circular, rectangular, or square recharge area. This model can be used to predict water table rises, with these rises being of particular importance in areas with shallow ground water. This model does not address quality considerations. The basic site and source information needed for use of this model for septic tank system areas should be fairly readily available, or it can be based on defensible assumptions.

(3) A solute-transport model developed by Konikow and Bredehoeft for addressing ground water flow and pollutant transport in the subsurface environment. The K-B model is mathematically sophisticated and, although it is available in a packaged computer program, it requires extensive field information, both current and historical, for calibration and subsequent usage. An attempt was made to use the K-B model in a septic tank system study area near Edmond, Oklahoma; however, necessary input data was simply unavailable. This suggests that special field studies will be necessary in order to gather the input data necessary for use of solute-transport models for evaluation of septic tank systems or system areas.

An hierarchical structure for usage of the three models (types of models) is in Table 53. The potential usage is shown at three levels: (1) a septic tank system serving an individual home; (2) several hundred individual septic tank systems being used in a subdivision; and (3) a large scale septic tank system serving several hundred homes, with the daily wastewater flow being upwards of 100,000 gallons. The empirical assessment model could be used as part of the permitting procedure for all three levels; however, its greatest usage should probably be for the first two levels. The analytical model could be used for subdivisions and large scale systems, with the greatest usage probably associated with the former. Finally, the solute-transport model should be used for large scale systems since their potential for ground water pollution could justify the conduction of the necessary field studies to gather appropriate input data.

In addition to using the structure suggested in Table 53 as part of the permitting process, the models could also be used to evaluate existing septic tank systems or system areas. This evaluation can be useful in ground water pollution identification, monitoring planning, and development of ground water quality management strategies.

SELECTED REFERENCES

Anderson, M.P., "Using Models to Simulate the Movement of Contaminants Through Ground Water Flow Systems", Critical Reviews in Environmental Control, Vol. 91, Nov. 1979, pp. 97-156.

Table 53: Hierarchical Structure for Septic Tank System Modeling

Model	Key Characteristics of Model	Usage		
		Individual Home Septic Tank System	Subdivision with Several Hundred Septic Tank Systems	Large-Scale Septic Tank System
Empirical Assessment*	Provides ground water pollution potential index; data needs minimal	X	X	X
Analytical	Predicts water level rise; data needs minimal		X	X
Solute Transport	Predicts ground water flow and concentrations of pollutants; need field studies to get input data			X

*Either the adjusted SIA model or a new model specifically developed for septic tank systems

Appel, C.A. and Bredehoeft, J.D., "Status of Ground Water Modeling in the U.S. Geological Survey", Circular No. 737, 1976, U.S. Geological Survey, Washington, D.C.

Association of Central Oklahoma Governments (ACOG), Oklahoma City, Oklahoma, 1980.

Bachmat, Y. et al., "Utilization of Numerical Groundwater Models for Water Resource Management", EPA-600/8-78/012, June 1978, U.S. Environmental Protection Agency, Ada, Oklahoma.

Bauer, D.H., Conrad, E.T. and Sherman, D.G., "Evaluation of On-Site Wastewater Treatment and Disposal Options", Feb. 1979, U.S. Environmental Protection Agency, Cincinnati, Ohio.

Bingham, R.H. and Moore, R.L., "Reconnaissance of the Water Resources of the Oklahoma City Quadrangle, Central Oklahoma", Hydrologic Atlas 4, 1975, Oklahoma Geological Survey, Norman, Oklahoma.

Bredehoeft, J.D. and Pinder, G.F., "Mass Transport in Flowing Ground Water", Water Resources Research, Vol. 9, No. 1, Feb. 1971, pp. 194-210.

Burton, L.C. and Jacobsen, C.L., "Geologic Map of Cleveland and Oklahoma Counties, Oklahoma", 1967, Oklahoma Geological Survey, Norman, Oklahoma.

Canter, L.W., "Methods for the Assessment of Ground Water Pollution Potential", Oct. 1981, First International Conference on Ground Water Quality Research, Rice University, Houston, Texas, 51 pages.

Carr, J.E. and Marcher, M.V., "A Preliminary Appraisal of the Garber-Wellington Aquifer Southern Logan and Northern Oklahoma Counties Oklahoma", Open File Report 77-278, May 1977, U.S. Geological Survey, Oklahoma City, Oklahoma.

Carriere, G.D., "Priority Ranking of Septic Tank Systems in the Garber-Wellington Area", NCGWR 80-32, Nov. 1980, 148 pages.

Edmond Water Department, Edmond, Oklahoma, 1980.

Engineering Enterprises, Inc., Norman, Oklahoma, 1980.

Fisher, C.F. et al., "Soil Survey of Oklahoma County, Oklahoma", Feb. 1969, United States Department of Agriculture (Soil Conservation Service) in cooperation with Oklahoma Agricultural Experiment Station.

Freeze, R.A. and Cherry, J.A., Ground Water, Prentice-Hall, Inc., Englewood Cliffs, New Jersey, 1979.

Hantush, M.S., "Growth and Decay for Groundwater-Mounds in Response to Uniform Percolation", Water Resources Research, Vol. 3, 1967, p. 227.

Harkin, J.M. et al., "Evaluation of Mound Systems for Purification of Septic Tank Effluent", Technical Report WIS WRC 79-05, 1979, Madison, Wisconsin.

Khaleel, R. and Redell, D.L., "Simulation of Pollutant Movement in Ground Water Aquifers", OWRT-A-030-Tex(1), May 1977, 261 pp., Water Resources Institute, Texas A&M University, College Station, Texas.

Kincannon, D.F., Draft Final Report on Land Application of Wastewater, Oklahoma State University, Stillwater, Oklahoma, 1981.

Konikow, L.F. and Bredehoeft, J.D., "Computer Model of Two-Dimensional Solute Transport and Dispersion in Ground Water", Techniques of Water-Resources Investigations of the United States Geological Survey, Book 1, Chapter 2, 1978, U.S. Geological Survey, Washington, D.C.

Konikow, L.F. and Grove, D.B., "Derivation of Equations Describing Solute Transport in Ground Water", Water-Resources Investigations #77-19, 1977, U.S. Geological Survey, Washington, D.C.

LeGrand, H.E., "System of Reevaluation of Contamination Potential of Some Waste Disposal Sites", Journal American Water Works Association, Vol. 56, Aug. 1964, pp. 959-974.

Mercado, A., "Nitrate and Chloride Pollution of Aquifers: A Regional Study with the Aid of a Single-Cell Model", Water Resources Research, Vol. 12, No. 4, Aug. 1976, pp. 731-747.

Mercer, J.W. and Faust, C.R., "Ground Water Modeling: Mathematical Models", Ground Water, Vol. 18, No. 3, May-June 1980, pp. 212-227.

Mogg, J.L., Schoff, S.L. and Reed, E.W., "Ground Water of Canadian County", Bulletin 87, 1969, Oklahoma Geological Survey, Norman, Oklahoma.

Oklahoma Water Resources Board, Oklahoma City, Oklahoma, 1980.

Peavy, H.S., "Groundwater Pollution from Septic Tank Drainfields", June 1978, Montana State University, Montana.

Phillips, C.R., Nathwani, J.S. and Mooij, H., "Development of a Soil-Waste Interaction Matrix for Assessing Land Disposal of Industrial Waste", Water Research, Vol. 11, Nov. 1977, pp. 859-868.

Pickens, J.F. and Lennox, W.C., "Numerical Simulation of Waste Movement in Steady Ground Water Flow Systems", Water Resources Research, Vol. 12, No. 2, Apr. 1976, pp. 171-184.

Prickett, T.A. and Lonnquist, C.G., "Selected Digital Computer Techniques for Ground Water Resources Evaluation", Bulletin 55, 1971, Illinois Water Survey, Urbana, Illinois.

Prickett, T.A., "State-of-the-Art of Groundwater Modeling", Water Supply and Management, Vol. 3, No. 2, 1979, pp. 134-141.

Sohrabi, T.M., "Digital-Transport Model Study of Potential Nitrate Contamination from Septic Tank Systems Near Edmond, Oklahoma", NCGWR 80-38, Nov. 1980, 107 pages.

Sproul, O.J., "Virus Movement Into Ground Water from Septic Tank Systems", Paper No. 12, Sept. 1973, Rural Environmental Engineering Conference on Water Pollution Control in Low Density Areas Proceedings, University Press of New England, Hanover, New Hampshire.

Tanji, K.K. and Gupta, S.K., "Computer Simulation Modeling for Nitrogen in Irrigated Croplands", Nitrogen in the Environment, Vol. 1, Academic Press, Inc., New York, New York, 1978.

U.S. Environmental Protection Agency, "Quality Criteria for Water", July 1976, U.S. Government Printing Office, Washington, D.C.

U.S. Environmental Protection Agency, "A Manual for Evaluating Contamination Potential of Surface Impoundments", EPA 570/9-78-003, June 1978, Office of Drinking Water, Washington, D.C.

U.S. Environmental Protection Agency, "Management of Small Waste Flows", Report No. EPA-600/2-78-173, Sept. 1978, Cincinnati, Ohio.

Vilker, V.L., "An Adsorption Model for Prediction of Virus Breakthrough from Fixed Beds", Proceedings of International Symposium on Land Treatment of Wastewater, Vol. 2, 1978, U.S. Army Corps of Engineers Cold Regions Research and Engineering Laboratory, Hanover, New Hampshire, pp. 381-388.

Viraraghavan, T., "Septic Tank Efficiency", ASCE Journal Environmental Engineering Division, Vol. 102, No. EE2, Apr. 1976, pp. 505-508.

Wood, P.R. and Burton, L.C., "Ground-Water Resources Cleveland and Oklahoma Counties", Circular 71, 1968, Oklahoma Geological Survey, Oklahoma University, Norman, Oklahoma.

CHAPTER 5

SUMMARY

Septic tanks were introduced in the United States in 1884, and since then septic tank systems have become the most widely used method of on-site sewage disposal, with over 70 million people depending on them. Approximately 17 million housing units, or 1/3 of all housing units, dispose of domestic wastewater through these systems, and about 25 percent of all new homes being constructed are including them. The greatest densities of usage occur in the east and southeast as well as the northern tier and northwest portions of the United States. A septic tank system includes both the septic tank and the subsurface soil absorption system. Approximately 800 billion gallons of wastewater is discharged annually to the soil via tile fields following the 17 million septic tanks.

Septic tank systems that have been properly designed, constructed, and maintained are efficient and economical alternatives to public sewage disposal systems. However, due to poor locations for many septic tank systems, as well as poor designs and construction and maintenance practices, septic tank systems have polluted, or have the potential to pollute, underlying ground waters. A major concern in many locations is that the density of the septic tanks is greater than the natural ability of the subsurface environment to receive and purify system effluents prior to their movement into ground water. A related issue is that the design life of many septic tank systems is in the order of 10-15 years. Due to the rapid rate of placement of septic tank systems in the 1960's, the usable life of many of the systems is being exceeded, and ground water contamination is beginning to occur.

Septic tank systems are frequently reported sources of localized ground water pollution. Historical concerns have focused on bacterial and nitrate pollution; more recently, synthetic organic chemicals from septic tank cleaners have been identified in ground water. Regional ground water problems have also been recognized in areas of high septic tank system density. Within the United States there are four counties with more than 100,000 housing units served by septic tank systems and cesspools, and an additional 23 counties with more than 50,000 housing units served by these systems. Densities range from as low as 2 to greater than 346 per square mile based on assuming an even distribution of the septic tank systems and cesspools throughout the county. If they are localized in segments of the county the actual densities could be several times greater. An often-cited figure is that areas with more than 40 systems per square mile can be considered to have potential contamination problems.

Several types of institutional arrangements have been developed for regulating septic tank system design and installation, operation and maintenance, and failure detection and correction. Most of the regulatory activities are conducted by state and local governments. The U.S. Environmental Protection Agency can become a participant in the regulatory process based on the provision of funding for septic tank systems. Sections 201(h) and (j) of the Clean Water Act of 1977 (P.L. 95-217) authorized construction grants funding of privately owned treatment works serving individual housing units or groups of housing units (or small commercial establishments), provided that a public entity (which will ensure proper operation and maintenance) apply on behalf of a number of such individual systems.

The objective of this book has been to summarize existing literature relative to the types and mechanisms of ground water pollution from septic tank systems, and to provide information on technical methodologies for evaluating the ground water pollution potential of septic tank systems. The book includes a survey of published literature on the identification and evaluation of ground water pollution from septic tank systems; and the selection and evaluation of two empirical assessment methodologies, one numerical model, and one analytical model for their applicability to septic tank systems. Selection of the methodologies and models was based on considering their previous or potential use for septic tank systems; likely availability of required input data; resource requirements in terms of general personnel and technical specialists, computational equipment, and time or ease of implementation; understandability by nontechnical persons; and previous documentation for prediction of pollutant transport.

SEPTIC TANK SYSTEMS

The basic septic tank system consists of a buried tank where waterborne wastes are collected, and scum, grease and settleable solids are removed from the liquid by gravity separation; and a subsurface drain system where clarified effluent percolates into the soil. System performance is essentially a function of the design of the system components, construction techniques employed, characteristics of the wastes, rate of hydraulic loading, climate, areal geology and topography, physical and chemical composition of the soil mantle, and care given to periodic maintenance.

Design considerations related to a septic tank include determination of the appropriate volume, a choice between single and double compartments, selection of the construction material, and placement on the site. Placement of the septic tank on the site basically involves consideration of the site slope and minimum setback distances from various natural features or built structures. Soil absorption systems include the design and usage of trenches or beds, seepage pits, mounds, fills and artificially drained systems. Trench and bed systems are the most commonly used methods for on-site wastewater treatment and disposal. Site criteria which must be met for septic tank system approval include: a specified percolation rate, as determined by a percolation test; and a minimum 4-ft (1.2 m) separation between the bottom of the seepage system and the maximum seasonal elevation of ground water. In addition, there must be a reasonable thickness, again normally 4 ft, of relatively permeable soil between the seepage system and the top of a clay layer or impervious rock formation.

EFFLUENT QUALITY FROM SEPTIC TANK SYSTEMS

One of the key concerns associated with the design and usage of septic tank systems is the potential for inadvertently polluting ground water. This concern is increased when considering systems serving multiple housing units. Potential ground water pollutants from septic tank systems are primarily those associated with domestic wastewater, unless the systems receive industrial wastes. Contaminants originating from system cleaning can also contribute to the ground water pollution potential of septic tank systems. The typical wastewater flow from a household unit is about 150 to 170 liters/day/person. Typical sources of household wastewater, expressed on a percentage basis, are: toilet(s)--22 to 45 percent; laundry--4 to 26 percent; bath(s)--18 to 37 percent; kitchen--6 to 13 percent; and other--0 to 14 percent. Of concern in terms of ground water pollution is the quality of the effluent from the septic tank portion of the system, and the efficiency of constituent removal in the soil underlying the soil absorption system.

Based on a number of studies, the following represent typical physical and chemical parameter effluent concentrations from septic tanks: suspended solids--75 mg/l; BOD_5--140 mg/l; COD--300 mg/l; total nitrogen--40 mg/l; and total phosphorus--15 mg/l. Studies of the efficiency of soil absorption systems have indicated the following typical concentrations entering ground water: suspended solids--18-53 mg/l; BOD--28-84 mg/l; COD--57-142 mg/l; ammonia nitrogen--10-78 mg/l; and total phosphates--6-9 mg/l. In addition, other wastewater constituents of concern include bacteria, viruses, nitrates, synthetic organic contaminants such as trichloroethylene, metals (lead, tin, zinc, copper, iron, cadmium, and arsenic), and inorganic contaminants (sodium, chlorides, potassium, calcium, magnesium, and sulfates).

GROUND WATER POLLUTION FROM SEPTIC TANK SYSTEMS

Ground water degradation has occurred in many areas having high densities of septic tank systems, with the degradation exemplified by high concentrations of nitrates and bacteria in addition to potentially significant amounts of organic contaminants. One common reason for degradation is that the capacity of the soil to absorb effluent from the tank has been exceeded, and the waste added to the system moves to the soil surface above the lateral lines. Another reason of greater significance to ground water is when pollutants move too rapidly through soils. Many soils with high hydraulic absorptive capacity (permeability) can be rapidly overloaded with organic and inorganic chemicals and microorganisms, thus permitting rapid movement of contaminants from the lateral field to the ground water zone. In considering ground water contamination from septic tank systems, attention must be directed to the transport and fate of pollutants from the soil absorption system through underlying soils and into ground water.

Physical, chemical and biological removal mechanisms may occur in both the soil and ground water systems. Transport and fate issues must be considered in terms of biological contaminants (bacteria and viruses), inorganic contaminants (phosphorus, nitrogen, and metals), and organic contaminants (synthetic organics and pesticides).

Biological contaminants (pathogens) have a wide variety of physical and biological characteristics, including wide ranges in size, shape, surface properties, and die-away rates. The distance of travel of bacteria through soil is of considerable significance since contamination of ground supplies may present a health hazard. A number of environmental factors can influence the transport rate, including rainfall; soil moisture, temperature, and pH; and availability of organic matter. Environmental factors affecting the survival of enteric bacteria in soil include soil moisture content and holding capacity, temperature, pH, sunlight, organic matter, and antagonism from soil microflora. The physical process of straining (chance contact) and the chemical process of adsorption (bonding and chemical interaction) appear to be the most significant mechanisms in bacterial removal from water percolating through soil. Factors influencing the removal efficiency of viruses by soil include flow rate, cation concentrations, clays, soluble organics concentrations, pH, isoelectric point of the viruses, and general chemical composition of the soil. The most important mechanism of virus removal in soil is by adsorption of viruses onto soil particles.

While phosphorus can move through soils underlying soil absorption systems and reach ground water, this has not been a major concern since phosphorus can be easily retained in the underlying soils due to chemical changes and adsorption. Phosphate ions become chemisorbed on the surfaces of Fe and Al minerals in strongly acid to neutral systems and on Ca minerals in neutral to alkaline systems. In the pH range encountered in septic tank seepage fields, hydroxyapatite is the stable calcium phosphate precipitate. However, at relatively high phosphorus concentrations similar to those found in septic tank effluents, dicalcium phosphate or octacalcium phosphate are formed initially, followed by a slow conversion to hydroxyapatite. Ammonium ions can be discharged into the subsurface environment, or they can be generated within the upper layers of soil from the ammonification process (conversion of organic nitrogen to ammonia nitrogen). The transport and fate of ammonium ions may involve adsorption, cation exchange, incorporation into microbial biomass, or release to the atmosphere in the gaseous form. Adsorption is probably the major mechanism of removal in the subsurface environment. Nitrate ions can also be discharged directly or generated within the upper layers of soil. The transport and fate of nitrate ions may involve movement with the water phase, uptake in plants or crops, or denitrification. Nitrates can move with ground water with minimal transformation.

The four major reactions that metals may be involved in with soils are adsorption, ion exchange, chemical precipitation and complexation with organic substances. Of these four, adsorption seems to be the most important for the fixation of heavy metals. Ion exchange is thought to provide only a temporary or transitory mechanism for the retention of trace and heavy metals. Precipitation reactions are greatly influenced by pH and concentration, with precipitation predominantly occurring at neutral to high pH values and in macroconcentrations. Organic materials in soils may immobilize metals by complexation reactions or cation exchange. Fixation of heavy metals by soils by either of these four mechanisms is dependent on a number of factors including soil composition, soil texture, pH and the oxidation-reduction potential of the soil and associated ions.

The transport and fate of organic contaminants in the subsurface environment is a relatively new topical area of concern, thus the published literature is sparse. A variety of possibilities exist for the movement of organics, including transport with the water phase, volatilization and loss from the soil system, retention on the soil due to adsorption, incorporation into microbial or plant biomass, and bacterial degradation. The relative importance of these possibilities in a given situation is dependent upon the characteristics of the organic, the soil types and characteristics, and the subsurface environmental conditions. This very complicated topical area is being actively researched at this time. Several studies have been conducted on the movement and biodegradation of large concentrations of pesticides in soils.

EVALUATION OF THE GROUND WATER POLLUTION POTENTIAL OF SEPTIC TANK SYSTEMS

Several technical methodologies are available for evaluating the ground water pollution potential of septic tank systems. Technical methodologies range from empirical index approaches to sophisticated mathematical models. Models can range from analytical approaches addressing ground water flow to numerical approaches which aggregate both flow and solute transport considerations. Septic tank systems can be considered as area sources of ground water pollution, with the rectangular dimensions of the drainage field representing the source boundaries. Waste stabilization ponds (surface impoundments), and sanitary and chemical landfills also can be considered as potential area sources of ground water pollution. Empirical assessment methodologies refer to simple approaches for development of numerical indices of the ground water pollution potential of man's activities. Several methodologies have been developed for evaluating the ground water pollution potential of wastewater ponds and sanitary and chemical landfills. Methodologies typically contain several factors for evaluation, with the number and type, and importance weighting, varying from methodology to methodology.

Ground water models can be classified into flow models and solute transport models. Of interest herein are analytical models and numerical models. Analytical models include those where the behavior of an aquifer is described by differential equations which are derived from basic principles such as the laws of continuity and conservation of energy. Numerical models are actually analytical models that are so large they require the use of digital computers, capable of multiple iterations, to converge on a solution. The applicability of ground water models has been the subject of a number of studies. Prediction of the movement of contaminants in ground water systems through the use of models has been given increased emphasis in recent years because of the growing trend toward subsurface disposal of wastes.

Ground water modeling can be useful for evaluation of specific sites for systems, or even larger geographical areas that may be served by hundreds of systems. Modeling could be used to exclude septic tank system location on specific sites or in larger geographical areas. In addition, modeling can be useful in planning ground water monitoring programs for specific sites or geographical areas. As noted earlier, available technical methodologies for addressing the ground water effects of septic tank systems range from empirical assessment approaches to ground water flow and solute transport models. These methodologies differ in their input requirements, output characteristics, and general useability. Accordingly, certain selection criteria were identified as basic to the selection of technical methodologies (TM); the criteria statements were as follows:

1. The TM should have been previously used for evaluation of septic tank systems.

2. The TM should be potentially useable, or adaptable for use, for evaluation of septic tank systems.

3. If the TM needs to be calibrated prior to use, the necessary data for calibration should be readily available.

4. The input data required for the TM should be readily available, thus the use of the TM could be easily implemented.

5. The resource requirements for use of the TM should be minimal (resource requirements refer to personnel needs and personnel qualifications, computer needs, and the time necessary for TM calibration and usage).

6. Usage of the TM for prediction of pollutant transport in the subsurface environment should have been previously documented.

7. The conceptual framework of the TM as well as its output should be understandable by non-ground water modeling specialists.

No single technical methodology (TM) which met all seven criteria was identified. However, two empirical assessment methodologies (Surface Impoundment Assessment and Waste-Soil-Site Interaction Matrix), one analytical model (Hantush), and one solute-transport model (Konikow and Bredehoeft) was chosen for examination. The two empirical methodologies were used to determine the ground water pollution potential of 13 septic tank system areas in central Oklahoma. The rank order of the ground water pollution potential of the 13 areas was determined by the two methodologies adjusted by considering the annual wastewater flows in the areas. The two adjusted methodologies provided similar rank orderings of the 13 septic tank system areas. Key findings from this case study were:

(1) The final ranking of the 13 septic tank system areas was largely dependent upon the annual wastewater flow in the area, and this is directly related to the number of persons and septic tank systems in the area.

(2) Both the surface impoundment assessment method and the waste-soil-site interaction matrix can be used to develop a priority ranking of existing or planned septic tank system areas. Since the surface impoundment assessment method has 6 items of needed information versus 17 items in the interaction matrix, the SIA method is easier to use. However, it should be noted that neither methodology accounts for wastewater flow, and this is an important factor which should be given consideration in the use of either method for septic tank system areas.

The Hantush analytical model was developed to determine the rise and fall of the water table under circular, rectangular, or square recharge areas; it does not address ground water quality. This model was applied to a mound type septic tank system analogous to those used in Wisconsin, and it was determined that the water table rise only approaches a maximum of 8 inches; however, this could be a significant rise in view of the fact that mound systems are used in areas of high water tables. Actual loadings from septic tank systems will be intermittent and this will decrease the actual rise of the water table; however, increases in loading rates (either by malfunctioning or overloaded systems) could increase the water table rise.

The Konikow-Bredehoeft (K-B) numerical model was applied to a septic tank system study area near Edmond, Oklahoma, to determine its usefulness in predicting nitrate concentrations in ground water from this source type. The K-B model is a two-dimensional solute transport model which has been used in the analysis of ground water pollution from a variety of source types. The objective was to determine the feasibility of modeling the effects of septic tank systems on ground water quality by direct application of the K-B solute transport model computer package to an existing situation. The results of analysis by the K-B model of the Edmond study area must be classified as disappointing and frustrating. Disappointment stems from the fact that the model was unable to be calibrated, even for ground water flow (water levels). Frustration stems from the fact that the difficulties encountered with the model were due solely to the lack of and questionable validity of input data. The only conclusion to be drawn concerning the applicability of sophisticated ground water models to the problem of septic tank systems is that the utility of the models may be outweighed by their significant data requirements. This suggests that special field studies will be necessary in order to gather the input data necessary for use of solute-transport models for evaluation of septic tank systems or system areas.

Based on the results of the case studies, an hierarchical structure for usage of the three types of technical methodologies has been developed. Potential usage can be considered for three types of septic tank systems: (1) a septic tank system serving an individual home; (2) several hundred individual septic tank systems being used in a subdivision; and (3) a large scale septic tank system serving several hundred homes, with the daily wastewater flow being upwards of 100,000 gallons. The empirical assessment methodology (adjusted SIA method) could be used as part of the permitting procedure for all three types; however, its greatest usage should probably be for the first two types. The analytical model could be used for subdivisions and large scale systems, with the greatest usage probably associated with the former. Finally, the solute-transport model should be used for large scale systems since their potential for ground water pollution could justify the conduction of the necessary field studies to gather appropriate input data.

APPENDIX A

ANNOTATED BIBLIOGRAPHY

Anderson, M.P., "Using Models to Simulate the Movement of Contaminants Through Ground Water Flow Systems", Critical Reviews in Environmental Control, Vol. 91, Nov. 1979, pp. 97-156.

Prediction of the movement of contaminants in ground water systems through the use of models has been given increased emphasis in recent years because of the growing trend toward subsurface disposal of wastes. Prediction is especially critical when nuclear wastes are involved. Contaminant transport models which include the effects of dispersion have been applied to several field studies. Regional size models which limit the effects of dispersion have had limited success because of the scarcity and poor quality of field data. Another difficulty in the development of contaminant transport models is the current lack of knowledge regarding the quantification of chemical reaction terms. This review examines the formulation of contaminant transport models, application to field problems, difficulties involved in obtaining input data, and current status of modeling efforts.

Andreoli, A. et al., "Nitrogen Removal in a Subsurface Disposal System", Journal of the Water Pollution Control Federation, Vol. 51, No. 4, Apr. 1979, pp. 841-855.

The effects of subsurface waste disposal on ground water quality in Long Island, N.Y., are assessed. Residents of Long Island depend on ground water for their entire water supply. The geology of Long Island is reviewed. Described are the design, construction, and operation of a full scale system consisting of a conventional septic tank-leaching field wastewater disposal system combined with a subsurface system using natural soil treatment mechanisms for nitrogen removal. The septic tank reduces the inorganic nitrogen concentration of raw wastewater by 20 percent. About 36 percent of the total nitrogen applied to the soil is removed after 2 ft of travel through the soil. Nitrification occurs within 2-4 ft of vertical travel in Long Island soil.

Andrew, W.F., "Soil as a Media (sic) for Sewage Treatment", Third Annual Illinois Private Sewage Disposal Symposium, Feb. 1978, pp. 18-20.

In the selection of a site for a septic tank absorption field, the pollution abatement potential must be considered. Compared to air and water, soil is a very good medium for the treatment of septic tank effluent. Ideally, a soil should be able to convert a pollutant to an unpolluted state at a rate equal to or greater than the rate at which it is added to the soil. Several soil characteristics affect the soil's pollutant abatement potential. Septic tank installation is not recommended in soil subject to flooding. Where soil is shallow to bedrock or cemented pan, the volume of absorptive soil is reduced; the only alternative is to increase the size of the field. A high water table reduces the open pore space of the soil, reducing its absorptive capacity. It also reduces the O_2 in the soil and, consequently, the microbial capacity. Permeability of the soil reflects the ability of air and water to move through it. If movement is too slow, the field needs to be enlarged; if too rapid, there is danger of ground water contamination. Where soil is sloping, there is a danger of uneven distribution of the effluent, and the possibility of some coming to the surface. Coarse fragments cause installation problems and reduce the overall volume per area. Subsidence can be a severe problem. Filter lines may shift, causing blockage and excessive concentration of effluent. Depressions may occur over the lines and cause surface water to accumulate.

Appel, C.A. and Bredehoeft, J.D., "Status of Ground Water Modeling in the U.S. Geological Survey", Circular No. 737, 1976, U.S. Geological Survey, Washington, D.C.

The types of problems for which models have been, or are being, developed include: ground water flow in saturated or partially unsaturated material; land subsidence resulting from ground water extraction; flow in coupled ground water-stream systems; coupling of rainfall-runoff basin models with soil moisture-accounting aquifer flow models; interaction of economic and hydrologic considerations; predicting the transport of contaminants in an aquifer; and estimating the effects of proposed development schemes for geothermal systems. The status of modeling activity for various models is reported as being in a developmental, verification, operational, or continued improvement phase.

Bachmat, Y. et al., "Utilization of Numerical Groundwater Models for Water Resource Management", EPA-600/8-78/012, June 1973, U.S. Environmental Protection Agency, Ada, Oklahoma.

The study assesses the present status of international numerical models as a tool for ground water related water resource management. Among the problem areas considered are: the accessibility of models to users; communications between managers and technical personnel; inadequacies of data; and inadequacies in modeling. The report, which is directed toward the nontechnical reader, describes 250 models. These are categorized as prediction, management, identification, and data management models.

Brown, K.W. et al., "The Movement of Fecal Coliforms and Coliphages Below Septic Lines", *Journal of Environmental Quality*, Vol. 8, No. 1, 1979, pp. 121-125.

A two-year lysimetric study utilizing three undisturbed soils was conducted to investigate the movement of fecal coliforms and coliphages to the ground water. Septic tank effluent was applied to each of the three soils at appropriate design rates via subsurface septic lines. The soils included had sand contents of 80, 41 and 7.6 percent. Indigenous concentrations of fecal coliforms in the effluent were more than sufficient to assure detectability. During the winter the levels of indigenous coliphages decreased, and on several occasions the septic effluent was spiked with cultured coliphages. The remainder of the year, indigenous levels were sufficient to allow adequate detection. Leachate samples were analyzed on a continuous basis, and at the end of the study the soils below the septic lines were dissected and sampled on a grid pattern. They were analyzed for both fecal coliforms and coliphages. On only a few occasions were fecal coliforms present in leachate collected 120 cm below the septic lines. Subsequent samples from the same locations did not indicate the presence of fecal coliforms so that the few samples that were collected shortly after application began may have been a result of contamination, or they may be indicative of greater mobility before organic residue built up in the soil. Soil samples taken 1 and 2 years after application began indicated limited mobility and survival of fecal coliforms in all three soils. Coliphages were present in the leachate only in very low concentrations immediately after spiking of the applied sewage with 10^3 times more organisms than were applied. Soil samples also confirmed the limited mobility of coliphages. Thus, 120 cm of any of the soils tested appeared to be sufficient to minimize the possibility of ground water pollution by fecal coliform or coliphages from septic effluent disposal.

Brown, R.J., "Septic Tank and Household Sewage Systems Design and Use: (Citations from the Engineering Index Data Base), 1970-1979", NTIS/PS-79/0458, 1979, National Technical Information Service, U.S. Department of Commerce, Springfield, Virginia.

The bibliography provides worldwide research reports on septic tanks and other sewage treatment units used for household sewage systems. Construction materials, design, service life, and a comparison of systems are described. The suitability of soils for drainage and adsorption to prevent pollution of ground water from bacteria and viruses are discussed. Purification processes and the environmental constraints of disposal systems are included. This bibliography contains 159 abstracts.

Carlile, B.L., Stewart, L.W. and Sobsey, M.D., "Status of Alternative Systems for Septic Wastes Disposal in North Carolina", 1977, Proceedings of the Second Annual Illinois Private Sewage Disposal Symposium, Champaign, Illinois.

Dye studies indicated that septic tank systems in the study area contribute significant contamination to nearby shellfish harvesting waters via surface and subsurface flow. Surface ponding of septic tank effluent during periods of rainfall constitute a potential health hazard through possible direct contact with these wastes. Continued dependence on conventional septic tank systems for area waste treatment will result in further degradation of area water resources. Studies such as these and from evidence of vast acres of shellfish waters closed, provide convincing evidence that the "carrying capacity" or use potential of land sites have already been exceeded in many coastal areas of the state. If septic tanks are indiscriminately installed in the area, then a reasonable estimate is that approximately 90 percent will not function properly and will fail to some degree within the first year's use. Ultimately, a research goal is to define the carrying capacity of soil types to identify the basic soil limitations in determining loading intensities for conventional and alternative systems of septic waste disposal which would allow developments to proceed without creating additional pollution loads on surface and ground waters.

Childs, K.E., Upchurch, S.B. and Ellis, B., "Sampling of Variable Waste-Migration Patterns in Ground Water", Ground Water, Vol. 12, No. 6, Nov.-Dec. 1974, pp. 369-371.

A survey of waste-migration patterns from septic tank/tile field systems surrounding Houghton Lake, Michigan, indicates that sampling plans designed to detect and quantify waste migration in ground water should be predicated on the concept that the waste plume may be complex and that the plume may not follow regional, ground water flow. The waste-migration plumes at Houghton Lake range from simple, multichemical plumes that move with regional flow to complex plumes that bifurcate, that show different migration patterns for different chemicals, and that move up the regional gradient for short distances.

Cotteral, J.A. and Norris, D.P., "Septic Tank Systems", American Society of Civil Engineering, Journal of Sanitary Engineering Division, Vol. 95, No. SA4, Aug. 1969, pp. 715-746.

This paper reviews the history and theory of septic tank systems as a basis for the establishment of guidelines for the design and construction of satisfactory and economical systems. While survival curves show that system life is usually short, proper design and construction supplemented by regular inspection and maintenance can adequately extend the expected life. It is recommended that a single drainfield design loading rate be applied to all installations meeting minimum topographical and geological requirements. Control should be exercised by a county regulatory agency and should be based upon engineering control of design in lieu of a codified approach. Periodic county inspection and regular maintenance by homeowners is essential, and can be implemented by an enforcement program based upon annually renewable septic tank use permits. An adequately designed, constructed and maintained septic tank system is more expensive than complete community sewerage, but is nevertheless economically feasible.

Crosby, J.W., III et al., "Investigation of Techniques to Provide Advance Warning of Ground Water Pollution Hazards with Special Reference to Aquifers in Glacial Outwash", NTIS No. PB-203 748, Aug. 1971, Washington State University, Pullman, Washington.

Findings are recorded of a six-year investigation of pollution hazards involved with the use of septic tanks and drainfields in the Spokane Valley of eastern Washington. The geological setting of the study area was investigated by gravimetric and refraction seismic methods. The results of these studies indicated a generally simple, Ushaped valley incised in ancient granitic and metamorphic rocks. Valley fill materials appears to be almost entirely glaciofluvial sands and gravels. Previously postulated basalt flows and Latah clays are probably not present in significant amounts. Drilling and sampling of local drainfields revealed that the upper moist and wet valley fill materials pass into dry sands and gravels at depth. This phenomenon prompted a postulate that drainfield fluids must be moving laterally rather than vertically. Confirmatory laboratory measurements of soil moisture tension showed all of the soils, at depth, to be in a state of high moisture deficiency. Routine geophysical logging of monitoring wells indicated that moisture movement and variations were confined to upper soil layers. Infiltration tests substantiated other findings concerning the movement of soil moisture. Extensive sampling and analysis of surface and ground waters revealed no evidence of ground water contamination. Surface waters are generally of good quality east of Spokane but are seriously degraded in the immediate Spokane area.

Drewry, W.A., "Virus Movement in Ground Water Systems", OWRR-A-005-ARK(2), 1969, Water Resources Research Center, University of Arkansas, Fayetteville, Arkansas.

This study investigated the extent to which soil acts as an agent in the transmission of waterborne viruses. Since many waterborne outbreaks of viral diseases have involved small well water supplies contaminated by effluents from subsurface wastewater disposal systems, there is a great need for such information. Results show that virus adsorption by soils is greatly affected by the pH, ionic strength, and soil-water ratio of the soil-water system and various soil properties. It is shown that one cannot predict the relative virus adsorbing ability of a particular soil based on the various tests normally used to characterize a soil. It is shown that virus movement through a continuous stratum of common soil under gravity flow conditions and with intermittent dosing should present no health hazard if usual public health practices relating to locating water supply wells are followed. Test results also indicate no greater or lesser movement of virus through soils with a highly polluted water than with a nonpolluted water.

Ettesvold, W.L., "On-Site Wastewater Treatment Versus Collection Sewers: A Local Health Department Viewpoint", Journal of Environmental Health, Vol. 41, No. 6, May-June 1979, pp. 321-324.

The cost effectiveness of retaining, repairing, and improving on-site wastewater treatment units and collector sewers is compared. Septic tanks and drainfields are not likely to be cost effective in densely developed areas if the costs of maintenance, inspection, pumping, sludge disposal, and ground water protection are considered. It is suggested that one large septic tank and drainfield be constructed on suitable soil in an area distant from lakes, streams, or wells.

Fetter, C.W., Jr., Sloey, W.E. and Spangler, F.L., "Potential Replacement of Septic Tank Drain Fields by Artificial Marsh Waste Water Treatment Systems", Ground Water, Vol. 14, No. 6, Nov.-Dec. 1976, pp. 396-403.

Use of emergent marsh vegetation planted in a gravel substrate in a plastic-lined trench to treat septic tank effluent is demonstrated in this paper. Treatment of unchlorinated primary municipal effluent reduces BOD_5 by 77 percent, COD by 71 percent, orthophosphate by 35 percent, total phosphorus by 37 percent, nitrate by 22 percent, and coliform bacteria by 99.9 percent. The method is useful at summer cottages, camping areas, resorts, and roadside rest areas. Marsh treatment systems are inexpensive to operate and virtually automatic.

Fey, R.T., "Cost-Minded Community Chooses Small Diameter Gravity System", Water and Sewage Works, Vol. 125, No. 6, June 1978, pp. 58-62.

In 1975, a small town in Wisconsin decided to install a small diameter gravity system for wastewater treatment because of the system's environmental compatibility and cost effectiveness. The system has a small scale force main and septic tanks, which discharge into a common absorption field. Typical plugging ingredients are eliminated in the system, and the septic tanks retain the solids. Monitoring wells in and around the absorption field are used to determine any change in ground water quality attributable to the septic effluent.

General Accounting Office, "Community-Managed Septic Systems - A Viable Alternative to Sewage Treatment Plants", CED-78-168, Nov. 1978, Community and Economic Development Division, Washington, D.C.

This report discusses the benefits and obstacles concerning septic systems as viable waste water treatment alternatives to central treatment processes. Properly operating septic systems can be as permanent and effective as central treatment facilities, at considerably less cost.

Goldstein, S.M. et al., "A Study of Selected Economic and Environmental Aspects of Individual Home Wastewater Treatment Systems", Report No. M72-45, Mar. 1972, Mitre Corp., McLean, Virginia.

This report evaluates the potential effectiveness of individual home waste treatment systems and estimates the cost implications of increased use of individual systems. A review of previous research into septic tank system failures is summarized. Economic factors which can govern the choice between individual and collective systems are reviewed. The results of several economic analyses of the problem are discussed. A MITRE-developed economic model is used to generate both the time stream and the total present value of future costs of sewage treatment on a national basis for projected new individual homes. Simultaneous consideration is given to individual and central systems for a variety of independently specified parameters.

Hagedorn, C. et al., "Survival and Movement of Fecal Indicator Bacteria in Soil Under Conditions of Saturated Flow", Journal of Environmental Quality, Vol. 7, No. 1, Jan.-Mar. 1978, pp. 55-59.

Antibiotic resistant fecal bacteria were used to monitor the degree of movement and subsequent ground water contamination by septic tank effluent discharged into a drainfield under saturated conditions. Two pits of different depths were constructed to simulate drainfield beds, and ground water samples were removed during 32-day sampling intervals from sampling wells installed at set distances from each inoculation pit. The bacteria added to the deep pit were released into a B2t horizon which contained a higher clay content than the A horizon in which the shallower pit was installed. Streptomycin resistant strains of Escherichia coli and Streptococcus faecalis amended to each pit site moved in a directional manner, required more time to reach sampling wells when inoculated into the deeper of the two pits, and moved relatively long distances when considering that the area where the sites were located had only a 2 percent slope. Bacterial numbers peaked in the sampling wells in association with major rainfall patterns and the populations required longer periods to peak in the wells furthest from the inoculation pits. The results indicated that antibiotic resistant bacteria eliminated the problem of differentiating between the amended bacteria and those nonresistant strains already in the soil, and the potential is excellent for including this type of microbiological procedure for assessing the suitability of a soil site for septic tank and wastewater drainfield installations.

Healy, K.A. and Laak, R., "Site Evaluation and Design of Seepage Fields", American Society of Civil Engineers Journal of Environmental Engineering Division, Vol. 100, No. 5, Oct. 1974, pp. 1133-1146.

A reevaluation of previous work by others indicated that soil can absorb septic tank effluent indefinitely if the application rate is kept below a certain level, which is a function of soil permeability. This long term acceptance rate is independent of whether the soil is continuously or intermittently flooded, and varies from approxiamtely 0.3 gpd/sq ft (0.01 m/day) for clay loam and silt to approximately 0.8 gpd/sq ft (0.03 m/day) for sand. A study of the ground water flow pattern below a seepage field showed that it is, in many cases, the hydraulic conductivity of the ground surrounding the field, as determined by the external water table, soil permeability, and impervious strata, that controls the size of the field required. Reliable techniques for site evaluation of soil permeability, depth to water table, and depth to any impervious strata are presented, and a chart is given for designing a seepage field based on this information. Design examples are included in this paper.

Holzer, T.L., "Limits to Growth and Septic Tanks", Proceedings of the Rural Environmental Engineering Conference on Water Pollution Control in Low Density Areas, 1975, University Press of New England, Hanover, New Hampshire, pp. 65-87.

Appraisal of the potential pollution of ground water by septic tank systems requires an understanding of the ground water system into which the effluent is discharged. Flow paths of ground water and recharge areas must be delineated. The quantity of ground water recharge must be estimated because the recharge is a measure of the amount of water available for dilution of the effluent. The capability of the natural system to renovate effluent from septic tank systems must be known. Data are presented in graphical and tabular form in this paper.

Jaovich, B.A. and Couillard, D., "Septic Tanks: Considerations About Drainage", Eau du Quebec, Vol. 11, No. 2, Apr. 1978, pp. 77-80.

Percolation or water infiltration tests are normally performed on soils intended for septic tank installations, but these tests present problems of reproducibility and representativity because of their empirical nature and the phenomenon of clogging. Clogging layers often develop in septic tanks within 10 months of use. Other tests--tensiometry and electric resistivity--have been developed which take into account the factors of clogging and soil water content. In general, it is best to use a clogging test in conjunction with some sort of infiltration test. The infiltration and purification capacity of soils can be improved by intermittent addition and good distribution of effluents over the entire receiving surface.

Jones, E.E., "Improving Subsurface Disposal System Performance", Journal of Environmental Health, Vol. 40, No. 4, Jan.-Feb. 1978, pp. 186-191.

Onsite domestic waste disposal facilities can have nearly infinite life at reasonable cost. It is economically prohibitive to install sewers for low density populations, so improved design and maintenance of septic tanks can be as valuable as public sewerage systems. Current technologies are reviewed in this paper. Domestic waste disposal facilities can be engineered to: reduce ground water pollution; provide greater service life; lower annual costs; and make more beneficial use of effluent water and nutrients. An essential factor is soil aeration or oxidation potential, which is required by certain organic compounds for decomposition. Most new management systems need adequate soil drainage for proper functioning. Service life figures for five eastern soils are included.

Jones, R.A. and Lee, G.F., "Septic Tank Wastewater Disposal Systems as Phosphorus Sources for Surface Waters", Journal of the Water Pollution Control Federation, Vol. 51, No. 11, Nov. 1979, pp. 2764-2775.

A 4-year ground water monitoring study was conducted in the immediate vicinity of an active septic tank system in northwestern Wisconsin to determine the potential for septic tank effluent to contribute to the excessive fertilization of area surface waters. During the course of this study, movement of septic tank effluent in the ground water was indicated by measured values of several conservative parameters. However, there was no evidence of the transport of the phosphate from septic tank effluent through the ground water even at the monitoring point closest to the tile field (about 15 m down ground water gradient from the tile field). The results of this confirmed the conclusions drawn from similar studies in other areas reported in the literature, namely, that phosphorus from septic tank waste water disposal system effluent is usually not readily transported through the ground water.

Khaleel, R. and Redell, D.L., "Simulation of Pollutant Movement in Groundwater Aquifers", OWRT-A-030-Tex (1), May 1977, Water Resources Institute, Texas A&M University, College Station, Texas.

A three-dimensional model describing the two-phase (air-water) fluid flow equations in an integrated saturated-unsaturated porous medium was developed. Also, a three-dimensional convective-dispersion equation describing the movement of a conservative, noninteracting tracer in a nonhomogeneous, anisotropic porous medium was developed. Finite difference forms of these two equations were solved using an implicit scheme to solve for water or air pressures, an explicit scheme to solve for water and air saturations, and the method of characteristics with a numerical tensor transformation to solve the convective-dispersion equations. The inclusion of air as a second fluid phase caused the infiltration rate to decrease rapidly to a value well below the saturated hydraulic conductivity when the air became compressed. This is in contrast to one-phase fluid flow problems in which the saturated hydraulic conductivity is considered to be the lower bound for the infiltration rate. A typical two-dimensional drainage problem in agriculture was solved in a nonhomogeneous, integrated saturated-unsaturated medium using the total simulator of fluid flow and convective-dispersion equations. A variety of outputs, such as an equipotential map or a solute concentration map, were obtained at selected time steps. A field-size problem describing the migration of septic tank wastes around the perimeter of a lake was also considered and solved using the total simulator.

Klein, S.A., "NTA Removal in Septic Tank and Oxidation Pond Systems", Journal of Water Pollution Control Federation, Vol. 46, No. 1, Jan. 1974, pp. 78-88.

Four pilot-scale septic tank and leaching field systems were used to determine the survival of trisodium nitrilotriacetate (NTA) in ground waters and its removal by household treatment systems. Results indicated a 20 percent removal of NTA in the septic tank and complete removal for the total system when the percolation fields were aerobic.

Konikow, L.F. and Bredehoeft, J.D., "Computer Model of Two-Dimensional Solute Transport and Dispersion in Ground Water", Techniques of Water Resources Investigations of the United States Geological Survey, Book 1, Chapter 2, 1978, U.S. Geological Survey, Washington, D.C.

This report presents a model that simulates solute transport in flowing ground water. The model is both general and flexible in that it can be applied to a wide range of problem types. It is applicable to one-or two-dimensional problems having steady-state or transient flow. The model computes changes in concentration over time caused by the processes of convective transport, hydrodynamic dispersion, and mixing (or dilution) from fluid sources. The model assumes that the solute is nonreactive and that gradients of fluid density, viscosity, and temperature do not affect the velocity distribution. However, the aquifer may be heterogeneous and (or) anisotropic. The model couples the ground water flow equation with the solute-transport equation. The digital computer program uses an alternating-direction implicit procedure to solve a finite-difference approximation to the ground water flow equation, and it uses the method of characteristics to solve the solute-transport equation. The report includes a listing of the computer programs, which is written in FORTRAN IV and contains about 2,000 lines. The model is based on a rectangular, block-centered, finite-difference grid. It allows the specification of any number of injection or withdrawal wells and of spatially varying diffuse recharge of discharge, saturated thickness, transmissivity, boundary conditions, and initial heads and concentrations. The accuracy of the model was evaluated for two idealized problems for which analytical solutions could be obtained. In the case of one-dimensional flow the agreement was nearly exact, but in the case of plane radial flow a small amount of numerical dispersion occurred. An analysis of several test problems indicates that the error in the mass balance will be generally less than 10 percent. The test problems demonstrated that the accuracy and precision of the numerical solution is sensitive to the initial number of particles placed in each cell and to the size of the time increment, as determined by the stability criteria. Mass balance errors are commonly the greatest during the first several time increments, but tend to decrease and stabilize with time.

Kreissl, J.F., "Status of Pressure Sewer Technology", EPA Technology Transfer Report, Mar. 1977, U.S. Environmental Protection Agency, Cincinnati, Ohio.

Although sewage pumping has been practiced for years in municipal systems in the form of lift stations and force mains to avoid excessive depths of cut, and in many individual homes in the form of ejector or sump pumps, the wholesale use of small diameter pressure collection systems did not emerge until the later part of the 1960's. Pressure sewer systems are a viable alternative technology and should be considered in any cost-effective analysis of alternative wastewater management systems in rural communities. Pressure sewers offer many advantages in areas where population density is low, severe rock conditions exist, high ground water or unstable soils prevail, or undulating terrain predominates. The most serious impediment to wider adoption of pressure sewer technology is the lack of comprehensive long term operation and maintenance data and treatment information. The two types of pressure sewer system designs--grinder-pump systems and septic tank effluent pumping systems--are detailed.

Lotse, E.G., "Septic Tank Effluent Movement Through Soil", NTIS No. PB-261 368/5ST, June 1976, University of Maine at Orono, Orono, Maine.

The rate and extent of phosphorus and nitrogen movement in selected Maine soils were studied under continuous and intermittent loading. Conditions approximating those of septic tank absorption fields were simulated. For intermittently operated columns, there was no breakthrough of phosphorus when 8.0, 12.4, and 15.0 pore volumes of effluent, respectively, had been collected. For continuously operated columns, however, breakthrough occurred at 10.8 and 11.2 pore volumes, respectively. The greater the hydraulic loading, the greater was the rate of phosphorus movement through a given soil. Septic tank absorption field systems should have several trenches and large total length of trench in order to minimize the movement of phosphorus and contamination of ground waters.

McGrail, J.W. et al., "A Cost Comparison of Underground Disposal of Wastewater Versus Public Sewerage for Rural and Suburban Towns", New England Water Pollution Control Association, Vol. 12, No. 1, Apr. 1978, pp. 4-19.

A nonpoint source water quality model was developed, applied, and verified. Private sewage disposal may have only a minor effect on the trophic condition of many lakes, especially those draining large watersheds. Inadequate private sewage disposal may cause pathogen contamination in lakes and streams, and high P concentrations in nearshore lake waters. The installation of interceptor sewers along rural shorelines is often not cost-effective. The interceptor induced shoreline development may result in a net increase in P loading to the lake from increased runoff. Where technically feasible, a program to inspect, correct, and maintain rural private sewage disposal systems can minimize the potential for pathogen- or P-related septic tank problems at a cost lower than that for sewers. A proposed on-lot disposal system control program represents a new level of local government involvement in wastewater disposal. Communities which choose a private sewage disposal control program over public sewerage should be considered for federal and state funding.

Mellen, W.L., "Site Evaluation for Seepage Fields", Third Annual Illinois Private Sewage Disposal Symposium, 1978, Lake County Health Department, Waukegan, Illinois, pp. 1-8.

Before designing an individual sewage disposal system, it is necessary to determine if the soil is suitable for the absorption of septic tank effluent. Several conditions must be met. The maximum seasonal elevation of the ground water table should be 2 ft below the bottom of the trench. The most important clue to seasonal high water table is the color of the soil. If it has a uniform reddish-brown to yellow color, due to oxidation of Fe compounds, it indicates free alternate movements of air and water in and through the soil. Such a soil has desirable absorption characteristics. To determine impervious stratas, it may be necessary to run percolation tests in the routine manner and also at a depth of 12 in lower than the original test. The ground slope has an effect on the site's suitability and the type of distribution. Soils in humid areas of the country should have a one ft fall within the septic field area, and serial distribution or a dropbox system should be used. Level areas are subject to supersaturation from building runoff and sump pumps during heavy spring rains.

Mercer, J.W. and Faust, C.R., "Ground Water Modeling: Mathematical Models", Ground Water, Vol. 18, No. 3, May-June 1980, pp. 212-227.

Ground water modeling begins with a conceptual understanding of the physical problem. The next step in modeling is translating the physical system into mathematical terms. In general, the final results are the familiar ground water flow equation and transport equations. These equations, however, are often simplified, using site specific assumptions, to form a variety of equation subsets. An understanding of these equations and their associated boundary and initial conditions is necessary before a modeling problem can be formulated.

Otis, R.J. et al., "On-Site Disposal of Small Wastewater Flows", EPA Technology Transfer Report, 1977, U.S. Environmental Protection Agency, Washington, D.C.

A noncentral facility consisting of several treatment and disposal systems serving isolated individual residences or clusters of residences in rural areas offers an economical solution to the problem of waste disposal in those areas where conventional central facilities are impractical. Individual or shared septic tank systems could provide onsite treatment and disposal where wastes are generated. Noncentral facilities are also more ecologically sound than centralized systems, since the dispersed systems dispose of wastes over wider areas. Through this practice, the environment is able to assimilate the waste discharge more readily, thereby reducing the need for mechanical treatment and the associated energy consumption. Various treatment and disposal systems that would be applicable to the noncentralized theory are described, and potential problems associated with each system are reviewed.

Otis, R.J., Plews, G.D. and Patterson, D.H., "Design of Conventional Soil Absorption Trenches and Beds", Third Annual Illinois Private Sewage Disposal Symposium, 1978, Toledo Area Council of Governments, Toledo, Ohio, pp. 52-66.

A good soil absorption system should absorb all effluent generated, provide a high level of treatment before the effluent reaches the ground water, and have a long, useful life. To meet these goals, proper site selection is necessary. Factors to be considered include the hydraulic conductivity characteristics of the soil, the unsaturated depth of the soil, the distance to bedrock, characteristics of the bedrock, the landscape position, slope of the land, and proximity to surface waters, wells, road cuts, buildings, etc. Trench and bed designs are discussed in detail with reference to the "Manual of Septic Tank Practice" of the USPHS. Probably the most frequent cause of early failure of properly designed systems is poor construction. Absorption of waste effluent requires that soil pores remain open. If these are sealed during construction by compaction, smearing, or puddling, the system may be rendered useless. Careful construction techniques will minimize these causes of soil clogging.

Pickens, J.F. and Lennox, W.C., "Numerical Simulation of Waste Movement in Steady Ground Water Flow Systems", Water Resources Research, Vol. 12, No. 2, Apr. 1976, pp. 171-184.

The finite element method based on a Galerkin technique is used to formulate the problem of simulating the two-dimensional transient movement of conservative or nonconservative wastes in a steady state saturated ground water flow system. The convection-dispersion equation is solved in two ways: in the conventional Cartesian coordinate system, and in a transformed coordinate system equivalent to the orthogonal curvilinear coordinate system of streamlines and normals to those lines. The two formulations produce identical results. A sensitivity analysis on the dispersion parameter "dispersivity" is performed, establishing its importance in convection-dispersion problems. Examples involving the movement of nonconservative contaminants described by distribution coefficients and examples with variable input concentration are also given. The model can be applied to environmental problems related to ground water contamination from waste disposal sites.

Pitt, W.A.J., Jr., "Effects of Septic Tank Effluent on Ground Water Quality, Dade County, Florida: An Interim Report", Ground Water, Vol. 12, No. 6, Nov.-Dec. 1974, pp. 353-355.

At each of five sites, where individual (residence) septic tanks have been in operation for at least 15 years and where septic tank concentration is less than 5 per acre, a drainfield site was selected for investigation to determine the effects of septic tank effluent on the quality of the water in the Biscayne Aquifer. At each site two sets of multiple depth wells were drilled. The upgradient wells adjacent to the drainfields in most places, were constructed so that the aquifer could be sampled at 10, 30, 40, and 60 feet below the land surface.

Prickett, T.A. and Lonnquist, C.G., "Selected Digital Computer Techniques for Ground Water Resources Evaluation", Bulletin 55, 1971, Illinois Water Survey, Urbana, Illinois.

Generalized digital computer program listings are given that can simulate one-, two-, and three-dimensional nonsteady flow of ground water in heterogeneous aquifers under water table, nonleaky, and leaky artesian conditions. Programming techniques involving time varying pumpage from wells, natural or artificial recharge rates, the relationships of water exchange between surface waters and the ground water reservoir, the process of ground water evapotranspiration, and the mechanism of converting from artesian to water table conditions are also included. The discussion of the digital techniques includes the necessary mathematical background, documented program listings, theoretical versus computer comparisons, and field examples. Also presented are sample computer input data and explanations of job setup procedures. A finite difference approach is used to formulate the equations of ground water flow. A modified alternating direction implicit method is used to solve the set of resulting finite difference equations. The programs included are written in FORTRAN IV and will operate with any consistent set of units.

Prickett, T.A., "State-of-the-Art of Groundwater Modeling", Water Supply and Management, Vol. 3, No. 2, 1979, pp. 134-141.

An outline for ground water modeling techniques within the categories of mathematical, sand tank, analog and numerical models is presented. The models discussed are one of two types: ground water "flow" or "transport" models. The term "flow" refers to the type of model used to provide answers to the quantitative side of ground water problems; typical "flow" models would solve problems related to safe yields of well fields, interference effects of nearby wells and surface water-ground water relationships. The "transport" model is a water quality type; typical "transport" models might include energy considerations, dispersion and diffusion processes, chemical exchange reactions, or multiple fluid effects. Mathematical solutions are models of the particular conditions defined; their applications to field problems are therefore somewhat limited. Four main groups of flow models are discussed: (1) sand tank models which are a scaled down representation of an aquifer, including its boundary configuration and usually its hydraulic conductivity; (2) analog models, where the behavior of an aquifer is described by differential equations which are derived from basic principles such as the laws of continuity and conservation of energy; (3) analog models, which can be subdivided into the three major categories of viscous fluid models, electrical models, and miscellaneous models and techniques; and (4) numerical models, which have been powerful tools in aiding hydrologists in evaluating ground water resources. Development of the digital computer has made possible the practical use of the techniques in ground water flow modeling. Numerical models can be subdivided into four groups: finite-difference, finite-element variational, finite-element Galerkin, and miscellaneous.

Rahe, T.M. et al., "Transport of Antibiotic-Resistant Escherichia Coli Through Western Oregon Hillslope Soils Under Conditions of Saturated Flow", Journal of Environmental Quality, Vol. 7, No. 4, Oct.-Dec., 1978, pp. 487-494.

Field experiments using strains of antibiotic resistant Escherichia coli were conducted to evaluate the events which would occur when a septic tank drainfield became submerged in a perched water table and fecal bacteria were subsequently released into the ground water. Three separately distinguishable bacterial strains were inoculated into three horizontal lines installed in the A, B, and C horizons of two western Oregon hillslope soils. Movement was evaluated by collecting ground water samples from rows of modified piezometers (six piezometers/row) placed at various depths and distances downslope from the injection lines. Transport of E. coli differed at both sites with respect to movement rates, zones in the soil profiles through which major translocation occurred, and the relative numbers of cells transported over time. Movement rates of at least 1,500 cm/hour were observed in the B horizon at one site. The strains of E. coli survived in large numbers in the soils examined for at least 96 hours and appeared to be satisfactory as tracers of subsurface water flow. The concept of partial displacement (or turbulent flow through macropores) is discussed as an explanation of the rapid movement of substantial numbers of microbial cells through saturated profiles.

Rea, R.A. and Upchurch, S.B., "Influence of Regolith Properties on Migration of Septic Tank Effluent", Ground Water, Vol. 18, No. 2, Mar.-Apr. 1980, pp. 118-125.

An investigation of waste migration patterns from a septic tank system indicates that complex patterns result from minor variations in regolith adsorptive capacity and texture, local hydrology, and possibly soil microbiology. The existence of multichemical, bifurcating plumes suggest that monitor wells arranged up and downgradient and capable of multilevel sampling are essential to adequately delineate contaminant migration in ground water. The data also indicate that sampling for a single constituent could yield misleading information about the nature and distribution of other ground water contaminants. The ability for chemical removal by the regolith is in direct response to minor variations in silt-and clay-sized particle content and corresponds to Langmuir adsorption isotherms. Silt- and clay-sized particles are dominantly organic in origin. Minor iron and aluminum hydroxyoxides and clays are present. Substrate samples, when collected at regular intervals and analyzed for adsorbed constituents and textural variability, provide an integrated picture of the distribution of waste chemicals through time. Such samples also provide insight into the mechanics of plume configuration and flow characteristics within the regolith. The study shows that regolith adsorption data are essential to the determination of life expectancy of the regolith as a contaminant treatment system.

Reneau, R.B., "Changes in Concentrations of Selected Chemical Pollutants in Wet, Tile-Drained Soil Systems as Influenced by Disposal of Septic Tank Effluents", Journal of Environmental Quality, Vol. 8, No. 2, Apr.-Jun. 1979, pp. 189-197.

Investigations at three Virginia locations determined onsite changes in several chemical constituents of septic tank effluent in shallow ground waters and soils. Changes were related to distance traveled, soil properties, and seasonal variation between subsurface absorption fields and a subsurface tile drainage system. Fluctuations in phosphorus, ammonium, nitrate, nitrogen dioxide, chlorine, pH, and methylene blue active substances were measured. Most of the chemical constituents monitored had been lowered to acceptable levels by the time effluent was intercepted by the tile drain. Effective soil volume, as determined by distance to and depth of drainage system and by soil type, was directly related to effluent purification. Techniques for improving water quality are suggested.

Reneau, R.B. et al., "Distribution of Total and Fecal Coliform Organisms from Septic Effluent in Selected Coastal Plains", Public Health Reports, Vol. 92, No. 3, May-June 1977, pp. 251-260.

Distribution of total and fecal coliform bacteria in three Atlantic Coastal Plain soils was monitored in situ for three years. The soils studied were Varina, Goldsboro, and Beltsville sandy loams. These soils are found extensively in the populous U.S. Atlantic seaboard, which is considered only marginally suitable for septic tank installation because the restricting soil layers result in the subsequent development of seasonal perched water tables. As distance from the drainfield increased, large reductions in total and fecal coliform bacteria were noted in the perched ground waters above the restricting layers. These restricting soil layers appear to be effective barriers to the vertical movement of indicator organisms. The reduction in the density of the coliform bacteria above the restriction soil layers probably can be attributed to dilution, filtration, and dieoff as the bacteria move through the natural soil system.

Reneau, R.B. and Pettry, D.E., "Phosphorus Distribution from Septic Tank Effluent in Coastal Plain Soils", Journal of Environmental Quality, Vol. 5, No. 1, Jan.-Mar. 1976, pp. 34-40.

Phosphorus concentrations in perched ground waters around septic tank drainfields are determined. The influence of disposal of septic tank effluent on soil phosphorus fractions and their distribution in natural soil systems is described. Contamination of a permanent ground water table via vertical movement is a limited possibility at Varina and Goldsboro soil locations.

Reneau, R.B., Jr., "Changes in Inorganic Nitrogeneous Compounds from Septic Tank Effluent in Soil with a Fluctuating Water Table", Journal of Environmental Quality, Vol. 6, No. 2, Apr.-Jun. 1977, pp. 173-178.

Changes in ammonia and nitrates were monitored in situ during 1972, 1973, 1974, and 1975 in a Virginia Coastal Plain soil with a fluctuating water table. Samples of soil solution above and in a very slowly permeable plinthic horizon were analyzed for the above-mentioned inorganic N fractions. Ammonium-N in solution above the plinthic horizon decreased with increased distance from the drainfield in the direction of ground water flow. Decreases were attributed to the processes of adsorption and nitrification. Nitrite and nitrate concentrations did not change significantly with distance above the plinthic horizon, but did accumulate in the plinthic material beginning at a 1.27-m distance from the drainfield.

Reneau, R.B., Jr., "Influence of Artificial Drainage on Penetration of Coliform Bacteria from Septic Tank Effluents into Net Tile Drained Soils", Journal of Environmental Quality, Vol. 7, No. 1, Jan.-Mar. 1978, pp. 23-30.

The bacteria were monitored in ground waters at selected distances from the septic tank drainage fields in the direction of ground water flow and were compared to coliform densities in control wells and in tile outfalls. Fecal coliform densities were approximately 105/100 ml in ground waters adjacent to the disposal area as compared to 101-103/100 ml 152 cm from the agricultural tile and less than 3.0/100 ml in control wells. The outfall from the study area was normally less than 200 fecals/100 ml compared to less than 3.0/100 ml in outfall waters from a control area. Fecal coliform densities of the outfall from the study area were some tenfold less than the bacterial quality of the receiving stream. Coliform densities in ground waters decreased as a logarithmic function of distance. In these soils, artificial drainage systems apparently lowered the seasonal fluctuating water tables to such a degree that individual wastewater treatment systems did not fail as a result of untreated or partially treated effluent coming to the surface. It is more difficult to assess the adequacy of artificial drainage with respect to penetration of coliform organisms present in the wastewaters.

Russelman, H.B. and Turn, M.P., "Management of Septic Tank Solids", Third Annual Illinois Private Sewage Disposal Symposium, 1978, Toledo Area Council of Governments, Toledo, Ohio, pp. 9-17.

Septage from septic tanks is a biodegradable waste capable of affecting the environment through water and air pollution. Proper control of its disposal requires knowing the number of tanks installed and the rate of new installations; the quantity of septage being hauled and by whom; the generally used disposal practices and problems associated with them; and the regulatory framework controlling the disposal. A private sewage disposal program which attempts only to assure proper effluent disposal does not adequately address problems inherent in the ultimate disposal of the residue. To assure a more complete role in the implementation of a disposal control program, a community should establish a licensing fee consistent with administrative costs. This would help prevent irresponsible scavengers from operating and also provide revenue to offset costs of inspection. It should require periodic inspection of all hauling vehicles and permit the use only of disposal sites found adequate to prevent surface and ground water pollution. It can permit the utilization of experimental sites if the absence of public health hazards is determined from monitoring operations. Since land application is the most common and economical method available to contractors, septage application rates should be studied and contractors should be guided in achieving the most cost-effective methods consistent with public health criteria.

Sawhney, B.L., "Predicting Phosphate Movement through Soil Columns", Journal of Environmental Quality, Vol. 6, No. 1, Jan.-Mar. 1977, pp. 86-89.

To assess the potential pollution of ground water with P from septic tank drainfields, sorption capacities of various soils were determined over an extended period of time and related to P movement through soil columns using solutions having P concentrations similar to waste waters. The amounts of P sorbed by fine sandy loam (fsl) and silt loam (sil) soil columns before breakthrough occurred were approximately equal to the sorption capacities determined from isotherms obtained over a sufficiently long reaction time of about 200 hours. In Merrimac fsl, breakthrough occurred after about 50 pore volumes of waste water had passed through the column, while about 100 pore volumes passed through Buxton sil before the breakthrough occurred.

Sawhney, B.L. and Starr, J.L., "Movement of Phosphorus from a Septic System Drainfield", Journal Water Pollution Control Federation, Vol. 49, No. 11, Nov. 1977, pp. 2238-2242.

Movement of phosphorus (P) from a septic tank drainfield through the surrounding soil to ground water and its eventual discharge to surface waters was investigated. Suction probes and tensiometers were intalled at various distances below and beside the drainfield to obtain effluent solutions and moisture distribution. Soon after the septic tank was put into use, ponding of the effluent in the trench began. Movement of P from the trench occurred in both downward and the horizontal directions.

Scalf, M.R., Dunlap, W.J. and Kreissl, J.F., "Environmental Effects of Septic Tank Systems", Report No. EPA/600/3-77/096, Aug. 1977, Robert S. Kerr Environmental Research Laboratory, U.S. Environmental Protection Agency, Ada, Oklahoma.

Septic tank soil absorption systems are the most widely used method of onsite domestic waste disposal. Almost one-third of the United States population depends on such systems. Although the percentage of newly constructed homes utilizing septic tanks is decreasing, the total number continues to increase. Properly designed, constructed, and operated septic tank systems have been demonstrated as an efficient and economical alternative to public sewer systems, particularly in rural and sparsely developed suburban areas. However, because of their widespread use in unsuitable situations, they have also demonstrated the potential for contamination of ground and surface waters.

Shoemaker, C.A. and Porter, K.S., "Recharge and Nitrogen Transport Models for Nassau and Suffolk Counties, New York", NTIS PB-276 906/5ST, Jan. 1978, Cornell University, Ithaca, New York.

Ground water aquifers underlying Long Island are the only source of drinking water for more than 2.5 million people in Nassau and

Suffolk Counties in Long Island, New York. Due to residential and agricultural land use, the ground water is being contaminated by nitrogen. In order to quantify both the amount of recharge water and the nitrogen concentration in the recharge, a simulation model has been developed. The model calculates a mass balance of water and nitrogen on 762 cells, each of which is 1.5 miles square. The calculations which are computed daily or monthly are based upon land use, soil type, temperature, precipitation and sewerage in each grid. Detailed soil moisture data were collected at several sites. Data from the early part of the year were used to calibrate the model. Validation was achieved by comparison with independent data collected in the late part of the year. The average recharge of precipitation for Nassau and Suffolk Counties was estimated by the model to be 1140 mgpd or 20.5 in per year. Lawn fertilizer and septic systems were the major sources of nitrogen in the recharge water.

Sproul, O.J., "Virus Movement Into Ground Water from Septic Tank Systems", Rural Environmental Engineering Conference on Water Pollution Control in Low Density Areas Proceedings, Paper No. 12, Sept. 1973, University Press of New England, Hanover, New Hampshire.

Viruses can be recovered from any water that has been subjected to viral contamination. In situations where wastewater is to be discharged to the local environment, e.g., one's backyard, as with the septic tank system, the concern of the homeowner should be obvious, especially if his water supply is a private well only a few feet from the septic tank system. Viruses from these supplies are routinely involved in outbreaks of infectious hepatitis and gastroenteritis. Methods of predicting the capacity of a septic tank soil absorption system to remove viruses and to develop criteria to assess this capacity are discussed.

Troyan, J.J. and Norris, D.P., "Cost-Effectiveness Analysis of Alternatives for Small Wastewater Treatment Systems", EPA Technology Transfer Report, Mar. 1977, U.S. Environmental Protection Agency, Washington, D.C.

Information pertinent to the cost-effectiveness analysis of sewerage systems for both small communities and rural residential areas is presented. Procedures for use in determining the feasibility and desirability of employing four onsite systems and four types of community collection systems are described. Major objectives of the study include: identifying the problem conditions that must be considered in selecting sewerage alternatives; outlining the advantages, drawbacks, and limitations of the onsite and community collection alternatives presented; reviewing a procedure for screening and analyzing costs of alternatives for individual homes; and examining a set of case histories taken from recent sewerage reports and facilities plans.

Uttormark, P.D., Chapin, J.D. and Green, K.M., "Estimating Nutrient Loadings of Lakes from Non-Point Sources", EPA 660/3-74-020, Aug. 1974, University of Wisconsin, Madison, Wisconsin.

Data describing nutrient contributions from nonpoint sources were compiled from the literature, converted to kg/ha/yr, and tabulated in a format convenient for estimating nutrient loadings of lakes. Contributing areas are subdivided according to general use categories, including agricultural, urban, forested, and wetland. Data describing nutrient transport by ground water seepage and bulk precipitation are given along with data for nutrient contributions from manure handling, septic tanks, and agricultural fertilizers.

Vilker, V.L., "An Adsorption Model for Prediction of Virus Breakthrough from Fixed Beds", Land Treatment of Waste Water International Symposium, Hanover, New Hampshire, Aug. 1978, pp. 381-389.

Laboratory and field studies have demonstrated the potential for biological and chemical contamination of U.S. ground water supplies by percolation from land application of untreated and treated wastewater, sludge land spreading, septic tanks, and landfill leachates. Experiments were conducted and mathematical models were developed to predict the breakthrough of low levels of virus from percolating columns under conditions of adsorption and elution. Breakthrough of viruses was illustrated by ion exchange/adsorption equations. Predictions were in qualitative agreement with observations from experiments that measured virus uptake by activated carbon or silty soil in columns.

Vilker, V.L. et al., "Water - 1977 (Application of Ion Exchange/Adsorption Models to Virus Transport in Percolating Beds)", AICHE Symposium Series, Vol. 74, No. 178, 1978, pp. 84-92.

Ground water currently constitutes 95 percent of the U.S. freshwater supply. This supply is subject to biological and chemical contamination by percolation from surface spreading of untreated and treated wastewater, sludge land spreading, septic tanks, and landfill leachate. Examined is the magnitude of the threat of virus contamination of ground water supplies that is presented by these waste disposal practices. Described are initial experimental and mathematical modeling efforts to predict breakthrough of low levels of virus from percolating columns under conditions of adsorption and elution. This breakthrough is described by the ion exchange/adsorption equations that include the effects of external mass transfer and nonlinear adsorption isotherms. Predictions qualitatively agree with reported observations from experiments that measured virus uptake by columns packed with activated carbon or a silty soil.

Viraraghavan, T., "Influence of Temperature on the Performance of Septic Tank Systems", Water, Air and Soil Pollution, Vol. 7, No. 1, Jan. 1977, pp. 103-110.

Air, liquid and soil temperatures are important environmental factors that influence the operation of septic tank soil absorption systems. An investigation conducted near Ottawa, Ont., on the efficiency of an experimental tile system did not show any specific trend between soil temperatures (depth dependent) and efficiency; this can be attributed to the fact that the depth factor carries with it other elements such as proximity to ground water table, and oxygen penetration that significantly influence the efficiency of the system.

Viraraghavan, T., "Travel of Microorganisms from a Septic Tile", Water, Air, and Soil Pollution, Vol. 9, No. 3, Apr. 1978, pp. 355-362.

An investigation was carried out to monitor the horizontal travel of indicator microorganisms from the end of a 7.93-m-long septic tile in the direction of ground water flow. Ground water samples were collected on two occasions at distances of 0, 2.10, 3.05, 9.15, 12.20, and 15.25 m from the end of the septic tile by putting down bores about 2 m deep, and analyzed for indicator organisms (coliforms, fecal coliforms, and fecal streptococci). The microorganism levels exhibited a declining trend with distance away from the tile end. Because the unsaturated depth of soil available for microorganisms vertical travel was limited, relatively high levels of organisms were found in the ground water even at a horizontal distance of 15.25 m from the end of the septic tile.

Viraraghavan, T. and Warnock, R.G., "Groundwater Pollution from a Septic Tile Field", Water, Air and Soil Pollution, Vol. 5, No. 3, Apr. 1976, pp. 281-287.

The characteristics of the ground water below an existing septic tile field were studied during the summer of 1973. The concentrations for chemical constituents were found to be much lower in the ground water compared to the septic tank effluent; however, these were quite high compared to background levels for the ground water in the area, indicating the pattern of pollution that is taking place.

Walker, W.G. et al., "Nitrogen Transformations During Subsurface Disposal of Septic Tank Effluent in Sands: II. Ground Water Quality", Journal of Environmental Quality, Vol. 2, No. 4, 1973, pp. 521-525.

Ground water observation wells were installed in the immediate vicinity of four septic tank effluent soil disposal systems. Potentiometric maps were constructed from measurements of the ground water level at each site to establish the direction of movement. Ground water samples were pumped from each well to establish patterns of N enrichment in the ground water around the seepage bank and to evaluate the performance of these disposal systems in sands in terms of N removal. Soil disposal systems of septic tank effluent in sands were found to add significant quantities of nitrate (NO_3-N), formed by nitrification of NH_4-N, the dominant N form in the effluent, to underlying ground water. The data obtained suggest that in sands, the only active mechanism of lowering the NO_3-N content is by dilution with uncontaminated ground water. Relatively large areas of 0.2 ha (0.5 acre) down gradient were needed in the studied systems before concentrations in the top layer of the ground water were lower than 10 mg/l. The average N input per person was 8 kg (10 lb) per year. Essentially complete nitrification in the soil results in addition of approximately 33 kg NO_3-N (73 lb) to the ground water per year for an average family of four.

Waltz, J.P., "A System for Geologic Evaluation of Pollution Potential at Mountain Dwelling Sites", NTIS PB-240 820/2ST, Jan. 1975, Colorado State University, Fort Collins, Colorado.

Development of mountain homesites is accelerating in the Rocky Mountains of central Colorado. These homesites often require individual water wells and sewage disposal systems. Unfortunately, the widely used septic tank leach field system generally is not suited for use in the mountainous terrain where soils are thin or missing. Although current federal regulations call for 6 ft or more of soil at the leach field site, many of the individual sewage disposal systems now in operation in the Rocky Mountain Region of Colorado fail to meet this requirement. Sewage effluent at these sites may directly enter bedrock fractures and travel large distances without being purified. As a consequence, contamination of streams, lakes, and ground water from these malfunctioning leach fields has become a problem of increasing magnitude. Investigations of geologic, topographic, and hydrologic conditions at over 100 homesites in the Rocky Mountains of north-central Colorado have resulted in the development of objective criteria for evaluating pollution potential at mountain homesites.

Weeter, D.W., "The Use of Evapotranspiration as a Means of Wastewater Disposal", Report No. 70, May 1979, National Technical Information Service, U.S. Department of Commerce, Springfield, Virginia.

A laboratory study, mathematical models, and a literature search were employed to determine the applicability of evapotranspiration to treat onsite disposal of septic tank effluent and aerobically treated effluent. The relative fate of some trace metals within the evapotranspiration rates (outflow) and the infiltration rates (inflow) of the proposed evapotranspiration bed. A literature search related soil-ground water parameters to the inflow-outflow rates and attempted to determine the effective life of the system. Results of the study show that evapotranspiration rates of aerobically digested water are equal to the rates for septic tank effluent; that evapotranspiration is independent of the dry plant matter produced; and the two feed solutions showed equal metal uptake rates. It is concluded that the cost of this method is economically justifiable in certain circumstances.

Willis, R. and Dracup, J.A., "Optimization of the Assimilative Waste Capacity of the Unsaturated and Saturated Zones of an Unconfined Aquifer System", Report No. UCLA-NEG-7394, Dec. 1973, School of Engineering and Applied Science, University of California, Los Angeles, California.

A mathematical model to optimize the assimilative waste capacity of unconfined aquifers is formulated. The aquifer is to be used conjunctively with surface sources as a source of water supply.

Waste waters may be introduced into the ground water aquifer system by either well injection or by basin spreading of waste waters. In the model, three treatment processes are available to reduce constituent concentrations present in waste waters: (1) dilution; (2) surface treatment of each constituent; and (3) the assimilative capacity of the unsaturated and saturated zones of the aquifer system. The total cost for supplying the dilution water and the cost for surface treatment of each constituent is minimized by the model.

APPENDIX B

CHARACTERISTICS OF SEPTIC TANK AREAS IN CENTRAL OKLAHOMA

Data Sheet

Site: Arcadia, Oklahoma County, Oklahoma

Permeability

(a) soil type	-	Darnell-Stephenville fine sandy loam
(b) in/hr	-	high percolation
Depth to water table (ft)	-	24
Land/water table gradient (slope - %)	-	3-12% (land)
Distance to Public/Private water source (ft)	-	< 100 (private wells)
Thickness of Porous Layer to Bedrock (ft)	-	estimate < 2
Population of Area	-	410
(a) year of census	-	1975
(b) Estimated Application rate (MG/Yr)	-	7.8 (52 gal/person-day)

Data Sheet

Site: Arrowhead Hills, Oklahoma County, Oklahoma

Permeability

- (a) soil type - Darnell-Stephenville fine sandy loam
- (b) in/hr - high percolation

Depth to water table (ft) - 50

Land/water table gradient (slope - %) - 3-12% (land)

Distance to Public/Private water source (ft) - < 100 (private wells)

Thickness of Porous Layer to Bedrock (ft) - < 1 (severely eroded)

Population of Area - 488

- (a) year of census - 1975
- (b) Estimated Application rate (MG/Yr) - 9.3 (52 gal/person-day)

Data Sheet

Site: Crutcho, Oklahoma County, Oklahoma

Permeability

(a) soil type	-	Stephenville fine sandy loam
(b) in/hr	-	high permeability
Depth to water table (ft)	-	34
Land/water table gradient (slope - %)	-	3-5% (land)
Distance to Public/Private water source (ft)	-	< 100 (private wells)
Thickness of Porous Layer to Bedrock (ft)	-	3-4 -
Population of Area	-	587 (3522/6)
(a) year of census	-	1977 (estimated)
(b) Estimated Application rate (MG/Yr)	-	11.1 (average 52 gal/person-day)

Data Sheet

Site: Del City, Oklahoma County, Oklahoma

Permeability		
(a) soil type	-	Renfrow clay loam
(b) in/hr	-	0.06
Depth to water table (ft)	-	140 to 180
Land/water table gradient (slope - %)	-	1-3% (land)
Distance to Public/Private water source (ft)	-	< 200 (public water wells in area)
Thickness of Porous Layer to Bedrock (ft)	-	1 to 4 -
Population of Area	-	246
(a) year of census	-	1975
(b) Estimated Application rate (MG/Yr)	-	4.7 (52 gal/person-day)

Data Sheet

Site: Forest Park, Lake Hiwassee, and Lake Alma, Oklahoma County, Oklahoma

Permeability		
(a) soil type	-	Darnell-Stephenville fine sandy loam
(b) in/hr	-	high percolation
Depth to water table (ft)	-	65-100
Land/water table gradient (slope - %)	-	3-12% (land)
Distance to Public/Private water source (ft)	-	< 100 (private wells)
Thickness of Porous Layer to Bedrock (ft)	-	1 -
Population of Area	-	1200
(a) year of census	-	1975
(b) Estimated Application rate (MG/Yr)	-	27.0 (52 gal/person-day)

Data Sheet

Site: Green Pastures, Oklahoma County, Oklahoma

Permeability

(a) soil type	-	Darnell-Stephenville fine sandy loam
(b) in/hr	-	very rapid percolation
Depth to water table (ft)	-	50 to 60
Land/water table gradient (slope - %)	-	3-12% (land)
Distance to Public/Private water source (ft)	-	< 100 (private wells)
Thickness of Porous Layer to Bedrock (ft)	-	1
Population of Area	-	2313
(a) year of census	-	1977 (estimate)
(b) Estimated Application rate (MG/Yr)	-	43.9 (52 gal/person-day)

Data Sheet

Site: Midwest City, Oklahoma County, Oklahoma

Permeability		
(a) soil type	-	Darnell-Stephenville fine sandy loam
(b) in/hr	-	high percolation
Depth to water table (ft)	-	36 to 44
Land/water table gradient (slope - %)	-	3-12% (land)
Distance to Public/Private water source (ft)	-	< 200 (public water wells in area)
Thickness of Porous Layer to Bedrock (ft)	-	1 to 4 -
Population of Area	-	12040
(a) year of census	-	1975
(b) Estimated Application rate (MG/Yr)	-	228.5 (52 gal/person-day)

Data Sheet

Site: Mustang, Canadian County, Oklahoma

Permeability		
(a) soil type	-	Binger fine sandy loam
(b) in/hr	-	high percolation
Depth to water table (ft)	-	20
Land/water table gradient (slope - %)	-	1-5 (land)
Distance to Public/Private water source (ft)	-	5 miles (Lake Overholser)
Thickness of Porous Layer to Bedrock (ft)	-	3-4
Population of Area	-	3550
(a) year of census	-	1975
(b) Estimated Application rate (MG/Yr)	-	67.4 (52 gal/person-day)

Data Sheet

Site: Nicoma Park, Oklahoma County, Oklahoma

Permeability		
(a) soil type	–	Darnell-Stephenville fine sandy loam
(b) in/hr	–	high percolation
Depth to water table (ft)	–	62 to 83
Land/water table gradient (slope – %)	–	3–12%
Distance to Public/Private water source (ft)	–	< 100 (private wells)
Thickness of Porous Layer to Bedrock (ft)	–	< 1 -
Population of Area	–	3000
(a) year of census	–	1975
(b) Estimated Application rate (MG/Yr)	–	57 (52 gal/person-day)

Data Sheet

Site: East Norman, (East of 24th Street), Cleveland County, Oklahoma

Permeability	
(a) soil type	Darnell-Stephenville fine sandy loam
(b) in/hr	high percolation
Depth to water table (ft)	145 to 185 (low due to water well drawdown)
Land/water table gradient (slope - %)	3-12% (land)
Distance to Public/Private water source (ft)	< 1/2 mile (public water well)
Thickness of Porous Layer to Bedrock (ft)	2 to 3 -
Population of Area	estimate 8000 (Koscinski, 1980)
(a) year of census	1980
(b) Estimated Application rate (MG/Yr)	151.8 (52 gal/person-day)

Data Sheet

Site: Seward Area (South of Guthrie to Oklahoma County Line) Logan County, Oklahoma

Permeability		
(a) soil type	-	Darnell-Stephenville fine sandy loam
(b) in/hr	-	high percolation
Depth to water table (ft)	-	13 to 31
Land/water table gradient (slope - %)	-	estimate 3-12 (land)
Distance to Public/Private water source (ft)	-	< 200 (private wells)
Thickness of Porous Layer to Bedrock (ft)	-	2 to 3-
Population of Area	-	2247 (Gaither, 1980)
(a) year of census	-	1980
(b) Estimated Application rate (MG/Yr)	-	174.6 (52 gal/person-day)

Data Sheet

Site: Silver Lake Estates, Oklahoma County, Oklahoma

Permeability		
(a) soil type	-	Vernon-Zaneis soil
(b) in/hr	-	slow percolation - 0.06
Depth to water table (ft)	-	12
Land/water table gradient (slope - %)	-	3-5%
Distance to Public/Private water source (ft)	-	0.25 miles to Lake Hefner
Thickness of Porous Layer to Bedrock (ft)	-	< 4 -
Population of Area	-	325
(a) year of census	-	1975
(b) Estimated Application rate (MG/Yr)	-	6.2 (52 gal/person-day)

Data Sheet

Site: Sunvalley Acres, Canadian County, Oklahoma

Permeability

- (a) soil type - Reinach fine sandy loam
- (b) in/hr - moderate to rapid percolation

Depth to water table (ft) - 7

Land/water table gradient (slope - %) - 1-3 (land)

Distance to Public/Private water source (ft) - < 100 (private wells)

Thickness of Porous Layer to Bedrock (ft) - estimate 3 to 4

Population of Area - 150

- (a) year of census - 1975
- (b) Estimated Application rate (MG/Yr) - 2.85 (52 gal/person-day)

APPENDIX C

PHILLIPS, NATHWANI AND MOOIJ ASSESSMENT MATRICES

Site: Arrowhead Hills, Oklahoma County, Oklahoma

WASTE \ Soil	P \ P	NP	NS	WT	G	I	D	T	TOTAL
		5.9	7.1	4.8	3	1	9.2	10	---
Ht	10	59	71	48	30	10	92	100	410
Gt	4.3	25	31	21	13	4	40	43	177
Dp	8.5	50	60	41	26	9	78	85	349
Cp	5	30	36	24	15	5	46	50	206
Bp	1.6	9	11	8	5	2	15	16	66
So	10	59	71	48	30	10	92	100	410
Vi	5	30	36	24	15	5	46	50	206
Sy	5	30	36	24	15	5	46	50	206
Ab	0	0	0	0	0	0	0	0	0
Ar	8.5	50	60	41	26	9	78	85	349
TOTAL	---	342	412	279	175	59	533	579	2379

P = normalized score

Site: Crutcho, Oklahoma County, Oklahoma

WASTE \ Soil	P \ P	NP	NS	WT	G	I	D	T	TOTAL
		5.9	7.1	7.1	3.75	1	9.2	10	---
Ht	10	59	71	71	38	10	92	100	441
Gt	4.3	25	31	31	16	4	40	43	190
Dp	8.5	50	60	60	32	9	78	85	374
Cp	5	30	36	36	19	5	46	50	222
Bp	1.6	9	11	11	6	2	15	16	70
So	10	59	71	71	38	10	92	100	441
Vi	5	30	36	36	19	5	46	50	222
Sy	5	30	36	36	19	5	46	50	222
Ab	0	0	0	0	0	0	0	0	0
Ar	8.5	50	60	60	32	9	78	85	374
TOTAL	---	342	412	412	219	59	533	579	2556

P = normalized score

Site: Del City, Oklahoma County, Oklahoma

WASTE \ Soil	P \ P	NP	NS	WT	G	I	D	T	TOTAL
		3.1	5	2.4	5.6	1	8.3	10	---
Ht	10	31	50	24	56	10	83	100	354
Gt	4.3	13	22	10	24	4	36	43	152
Dp	8.5	26	43	20	48	9	71	85	302
Cp	5	16	25	12	28	5	42	50	178
Bp	1.6	5	8	4	9	2	13	16	57
So	10	31	50	24	56	10	83	100	354
Vi	5	16	25	12	28	5	42	50	178
Sy	5	16	25	12	28	5	42	50	178
Ab	0	0	0	0	0	0	0	0	0
Ar	7.8	24	39	19	44	8	65	78	277
TOTAL	---	178	287	137	321	58	477	572	2030

P = normalized score

Site: Forest Park, Lake Hiwassee, and Lake Alma, Oklahoma County, Oklahoma

WASTE \ Soil		NP	NS	WT	G	I	D	T	TOTAL
	P \ P	5.9	7.1	3.6	3	1	9.2	10	---
Ht	10	59	71	36	30	10	92	100	398
Gt	4.3	25	31	16	13	4	40	43	172
Dp	8.5	50	60	31	26	9	78	85	339
Cp	5	30	36	18	15	5	46	50	200
Bp	1.6	9	11	6	5	2	15	16	64
So	10	59	71	36	30	10	92	100	398
Vi	5	30	36	18	15	5	46	50	200
Sy	5	30	36	18	15	5	46	50	200
Ab	0	0	0	0	0	0	0	0	0
Ar	8.5	50	60	31	26	9	78	85	339
TOTAL	---	342	412	210	175	59	533	579	2310

P = normalized score

Site: Green Pastures, Oklahoma County, Oklahoma

Soil / WASTE	P	Ht	Gt	Dp	Cp	Bp	So	Vi	Sy	Ab	Ar	TOTAL
P		10	4.3	8.5	5	1.6	10	5	5	0	8.5	---
NP	5.9	59	25	50	30	9	59	30	30	0	50	342
NS	7.1	71	31	60	36	11	71	36	36	0	60	412
WT	4.6	46	20	39	23	7	46	23	23	0	39	266
G	3.0	30	13	26	15	5	30	15	15	0	26	175
I	1	10	4	9	5	2	10	5	5	0	9	59
D	9.2	92	40	78	46	15	92	46	46	0	78	533
T	10	100	43	85	50	16	100	50	50	0	85	579
TOTAL	---	408	176	347	205	65	408	205	205	0	347	2366

P = normalized score

Site: Midwest City, Oklahoma County, Oklahoma

WASTE \ Soil	P \ P	NP	NS	WT	G	I	D	T	TOTAL
		5.9	7.1	5.4	3	1	8.3	10	---
Ht	10	59	71	54	30	10	83	100	407
Gt	4.3	25	31	23	13	4	36	43	175
Dp	8.5	50	60	46	26	9	71	85	347
Cp	5	30	36	27	15	5	42	50	205
Bp	1.6	9	11	9	5	2	13	16	65
So	10	59	71	54	30	10	83	100	407
Vi	5	30	36	27	15	5	42	50	205
Sy	5	30	36	27	15	5	42	50	205
Ab	0	0	0	0	0	0	0	0	0
Ar	8.5	50	60	46	26	9	71	85	347
TOTAL	---	342	412	313	175	59	483	579	2363

P = normalized score

Site: Mustang, Canadian County, Oklahoma

WASTE \ Soil	P \ P	NP	NS	WT	G	I	D	T	TOTAL
WASTE	P	5.9	7.1	7.9	5	1	2.5	10	---
Ht	10	59	71	79	50	10	25	100	394
Gt	4.3	25	31	34	22	4	11	43	170
Dp	8.5	50	60	67	43	9	21	85	335
Cp	5	30	36	40	25	5	13	50	199
Bp	1.6	9	11	13	8	2	4	16	63
So	10	59	71	79	50	10	25	100	394
Vi	5	30	36	40	25	5	13	50	199
Sy	5	30	36	40	25	5	13	50	199
Ab	0	0	0	0	0	0	0	0	0
Ar	8.5	50	60	67	43	9	21	85	335
TOTAL	---	342	412	459	291	59	146	579	2288

P = normalized score

Site: Nicoma Park, Oklahoma County, Oklahoma

WASTE \ Soil	P	NP	NS	WT	G	I	D	T	TOTAL
P		5.9	7.1	3.8	3	1	9.2	10	---
Ht	10	59	71	38	30	10	92	100	400
Gt	4.3	25	31	16	13	4	40	43	172
Dp	8.5	50	60	32	26	9	78	85	340
Cp	5	30	36	19	15	5	46	50	201
Bp	1.6	9	11	6	5	2	15	16	64
So	10	59	71	38	30	10	92	100	400
Vi	5	30	36	19	15	5	46	50	201
Sy	5	30	36	19	15	5	46	50	201
Ab	0	0	0	0	0	0	0	0	0
Ar	8.5	50	50	32	26	9	78	85	340
TOTAL	---	342	412	219	175	59	533	579	2319

P = normalized score

Site: East Norman (East of E. 24th Street), Cleveland County, Oklahoma

Soil / WASTE	P \ P	NP	NS	WT	G	I	D	T	TOTAL
WASTE		5.9	7.1	2.4	3	1	5.2	10	---
Ht	10	59	71	24	30	10	52	100	346
Gt	4.3	25	31	10	13	4	22	43	148
Dp	8.5	50	60	20	26	9	44	85	294
Cp	5	30	36	12	15	5	26	50	174
Bp	1.6	9	11	4	5	1	8	16	55
So	10	59	71	24	30	10	52	100	346
Vi	5	30	36	12	15	5	26	50	174
Sy	5	30	36	12	15	5	26	50	174
Ab	0	0	0	0	0	0	0	0	0
Ar	8.5	50	60	20	26	9	44	85	294
TOTAL	---	342	412	138	175	59	300	579	2005

P = normalized score

Site: Seward Area (South of Guthrie to Oklahoma County Line)
Logan County, Oklahoma

WASTE \ Soil	P \ P	NP	NS	WT	G	I	D	T	TOTAL
		5.9	7.1	7.4	3	1	8.3	10	---
Ht	10	59	71	74	30	10	83	100	427
Gt	4.3	25	31	32	13	4	36	43	184
Dp	8.5	50	60	63	26	9	71	85	364
Cp	5	30	36	37	15	5	42	50	215
Bp	1.6	9	11	12	5	2	13	16	68
So	10	59	71	74	30	10	83	100	427
Vi	5	30	36	37	15	5	42	50	215
Sy	5	30	36	37	15	5	42	50	215
Ab	0	0	0	0	0	0	0	0	0
Ar	8.5	50	60	63	26	9	71	85	364
TOTAL	---	342	412	429	175	59	483	576	2479

P = normalized score

Site: Silver Lake Estates, Oklahoma County, Oklahoma

WASTE \ Soil	P	NP	NS	WT	G	I	D	T	TOTAL
P		2.5	5.7	8.9	4.4	1	5.8	10	---
Ht	10	25	57	89	44	10	58	100	383
Gt	4.3	11	25	38	19	4	25	43	165
Dp	8.5	21	49	76	37	9	49	85	326
Cp	5	13	29	45	22	5	29	50	193
Bp	1.6	4	9	14	7	2	9	16	61
So	10	25	57	89	44	10	58	100	383
Vi	5	13	29	45	22	5	29	50	193
Sy	5	13	29	45	22	5	29	50	193
Ab	0	0	0	0	0	0	0	0	0
Ar	8	20	46	71	35	8	46	80	306
TOTAL	---	145	330	512	252	58	332	574	2203

P = normalized score

Site: Sunvalley Acres, Canadian County, Oklahoma

WASTE \ Soil	P \ P	NP	NS	WT	G	I	D	T	TOTAL
		5.9	5.7	9.5	5	1	9.2	10	---
Ht	10	59	57	95	50	10	92	100	463
Gt	4.3	25	25	41	22	4	40	43	200
Dp	8.5	50	49	81	43	9	78	85	395
Cp	5	22	29	48	25	5	46	50	225
Bp	1.6	9	9	15	8	2	15	16	74
So	10	59	57	95	50	10	92	100	463
Vi	5	22	29	48	25	5	46	50	225
Sy	5	22	29	48	25	5	46	50	225
Ab	0	0	0	0	0	0	0	0	0
Ar	8	47	46	76	40	8	74	80	371
TOTAL	---	315	330	547	288	58	529	574	2641

P = normalized score

APPENDIX D

ERROR FUNCTION IN HANTUSH ANALYTICAL MODEL

β =	0.005	0.010	0.020	0.030	0.040	0.050	0.060	0.070	0.080	0.090	0.100	0.110	0.120
α													
0.005	0.0007	0.0009	0.0014	0.0019	0.0023	0.0027	0.0030	0.0033	0.0036	0.0038	0.0040	0.0042	0.0043
0.010	0.0009	0.0014	0.0023	0.0032	0.0040	0.0047	0.0054	0.0060	0.0066	0.0071	0.0076	0.0080	0.0083
0.020	0.0014	0.0023	0.0041	0.0057	0.0072	0.0087	0.0101	0.0113	0.0125	0.0136	0.0146	0.0154	0.0162
0.030	0.0019	0.0032	0.0057	0.0081	0.0103	0.0125	0.0145	0.0164	0.0181	0.0197	0.0212	0.0226	0.0239
0.040	0.0023	0.0040	0.0072	0.0103	0.0133	0.0161	0.0187	0.0212	0.0235	0.0257	0.0277	0.0295	0.0312
0.050	0.0027	0.0047	0.0087	0.0125	0.0161	0.0195	0.0227	0.0258	0.0287	0.0313	0.0338	0.0362	0.0383
0.060	0.0030	0.0054	0.0101	0.0145	0.0187	0.0227	0.0266	0.0302	0.0336	0.0368	0.0398	0.0425	0.0451
0.070	0.0033	0.0060	0.0113	0.0164	0.0212	0.0258	0.0302	0.0343	0.0382	0.0419	0.0454	0.0486	0.0517
0.080	0.0036	0.0066	0.0125	0.0181	0.0235	0.0287	0.0336	0.0382	0.0427	0.0468	0.0508	0.0545	0.0579
0.090	0.0038	0.0071	0.0136	0.0197	0.0257	0.0313	0.0368	0.0419	0.0468	0.0515	0.0559	0.0600	0.0639
0.100	0.0040	0.0076	0.0146	0.0212	0.0277	0.0338	0.0398	0.0454	0.0508	0.0559	0.0608	0.0653	0.0697
0.110	0.0042	0.0080	0.0154	0.0226	0.0295	0.0362	0.0425	0.0486	0.0545	0.0600	0.0653	0.0704	0.0751
0.120	0.0045	0.0086	0.0165	0.0241	0.0315	0.0385	0.0453	0.0519	0.0581	0.0641	0.0698	0.0752	0.0804
0.130	0.0048	0.0090	0.0174	0.0255	0.0333	0.0408	0.0480	0.0550	0.0616	0.0680	0.0741	0.0799	0.0855
0.140	0.0050	0.0095	0.0183	0.0268	0.0350	0.0430	0.0506	0.0580	0.0650	0.0718	0.0783	0.0845	0.0905
0.150	0.0052	0.0099	0.0191	0.0281	0.0367	0.0451	0.0531	0.0609	0.0683	0.0755	0.0824	0.0890	0.0953
0.160	0.0054	0.0103	0.0200	0.0293	0.0384	0.0471	0.0555	0.0637	0.0715	0.0791	0.0863	0.0933	0.1000
0.170	0.0056	0.0107	0.0208	0.0305	0.0399	0.0491	0.0579	0.0664	0.0746	0.0825	0.0902	0.0975	0.1045
0.180	0.0058	0.0111	0.0216	0.0317	0.0415	0.0510	0.0602	0.0690	0.0776	0.0859	0.0939	0.1015	0.1089
0.190	0.0060	0.0115	0.0223	0.0328	0.0429	0.0528	0.0623	0.0716	0.0805	0.0891	0.0974	0.1054	0.1131
0.200	0.0062	0.0119	0.0230	0.0338	0.0443	0.0546	0.0644	0.0740	0.0833	0.0922	0.1008	0.1092	0.1172
0.210	0.0063	0.0122	0.0237	0.0349	0.0457	0.0562	0.0665	0.0764	0.0859	0.0952	0.1041	0.1128	0.1211
0.220	0.0065	0.0125	0.0243	0.0358	0.0470	0.0579	0.0684	0.0786	0.0885	0.0981	0.1073	0.1163	0.1249
0.230	0.0066	0.0128	0.0250	0.0368	0.0483	0.0594	0.0703	0.0808	0.0910	0.1008	0.1104	0.1196	0.1285
0.240	0.0068	0.0131	0.0256	0.0377	0.0494	0.0609	0.0720	0.0828	0.0933	0.1035	0.1133	0.1228	0.1320
0.250	0.0069	0.0134	0.0261	0.0385	0.0506	0.0623	0.0737	0.0848	0.0955	0.1060	0.1160	0.1258	0.1352

0.260	0.0070	0.0136	0.0267	0.0393	0.0517	0.0637	0.0754	0.0867	0.0977	0.1084	0.1188	0.1288	0.1385
0.270	0.0071	0.0139	0.0272	0.0401	0.0527	0.0650	0.0769	0.0886	0.0998	0.1108	0.1214	0.1317	0.1416
0.280	0.0073	0.0142	0.0277	0.0409	0.0538	0.0663	0.0785	0.0904	0.1019	0.1131	0.1240	0.1345	0.1447
0.290	0.0074	0.0144	0.0282	0.0417	0.0548	0.0676	0.0800	0.0921	0.1039	0.1154	0.1265	0.1373	0.1477
0.300	0.0075	0.0147	0.0287	0.0424	0.0558	0.0688	0.0815	0.0939	0.1059	0.1176	0.1289	0.1399	0.1506
0.310	0.0076	0.0149	0.0292	0.0431	0.0567	0.0700	0.0830	0.0956	0.1078	0.1197	0.1313	0.1425	0.1534
0.320	0.0077	0.0151	0.0297	0.0438	0.0577	0.0712	0.0844	0.0972	0.1097	0.1218	0.1336	0.1451	0.1562
0.330	0.0078	0.0153	0.0301	0.0445	0.0586	0.0723	0.0857	0.0988	0.1115	0.1238	0.1359	0.1475	0.1589
0.340	0.0079	0.0156	0.0306	0.0452	0.0595	0.0734	0.0871	0.1003	0.1132	0.1258	0.1381	0.1499	0.1615
0.350	0.0080	0.0158	0.0310	0.0458	0.0603	0.0745	0.0883	0.1018	0.1149	0.1277	0.1402	0.1523	0.1640
0.360	0.0082	0.0160	0.0314	0.0465	0.0612	0.0756	0.0896	0.1033	0.1166	0.1296	0.1422	0.1545	0.1665
0.370	0.0082	0.0162	0.0318	0.0471	0.0620	0.0766	0.0908	0.1047	0.1182	0.1314	0.1442	0.1567	0.1688
0.380	0.0084	0.0165	0.0325	0.0480	0.0633	0.0781	0.0927	0.1068	0.1207	0.1341	0.1473	0.1600	0.1725
0.390	0.0085	0.0167	0.0328	0.0485	0.0639	0.0789	0.0936	0.1079	0.1218	0.1355	0.1487	0.1616	0.1742
0.400	0.0086	0.0168	0.0331	0.0489	0.0645	0.0796	0.0945	0.1089	0.1230	0.1368	0.1502	0.1633	0.1760
0.410	0.0086	0.0170	0.0334	0.0494	0.0651	0.0804	0.0954	0.1100	0.1242	0.1381	0.1517	0.1649	0.1777
0.420	0.0087	0.0171	0.0337	0.0498	0.0657	0.0811	0.0963	0.1110	0.1254	0.1395	0.1532	0.1665	0.1795

β =	0.005	0.010	0.020	0.030	0.040	0.050	0.060	0.070	0.080	0.090	0.100	0.110	0.120
α													
0.430	0.0088	0.0173	0.0340	0.0503	0.0663	0.0819	0.0972	0.1121	0.1266	0.1408	0.1546	0.1681	0.1812
0.440	0.0088	0.0174	0.0343	0.0507	0.0669	0.0826	0.0981	0.1131	0.1278	0.1421	0.1561	0.1697	0.1830
0.450	0.0089	0.0175	0.0346	0.0512	0.0675	0.0834	0.0990	0.1141	0.1290	0.1435	0.1576	0.1713	0.1847
0.460	0.0090	0.0177	0.0349	0.0516	0.0681	0.0841	0.0999	0.1152	0.1302	0.1448	0.1591	0.1729	0.1865
0.470	0.0090	0.0178	0.0352	0.0521	0.0687	0.0849	0.1008	0.1162	0.1314	0.1461	0.1605	0.1746	0.1882
0.480	0.0091	0.0180	0.0355	0.0525	0.0693	0.0856	0.1017	0.1173	0.1326	0.1475	0.1620	0.1762	0.1900
0.490	0.0092	0.0181	0.0358	0.0530	0.0699	0.0864	0.1026	0.1183	0.1337	0.1488	0.1635	0.1778	0.1917
0.500	0.0093	0.0183	0.0361	0.0535	0.0705	0.0872	0.1035	0.1194	0.1349	0.1501	0.1650	0.1794	0.1935
0.520	0.0098	0.0188	0.0368	0.0543	0.0716	0.0885	0.1050	0.1212	0.1371	0.1526	0.1678	0.1826	0.1570
0.540	0.0098	0.0190	0.0372	0.0550	0.0725	0.0896	0.1064	0.1229	0.1389	0.1547	0.1701	0.1851	0.1998
0.560	0.0098	0.0191	0.0376	0.0557	0.0734	0.0908	0.1078	0.1244	0.1407	0.1567	0.1723	0.1875	0.2023
0.580	0.0098	0.0193	0.0380	0.0563	0.0742	0.0918	0.1091	0.1259	0.1424	0.1586	0.1744	0.1898	0.2048
0.600	0.0098	0.0194	0.0383	0.0569	0.0750	0.0928	0.1103	0.1273	0.1440	0.1604	0.1764	0.1920	0.2072
0.620	0.0099	0.0196	0.0387	0.0574	0.0758	0.0938	0.1114	0.1287	0.1456	0.1621	0.1783	0.1940	0.2094
0.640	0.0099	0.0197	0.0390	0.0580	0.0765	0.0947	0.1125	0.1300	0.1470	0.1637	0.1801	0.1960	0.2116
0.660	0.0100	0.0199	0.0394	0.0585	0.0772	0.0956	0.1136	0.1312	0.1484	0.1653	0.1818	0.1979	0.2136
0.680	0.0100	0.0200	0.0397	0.0589	0.0778	0.0964	0.1145	0.1323	0.1497	0.1667	0.1834	0.1996	0.2155
0.700	0.0101	0.0202	0.0400	0.0594	0.0785	0.0971	0.1154	0.1333	0.1509	0.1681	0.1849	0.2013	0.2173
0.720	0.0102	0.0203	0.0403	0.0598	0.0790	0.0978	0.1163	0.1343	0.1520	0.1693	0.1863	0.2028	0.2190
0.740	0.0103	0.0205	0.0406	0.0602	0.0796	0.0985	0.1171	0.1352	0.1531	0.1705	0.1876	0.2042	0.2206
0.760	0.0104	0.0206	0.0408	0.0606	0.0800	0.0991	0.1178	0.1361	0.1540	0.1716	0.1888	0.2056	0.2220
0.780	0.0104	0.0207	0.0411	0.0610	0.0806	0.0998	0.1186	0.1370	0.1550	0.1726	0.1898	0.2067	0.2231
0.800	0.0105	0.0208	0.0413	0.0613	0.0810	0.1003	0.1192	0.1377	0.1558	0.1736	0.1910	0.2080	0.2246
0.820	0.0105	0.0209	0.0415	0.0616	0.0814	0.1007	0.1198	0.1384	0.1567	0.1745	0.1921	0.2092	0.2260
0.840	0.0106	0.0210	0.0416	0.0619	0.0817	0.1012	0.1203	0.1391	0.1574	0.1754	0.1930	0.2103	0.2272
0.860	0.0106	0.0211	0.0418	0.0621	0.0821	0.1016	0.1208	0.1397	0.1582	0.1763	0.1940	0.2113	0.2283

0.880	0.0107	0.0212	0.0420	0.0624	0.0824	0.1021	0.1214	0.1403	0.1588	0.1770	0.1948	0.2123	0.2294
0.900	0.0107	0.0213	0.0422	0.0626	0.0827	0.1025	0.1219	0.1409	0.1595	0.1778	0.1957	0.2132	0.2303
0.920	0.0108	0.0214	0.0423	0.0629	0.0831	0.1029	0.1223	0.1414	0.1601	0.1784	0.1964	0.2140	0.2312
0.940	0.0108	0.0215	0.0425	0.0631	0.0834	0.1032	0.1228	0.1419	0.1607	0.1791	0.1971	0.2148	0.2320
0.960	0.0109	0.0215	0.0426	0.0633	0.0836	0.1036	0.1232	0.1424	0.1612	0.1797	0.1977	0.2155	0.2328
0.980	0.0109	0.0216	0.0428	0.0635	0.0839	0.1039	0.1236	0.1428	0.1617	0.1802	0.1984	0.2161	0.2335
1.000	0.0109	0.0217	0.0429	0.0637	0.0842	0.1042	0.1239	0.1432	0.1622	0.1807	0.1989	0.2167	0.2342
1.200	0.0112	0.0221	0.0437	0.0649	0.0857	0.1062	0.1263	0.1460	0.1653	0.1843	0.2030	0.2212	0.2391
1.400	0.0113	0.0223	0.0440	0.0654	0.0864	0.1070	0.1273	0.1471	0.1666	0.1858	0.2045	0.2229	0.2409
1.600	0.0113	0.0224	0.0442	0.0657	0.0868	0.1076	0.1279	0.1479	0.1675	0.1867	0.2056	0.2240	0.2421
1.800	0.0113	0.0224	0.0444	0.0659	0.0871	0.1078	0.1283	0.1483	0.1680	0.1872	0.2062	0.2247	0.2429
2.000	0.0113	0.0224	0.0444	0.0659	0.0871	0.1079	0.1284	0.1484	0.1681	0.1875	0.2064	0.2250	0.2432
2.200	0.0113	0.0224	0.0444	0.0659	0.0871	0.1079	0.1284	0.1484	0.1681	0.1875	0.2065	0.2251	0.2433
2.400	0.0113	0.0224	0.0444	0.0659	0.0871	0.1079	0.1284	0.1484	0.1681	0.1875	0.2065	0.2251	0.2433
2.600	0.0113	0.0224	0.0444	0.0659	0.0871	0.1079	0.1284	0.1484	0.1681	0.1875	0.2065	0.2251	0.2433
2.800	0.0113	0.0224	0.0444	0.0659	0.0871	0.1079	0.1284	0.1484	0.1681	0.1875	0.2065	0.2251	0.2433
3.000	0.0113	0.0224	0.0444	0.0659	0.0871	0.1079	0.1284	0.1484	0.1681	0.1875	0.2065	0.2251	0.2433

β =	0.130	0.140	0.150	0.160	0.170	0.180	0.190	0.200	0.210	0.220	0.230	0.240	0.250
α													
0.005	0.0048	0.0050	0.0052	0.0054	0.0056	0.0058	0.0060	0.0062	0.0063	0.0065	0.0066	0.0068	0.0069
0.010	0.0090	0.0095	0.0099	0.0103	0.0107	0.0111	0.0115	0.0119	0.0122	0.0125	0.0128	0.0131	0.0134
0.020	0.0174	0.0183	0.0191	0.0200	0.0208	0.0216	0.0223	0.0230	0.0237	0.0243	0.0250	0.0256	0.0261
0.030	0.0255	0.0268	0.0281	0.0293	0.0305	0.0317	0.0328	0.0338	0.0349	0.0358	0.0368	0.0377	0.0385
0.040	0.0333	0.0350	0.0367	0.0384	0.0399	0.0415	0.0429	0.0443	0.0457	0.0470	0.0483	0.0494	0.0506
0.050	0.0408	0.0430	0.0451	0.0471	0.0491	0.0510	0.0528	0.0546	0.0562	0.0579	0.0594	0.0609	0.0623
0.060	0.0480	0.0506	0.0531	0.0555	0.0579	0.0602	0.0623	0.0644	0.0665	0.0684	0.0703	0.0720	0.0737
0.070	0.0550	0.0580	0.0609	0.0637	0.0664	0.0690	0.0716	0.0740	0.0764	0.0786	0.0808	0.0828	0.0848
0.080	0.0616	0.0650	0.0683	0.0715	0.0746	0.0776	0.0805	0.0833	0.0859	0.0885	0.0910	0.0933	0.0956
0.090	0.0680	0.0718	0.0755	0.0791	0.0825	0.0859	0.0891	0.0922	0.0952	0.0981	0.1008	0.1035	0.1060
0.100	0.0741	0.0783	0.0824	0.0863	0.0902	0.0939	0.0974	0.1008	0.1041	0.1073	0.1104	0.1133	0.1161
0.110	0.0799	0.0845	0.0890	0.0933	0.0975	0.1015	0.1054	0.1092	0.1128	0.1163	0.1196	0.1228	0.1259
0.120	0.0855	0.0905	0.0953	0.0999	0.1044	0.1088	0.1130	0.1171	0.1210	0.1248	0.1284	0.1319	0.1352
0.130	0.0910	0.0963	0.1015	0.1065	0.1113	0.1160	0.1205	0.1249	0.1291	0.1331	0.1370	0.1408	0.1444
0.140	0.0963	0.1020	0.1075	0.1128	0.1180	0.1230	0.1278	0.1325	0.1370	0.1413	0.1455	0.1495	0.1533
0.150	0.1015	0.1075	0.1133	0.1190	0.1244	0.1298	0.1349	0.1399	0.1446	0.1493	0.1537	0.1580	0.1621
0.160	0.1065	0.1128	0.1190	0.1249	0.1307	0.1363	0.1418	0.1470	0.1521	0.1570	0.1617	0.1662	0.1706
0.170	0.1113	0.1180	0.1244	0.1307	0.1368	0.1427	0.1485	0.1540	0.1594	0.1645	0.1695	0.1743	0.1789
0.180	0.1160	0.1230	0.1298	0.1363	0.1427	0.1490	0.1550	0.1608	0.1664	0.1718	0.1770	0.1821	0.1869
0.190	0.1205	0.1278	0.1349	0.1418	0.1485	0.1550	0.1612	0.1673	0.1732	0.1789	0.1844	0.1896	0.1947
0.200	0.1249	0.1325	0.1399	0.1470	0.1540	0.1608	0.1673	0.1737	0.1798	0.1858	0.1915	0.1970	0.2023
0.210	0.1291	0.1370	0.1446	0.1521	0.1594	0.1664	0.1732	0.1798	0.1862	0.1924	0.1984	0.2041	0.2097
0.220	0.1331	0.1413	0.1493	0.1570	0.1645	0.1718	0.1789	0.1858	0.1924	0.1988	0.2050	0.2110	0.2168
0.230	0.1370	0.1455	0.1537	0.1617	0.1695	0.1770	0.1844	0.1915	0.1984	0.2050	0.2115	0.2177	0.2237
0.240	0.1408	0.1495	0.1580	0.1662	0.1743	0.1821	0.1896	0.1970	0.2041	0.2110	0.2177	0.2241	0.2303
0.250	0.1443	0.1533	0.1620	0.1705	0.1788	0.1869	0.1947	0.2023	0.2096	0.2167	0.2236	0.2303	0.2367

0.260	0.1478	0.1570	0.1660	0.1748	0.1833	0.1916	0.1996	0.2074	0.2150	0.2223	0.2294	0.2363	0.2429
0.270	0.1512	0.1607	0.1699	0.1789	0.1876	0.1961	0.2044	0.2124	0.2202	0.2278	0.2351	0.2422	0.2490
0.280	0.1545	0.1642	0.1737	0.1829	0.1919	0.2006	0.2091	0.2173	0.2253	0.2331	0.2406	0.2479	0.2549
0.290	0.1577	0.1677	0.1774	0.1868	0.1960	0.2049	0.2136	0.2221	0.2303	0.2383	0.2460	0.2535	0.2607
0.300	0.1609	0.1710	0.1810	0.1906	0.2000	0.2092	0.2181	0.2268	0.2352	0.2433	0.2513	0.2589	0.2664
0.310	0.1640	0.1743	0.1845	0.1943	0.2039	0.2133	0.2224	0.2313	0.2399	0.2483	0.2564	0.2642	0.2718
0.320	0.1669	0.1775	0.1879	0.1979	0.2078	0.2173	0.2267	0.2357	0.2445	0.2531	0.2614	0.2694	0.2772
0.330	0.1698	0.1806	0.1912	0.2015	0.2115	0.2213	0.2308	0.2400	0.2490	0.2577	0.2662	0.2744	0.2824
0.340	0.1726	0.1836	0.1944	0.2049	0.2151	0.2251	0.2348	0.2442	0.2534	0.2623	0.2709	0.2793	0.2875
0.350	0.1754	0.1866	0.1975	0.2082	0.2186	0.2287	0.2386	0.2482	0.2576	0.2667	0.2755	0.2841	0.2924
0.360	0.1780	0.1894	0.2005	0.2114	0.2220	0.2323	0.2424	0.2522	0.2617	0.2710	0.2800	0.2887	0.2972
0.370	0.1806	0.1921	0.2035	0.2145	0.2253	0.2358	0.2460	0.2560	0.2657	0.2751	0.2843	0.2932	0.3018
0.380	0.1844	0.1963	0.2079	0.2192	0.2302	0.2410	0.2515	0.2617	0.2717	0.2814	0.2908	0.2999	0.3088
0.390	0.1864	0.1984	0.2101	0.2215	0.2327	0.2436	0.2542	0.2646	0.2747	0.2845	0.2940	0.3033	0.3123
0.400	0.1883	0.2004	0.2123	0.2239	0.2352	0.2462	0.2570	0.2675	0.2777	0.2876	0.2973	0.3067	0.3158
0.410	0.1902	0.2025	0.2145	0.2262	0.2377	0.2488	0.2597	0.2704	0.2807	0.2908	0.3006	0.3101	0.3193
0.420	0.1921	0.2045	0.2167	0.2285	0.2401	0.2515	0.2625	0.2732	0.2837	0.2939	0.3038	0.3135	0.3229

α \ β =	0.130	0.140	0.150	0.160	0.170	0.180	0.190	0.200	0.210	0.220	0.230	0.240	0.250
0.430	0.1940	0.2066	0.2189	0.2309	0.2426	0.2541	0.2652	0.2761	0.2867	0.2971	0.3071	0.3169	0.3264
0.440	0.1959	0.2086	0.2211	0.2332	0.2451	0.2567	0.2680	0.2790	0.2898	0.3002	0.3104	0.3203	0.3299
0.450	0.1978	0.2107	0.2233	0.2356	0.2476	0.2593	0.2707	0.2819	0.2928	0.3034	0.3137	0.3237	0.3334
0.460	0.1998	0.2128	0.2255	0.2379	0.2500	0.2619	0.2735	0.2848	0.2958	0.3065	0.3169	0.3271	0.3370
0.470	0.2017	0.2148	0.2277	0.2402	0.2525	0.2645	0.2762	0.2877	0.2988	0.3096	0.3202	0.3305	0.3405
0.480	0.2036	0.2169	0.2299	0.2426	0.2550	0.2671	0.2790	0.2905	0.3018	0.3128	0.3235	0.3339	0.3440
0.490	0.2055	0.2189	0.2321	0.2449	0.2575	0.2697	0.2817	0.2934	0.3048	0.3159	0.3268	0.3373	0.3475
0.500	0.2074	0.2210	0.2343	0.2472	0.2599	0.2724	0.2845	0.2963	0.3078	0.3191	0.3300	0.3407	0.3511
0.520	0.2111	0.2250	0.2385	0.2518	0.2648	0.2774	0.2898	0.3019	0.3137	0.3252	0.3363	0.3472	0.3578
0.540	0.2141	0.2281	0.2419	0.2554	0.2686	0.2814	0.2940	0.3063	0.3183	0.3300	0.3414	0.3525	0.3633
0.560	0.2169	0.2312	0.2451	0.2588	0.2722	0.2853	0.2981	0.3105	0.3227	0.3346	0.3462	0.3575	0.3685
0.580	0.2196	0.2341	0.2482	0.2621	0.2757	0.2890	0.3019	0.3146	0.3270	0.3391	0.3508	0.3623	0.3735
0.600	0.2221	0.2368	0.2512	0.2652	0.2790	0.2925	0.3056	0.3185	0.3310	0.3433	0.3552	0.3669	0.3782
0.620	0.2246	0.2394	0.2540	0.2682	0.2822	0.2958	0.3091	0.3221	0.3349	0.3473	0.3594	0.3712	0.3828
0.640	0.2269	0.2419	0.2566	0.2710	0.2852	0.2990	0.3125	0.3256	0.3385	0.3511	0.3634	0.3754	0.3871
0.660	0.2291	0.2443	0.2592	0.2737	0.2880	0.3020	0.3156	0.3290	0.3420	0.3547	0.3672	0.3793	0.3911
0.680	0.2312	0.2465	0.2615	0.2763	0.2907	0.3048	0.3186	0.3321	0.3453	0.3582	0.3707	0.3830	0.3950
0.700	0.2331	0.2486	0.2638	0.2786	0.2932	0.3074	0.3214	0.3350	0.3483	0.3614	0.3741	0.3865	0.3986
0.720	0.2349	0.2506	0.2659	0.2809	0.2955	0.3099	0.3240	0.3378	0.3512	0.3644	0.3772	0.3897	0.4020
0.740	0.2366	0.2524	0.2678	0.2829	0.2977	0.3123	0.3264	0.3403	0.3539	0.3672	0.3801	0.3928	0.4051
0.760	0.2382	0.2541	0.2696	0.2849	0.2998	0.3144	0.3287	0.3427	0.3564	0.3698	0.3828	0.3956	0.4080
0.780	0.2396	0.2556	0.2713	0.2866	0.3017	0.3164	0.3308	0.3450	0.3588	0.3723	0.3855	0.3984	0.4110
0.800	0.2411	0.2572	0.2729	0.2884	0.3035	0.3184	0.3329	0.3471	0.3611	0.3747	0.3880	0.4009	0.4136
0.820	0.2425	0.2586	0.2745	0.2901	0.3053	0.3203	0.3349	0.3492	0.3632	0.3769	0.3903	0.4033	0.4161
0.840	0.2437	0.2600	0.2760	0.2916	0.3070	0.3220	0.3367	0.3511	0.3652	0.3790	0.3924	0.4056	0.4184
0.860	0.2449	0.2613	0.2773	0.2931	0.3085	0.3236	0.3384	0.3529	0.3671	0.3809	0.3945	0.4077	0.4206

0.880	0.2461	0.2625	0.2786	0.2944	0.3099	0.3251	0.3400	0.3546	0.3688	0.3828	0.3964	0.4097	0.4227
0.900	0.2471	0.2636	0.2798	0.2957	0.3113	0.3266	0.3415	0.3561	0.3705	0.3845	0.3982	0.4115	0.4246
0.920	0.2481	0.2647	0.2809	0.2969	0.3125	0.3279	0.3429	0.3576	0.3720	0.3861	0.3998	0.4133	0.4264
0.940	0.2490	0.2656	0.2820	0.2980	0.3137	0.3291	0.3442	0.3590	0.3734	0.3876	0.4014	0.4149	0.4281
0.960	0.2498	0.2665	0.2829	0.2990	0.3148	0.3303	0.3454	0.3602	0.3748	0.3890	0.4028	0.4164	0.4297
0.980	0.2506	0.2674	0.2838	0.3000	0.3158	0.3314	0.3466	0.3614	0.3760	0.3903	0.4042	0.4178	0.4312
1.000	0.2513	0.2682	0.2847	0.3009	0.3168	0.3324	0.3476	0.3626	0.3772	0.3915	0.4055	0.4192	0.4325
1.200	0.2566	0.2738	0.2907	0.3073	0.3236	0.3396	0.3552	0.3705	0.3855	0.4002	0.4146	0.4286	0.4423
1.400	0.2586	0.2760	0.2931	0.3099	0.3263	0.3425	0.3583	0.3737	0.3889	0.4038	0.4183	0.4325	0.4464
1.600	0.2600	0.2775	0.2947	0.3115	0.3281	0.3443	0.3602	0.3758	0.3911	0.4060	0.4206	0.4350	0.4489
1.800	0.2608	0.2783	0.2956	0.3125	0.3291	0.3454	0.3613	0.3769	0.3923	0.4073	0.4219	0.4363	0.4503
2.000	0.2611	0.2786	0.2959	0.3128	0.3295	0.3458	0.3617	0.3774	0.3927	0.4077	0.4224	0.4368	0.4509
2.200	0.2611	0.2787	0.2959	0.3128	0.3295	0.3458	0.3617	0.3774	0.3927	0.4077	0.4224	0.4368	0.4509
2.400	0.2611	0.2787	0.2959	0.3128	0.3295	0.3458	0.3617	0.3774	0.3927	0.4077	0.4224	0.4368	0.4509
2.600	0.2611	0.2787	0.2959	0.3128	0.3295	0.3458	0.3617	0.3774	0.3927	0.4077	0.4224	0.4368	0.4509
2.800	0.2611	0.2787	0.2959	0.3128	0.3295	0.3458	0.3617	0.3774	0.3927	0.4077	0.4224	0.4368	0.4509
3.000	0.2611	0.2787	0.2959	0.3128	0.3295	0.3458	0.3617	0.3774	0.3927	0.4077	0.4224	0.4368	0.4509

β = / α	0.260	0.270	0.280	0.290	0.300	0.310	0.320	0.330	0.340	0.350	0.360	0.370	0.380
0.005	0.0070	0.0071	0.0073	0.0074	0.0075	0.0076	0.0077	0.0078	0.0079	0.0080	0.0082	0.0082	0.0083
0.010	0.0136	0.0139	0.0142	0.0144	0.0147	0.0149	0.0151	0.0153	0.0156	0.0158	0.0160	0.0162	0.0164
0.020	0.0267	0.0272	0.0277	0.0282	0.0287	0.0292	0.0297	0.0301	0.0306	0.0310	0.0314	0.0318	0.0322
0.030	0.0393	0.0401	0.0409	0.0417	0.0424	0.0431	0.0438	0.0445	0.0452	0.0458	0.0465	0.0471	0.0477
0.040	0.0517	0.0527	0.0538	0.0548	0.0558	0.0567	0.0577	0.0586	0.0595	0.0603	0.0612	0.0620	0.0628
0.050	0.0637	0.0650	0.0663	0.0676	0.0688	0.0700	0.0712	0.0723	0.0734	0.0745	0.0756	0.0766	0.0775
0.060	0.0754	0.0769	0.0785	0.0800	0.0815	0.0830	0.0844	0.0857	0.0871	0.0883	0.0896	0.0908	0.0920
0.070	0.0867	0.0886	0.0904	0.0921	0.0939	0.0956	0.0972	0.0988	0.1003	0.1018	0.1033	0.1047	0.1060
0.080	0.0977	0.0998	0.1019	0.1039	0.1059	0.1078	0.1097	0.1115	0.1132	0.1149	0.1166	0.1182	0.1197
0.090	0.1084	0.1108	0.1131	0.1154	0.1176	0.1197	0.1218	0.1238	0.1258	0.1277	0.1296	0.1314	0.1331
0.100	0.1188	0.1214	0.1240	0.1265	0.1289	0.1313	0.1336	0.1359	0.1381	0.1402	0.1422	0.1442	0.1461
0.110	0.1288	0.1317	0.1345	0.1373	0.1399	0.1425	0.1451	0.1475	0.1499	0.1523	0.1545	0.1567	0.1588
0.120	0.1384	0.1415	0.1446	0.1476	0.1505	0.1533	0.1561	0.1588	0.1614	0.1639	0.1663	0.1687	0.1710
0.130	0.1478	0.1512	0.1545	0.1577	0.1609	0.1640	0.1669	0.1698	0.1726	0.1754	0.1780	0.1806	0.1830
0.140	0.1570	0.1607	0.1642	0.1677	0.1710	0.1743	0.1775	0.1806	0.1836	0.1866	0.1894	0.1921	0.1948
0.150	0.1660	0.1699	0.1737	0.1774	0.1810	0.1845	0.1879	0.1912	0.1944	0.1975	0.2005	0.2035	0.2063
0.160	0.1748	0.1789	0.1829	0.1868	0.1906	0.1943	0.1979	0.2015	0.2049	0.2082	0.2114	0.2145	0.2175
0.170	0.1833	0.1876	0.1919	0.1960	0.2000	0.2039	0.2078	0.2115	0.2151	0.2186	0.2220	0.2253	0.2285
0.180	0.1916	0.1961	0.2006	0.2049	0.2092	0.2133	0.2173	0.2213	0.2251	0.2287	0.2323	0.2358	0.2391
0.190	0.1996	0.2044	0.2091	0.2136	0.2181	0.2224	0.2267	0.2308	0.2348	0.2386	0.2424	0.2460	0.2495
0.200	0.2074	0.2124	0.2173	0.2221	0.2268	0.2313	0.2357	0.2400	0.2442	0.2482	0.2522	0.2560	0.2597
0.210	0.2150	0.2202	0.2253	0.2303	0.2352	0.2399	0.2445	0.2490	0.2534	0.2576	0.2617	0.2657	0.2696
0.220	0.2223	0.2278	0.2331	0.2383	0.2433	0.2483	0.2531	0.2577	0.2623	0.2667	0.2710	0.2751	0.2791
0.230	0.2294	0.2351	0.2406	0.2460	0.2513	0.2564	0.2614	0.2662	0.2709	0.2755	0.2800	0.2843	0.2885
0.240	0.2363	0.2422	0.2479	0.2535	0.2589	0.2642	0.2694	0.2744	0.2793	0.2841	0.2887	0.2932	0.2975
0.250	0.2429	0.2490	0.2549	0.2607	0.2664	0.2719	0.2772	0.2824	0.2875	0.2924	0.2972	0.3018	0.3063

0.260	0.2494	0.2556	0.2617	0.2677	0.2735	0.2792	0.2847	0.2901	0.2954	0.3004	0.3054	0.3102	0.3148
0.270	0.2556	0.2621	0.2684	0.2745	0.2805	0.2864	0.2921	0.2976	0.3030	0.3083	0.3134	0.3183	0.3231
0.280	0.2617	0.2684	0.2749	0.2812	0.2874	0.2934	0.2993	0.3050	0.3105	0.3159	0.3212	0.3263	0.3312
0.290	0.2677	0.2745	0.2812	0.2877	0.2940	0.3002	0.3063	0.3121	0.3179	0.3234	0.3288	0.3341	0.3392
0.300	0.2735	0.2805	0.2874	0.2940	0.3006	0.3069	0.3131	0.3191	0.3250	0.3307	0.3363	0.3417	0.3469
0.310	0.2792	0.2864	0.2934	0.3002	0.3069	0.3134	0.3198	0.3260	0.3320	0.3379	0.3436	0.3491	0.3545
0.320	0.2847	0.2921	0.2993	0.3063	0.3131	0.3198	0.3263	0.3326	0.3388	0.3448	0.3507	0.3564	0.3619
0.330	0.2901	0.2976	0.3050	0.3121	0.3191	0.3260	0.3326	0.3391	0.3455	0.3516	0.3576	0.3635	0.3691
0.340	0.2954	0.3030	0.3105	0.3179	0.3250	0.3320	0.3388	0.3455	0.3520	0.3583	0.3644	0.3703	0.3761
0.350	0.3004	0.3083	0.3159	0.3234	0.3307	0.3379	0.3448	0.3516	0.3583	0.3647	0.3710	0.3770	0.3830
0.360	0.3054	0.3134	0.3212	0.3288	0.3363	0.3436	0.3507	0.3576	0.3644	0.3710	0.3774	0.3836	0.3896
0.370	0.3102	0.3183	0.3263	0.3341	0.3417	0.3491	0.3564	0.3635	0.3703	0.3770	0.3836	0.3899	0.3961
0.380	0.3174	0.3257	0.3340	0.3420	0.3498	0.3575	0.3649	0.3722	0.3793	0.3862	0.3929	0.3995	0.4058
0.390	0.3210	0.3295	0.3378	0.3460	0.3539	0.3617	0.3693	0.3767	0.3839	0.3909	0.3977	0.4043	0.4108
0.400	0.3247	0.3333	0.3417	0.3500	0.3581	0.3659	0.3736	0.3811	0.3884	0.3955	0.4025	0.4092	0.4157
0.410	0.3283	0.3371	0.3456	0.3540	0.3622	0.3702	0.3780	0.3856	0.3930	0.4002	0.4072	0.4141	0.4207
0.420	0.3320	0.3408	0.3495	0.3580	0.3663	0.3744	0.3823	0.3900	0.3976	0.4049	0.4120	0.4190	0.4257

β =	0.260	0.270	0.280	0.290	0.300	0.310	0.320	0.330	0.340	0.350	0.360	0.370	0.380
α													
0.430	0.3356	0.3446	0.3534	0.3620	0.3704	0.3786	0.3867	0.3945	0.4021	0.4095	0.4168	0.4238	0.4307
0.440	0.3393	0.3484	0.3573	0.3660	0.3745	0.3829	0.3910	0.3989	0.4067	0.4142	0.4216	0.4287	0.4357
0.450	0.3429	0.3521	0.3612	0.3700	0.3787	0.3871	0.3953	0.4034	0.4112	0.4189	0.4263	0.4336	0.4406
0.460	0.3466	0.3559	0.3651	0.3740	0.3828	0.3913	0.3997	0.4078	0.4158	0.4236	0.4311	0.4385	0.4456
0.470	0.3502	0.3597	0.3690	0.3780	0.3869	0.3956	0.4040	0.4123	0.4204	0.4282	0.4359	0.4433	0.4506
0.480	0.3539	0.3634	0.3728	0.3820	0.3910	0.3998	0.4084	0.4168	0.4249	0.4329	0.4407	0.4482	0.4556
0.490	0.3575	0.3672	0.3767	0.3860	0.3951	0.4040	0.4127	0.4212	0.4295	0.4376	0.4454	0.4531	0.4605
0.500	0.3612	0.3710	0.3806	0.3900	0.3993	0.4083	0.4171	0.4257	0.4341	0.4422	0.4502	0.4580	0.4655
0.520	0.3681	0.3782	0.3880	0.3976	0.4071	0.4163	0.4253	0.4341	0.4426	0.4510	0.4592	0.4671	0.4748
0.540	0.3738	0.3840	0.3941	0.4039	0.4135	0.4229	0.4321	0.4411	0.4498	0.4584	0.4667	0.4748	0.4827
0.560	0.3792	0.3896	0.3999	0.4099	0.4197	0.4292	0.4386	0.4478	0.4567	0.4654	0.4739	0.4822	0.4902
0.580	0.3844	0.3950	0.4054	0.4156	0.4256	0.4353	0.4448	0.4542	0.4633	0.4721	0.4808	0.4892	0.4974
0.600	0.3893	0.4001	0.4107	0.4210	0.4312	0.4411	0.4508	0.4603	0.4695	0.4785	0.4873	0.4959	0.5043
0.620	0.3940	0.4050	0.4157	0.4262	0.4365	0.4466	0.4564	0.4661	0.4755	0.4846	0.4936	0.5023	0.5108
0.640	0.3984	0.4096	0.4205	0.4311	0.4416	0.4518	0.4618	0.4716	0.4811	0.4904	0.4995	0.5084	0.5170
0.660	0.4027	0.4139	0.4250	0.4358	0.4464	0.4567	0.4669	0.4768	0.4865	0.4959	0.5051	0.5141	0.5228
0.680	0.4066	0.4180	0.4292	0.4402	0.4509	0.4614	0.4717	0.4817	0.4915	0.5011	0.5104	0.5195	0.5284
0.700	0.4104	0.4219	0.4332	0.4443	0.4552	0.4658	0.4761	0.4863	0.4962	0.5059	0.5153	0.5246	0.5335
0.720	0.4139	0.4255	0.4370	0.4482	0.4591	0.4699	0.4803	0.4906	0.5006	0.5104	0.5200	0.5293	0.5384
0.740	0.4172	0.4289	0.4405	0.4518	0.4628	0.4737	0.4843	0.4946	0.5048	0.5146	0.5243	0.5337	0.5429
0.760	0.4202	0.4321	0.4437	0.4551	0.4662	0.4772	0.4879	0.4983	0.5086	0.5185	0.5283	0.5378	0.5471
0.780	0.4233	0.4352	0.4469	0.4584	0.4697	0.4807	0.4915	0.5020	0.5123	0.5224	0.5323	0.5419	0.5513
0.800	0.4260	0.4380	0.4499	0.4614	0.4728	0.4839	0.4948	0.5054	0.5158	0.5260	0.5360	0.5457	0.5551
0.820	0.4286	0.4407	0.4526	0.4643	0.4758	0.4870	0.4979	0.5087	0.5192	0.5294	0.5394	0.5492	0.5588
0.840	0.4310	0.4432	0.4552	0.4670	0.4785	0.4898	0.5009	0.5117	0.5223	0.5326	0.5427	0.5525	0.5622
0.860	0.4332	0.4456	0.4577	0.4695	0.4811	0.4925	0.5036	0.5145	0.5252	0.5356	0.5457	0.5557	0.5653

0.880	0.4354	0.4478	0.4599	0.4719	0.4836	0.4950	0.5062	0.5172	0.5279	0.5384	0.5486	0.5586	0.5683
0.900	0.4374	0.4498	0.4621	0.4741	0.4859	0.4974	0.5086	0.5197	0.5305	0.5410	0.5513	0.5613	0.5712
0.920	0.4392	0.4518	0.4641	0.4762	0.4880	0.4996	0.5109	0.5220	0.5328	0.5434	0.5538	0.5639	0.5738
0.940	0.4410	0.4536	0.4660	0.4781	0.4900	0.5016	0.5130	0.5242	0.5351	0.5457	0.5562	0.5663	0.5763
0.960	0.4426	0.4553	0.4677	0.4799	0.4919	0.5036	0.5150	0.5262	0.5372	0.5479	0.5584	0.5686	0.5786
0.980	0.4442	0.4569	0.4694	0.4816	0.4936	0.5054	0.5169	0.5281	0.5392	0.5499	0.5604	0.5707	0.5807
1.000	0.4456	0.4584	0.4709	0.4832	0.4953	0.5071	0.5186	0.5299	0.5410	0.5518	0.5624	0.5727	0.5827
1.200	0.4558	0.4689	0.4818	0.4945	0.5069	0.5190	0.5309	0.5426	0.5540	0.5651	0.5760	0.5867	0.5971
1.400	0.4599	0.4732	0.4862	0.4990	0.5115	0.5238	0.5359	0.5476	0.5592	0.5705	0.5815	0.5923	0.6029
1.600	0.4626	0.4760	0.4891	0.5019	0.5146	0.5269	0.5391	0.5509	0.5626	0.5739	0.5851	0.5959	0.6066
1.800	0.4641	0.4775	0.4907	0.5036	0.5163	0.5287	0.5409	0.5528	0.5645	0.5759	0.5871	0.5980	0.6087
2.000	0.4646	0.4781	0.4913	0.5042	0.5169	0.5294	0.5416	0.5535	0.5652	0.5767	0.5879	0.5988	0.6095
2.200	0.4646	0.4781	0.4913	0.5042	0.5169	0.5294	0.5416	0.5536	0.5653	0.5767	0.5879	0.5988	0.6095
2.400	0.4646	0.4781	0.4913	0.5042	0.5169	0.5294	0.5416	0.5536	0.5653	0.5767	0.5879	0.5988	0.6095
2.600	0.4646	0.4781	0.4913	0.5042	0.5169	0.5294	0.5416	0.5536	0.5653	0.5767	0.5879	0.5988	0.6095
2.800	0.4646	0.4781	0.4913	0.5042	0.5169	0.5294	0.5416	0.5536	0.5653	0.5767	0.5879	0.5988	0.6095
3.000	0.4646	0.4781	0.4913	0.5042	0.5169	0.5294	0.5416	0.5536	0.5653	0.5767	0.5879	0.5988	0.6095

β = α	0.390	0.400	0.410	0.420	0.430	0.440	0.450	0.460	0.470	0.480	0.490	0.500	0.520
0.005	0.0085	0.0086	0.0086	0.0087	0.0088	0.0088	0.0089	0.0090	0.0090	0.0091	0.0092	0.0093	0.0094
0.010	0.0167	0.0168	0.0170	0.0171	0.0173	0.0174	0.0175	0.0177	0.0178	0.0180	0.0181	0.0183	0.0186
0.020	0.0328	0.0331	0.0334	0.0337	0.0340	0.0343	0.0346	0.0349	0.0352	0.0355	0.0358	0.0361	0.0367
0.030	0.0485	0.0489	0.0494	0.0498	0.0503	0.0507	0.0512	0.0516	0.0521	0.0525	0.0530	0.0535	0.0544
0.040	0.0639	0.0645	0.0651	0.0657	0.0663	0.0669	0.0675	0.0681	0.0687	0.0693	0.0699	0.0705	0.0717
0.050	0.0789	0.0796	0.0804	0.0811	0.0819	0.0826	0.0834	0.0841	0.0849	0.0856	0.0864	0.0872	0.0887
0.060	0.0936	0.0945	0.0954	0.0963	0.0972	0.0981	0.0990	0.0999	0.1008	0.1017	0.1026	0.1035	0.1053
0.070	0.1079	0.1089	0.1100	0.1110	0.1121	0.1131	0.1141	0.1152	0.1162	0.1173	0.1183	0.1194	0.1215
0.080	0.1218	0.1230	0.1242	0.1254	0.1266	0.1278	0.1290	0.1302	0.1314	0.1326	0.1337	0.1349	0.1373
0.090	0.1355	0.1368	0.1381	0.1395	0.1408	0.1421	0.1435	0.1448	0.1461	0.1475	0.1488	0.1501	0.1528
0.100	0.1487	0.1502	0.1517	0.1532	0.1546	0.1561	0.1576	0.1591	0.1605	0.1620	0.1635	0.1650	0.1679
0.110	0.1616	0.1633	0.1649	0.1665	0.1681	0.1697	0.1713	0.1729	0.1746	0.1762	0.1778	0.1794	0.1826
0.120	0.1741	0.1759	0.1776	0.1794	0.1812	0.1829	0.1847	0.1865	0.1882	0.1900	0.1918	0.1935	0.1971
0.130	0.1864	0.1883	0.1902	0.1921	0.1940	0.1959	0.1978	0.1998	0.2017	0.2036	0.2055	0.2074	0.2112
0.140	0.1984	0.2004	0.2025	0.2045	0.2066	0.2086	0.2107	0.2128	0.2148	0.2169	0.2189	0.2210	0.2251
0.150	0.2101	0.2123	0.2145	0.2167	0.2189	0.2211	0.2233	0.2255	0.2277	0.2299	0.2321	0.2343	0.2387
0.160	0.2215	0.2239	0.2262	0.2285	0.2309	0.2332	0.2356	0.2379	0.2402	0.2426	0.2449	0.2472	0.2519
0.170	0.2327	0.2352	0.2377	0.2401	0.2426	0.2451	0.2476	0.2500	0.2525	0.2550	0.2575	0.2599	0.2649
0.180	0.2436	0.2462	0.2488	0.2515	0.2541	0.2567	0.2593	0.2619	0.2645	0.2671	0.2697	0.2724	0.2776
0.190	0.2542	0.2570	0.2597	0.2625	0.2652	0.2680	0.2707	0.2735	0.2762	0.2790	0.2817	0.2845	0.2900
0.200	0.2646	0.2675	0.2704	0.2732	0.2761	0.2790	0.2819	0.2848	0.2877	0.2905	0.2934	0.2963	0.3021
0.210	0.2747	0.2777	0.2807	0.2837	0.2867	0.2898	0.2928	0.2958	0.2988	0.3018	0.3048	0.3078	0.3139
0.220	0.2845	0.2876	0.2908	0.2939	0.2971	0.3002	0.3034	0.3065	0.3096	0.3128	0.3159	0.3191	0.3254
0.230	0.2940	0.2973	0.3006	0.3038	0.3071	0.3104	0.3137	0.3169	0.3202	0.3235	0.3268	0.3300	0.3366
0.240	0.3033	0.3067	0.3101	0.3135	0.3169	0.3203	0.3237	0.3271	0.3305	0.3339	0.3373	0.3407	0.3475
0.250	0.3123	0.3158	0.3194	0.3229	0.3264	0.3299	0.3335	0.3370	0.3405	0.3440	0.3476	0.3511	0.3582

0.260	0.3210	0.3247	0.3283	0.3320	0.3356	0.3393	0.3429	0.3466	0.3502	0.3539	0.3575	0.3612	0.3685
0.270	0.3295	0.3333	0.3371	0.3408	0.3446	0.3484	0.3521	0.3559	0.3597	0.3634	0.3672	0.3710	0.3785
0.280	0.3378	0.3417	0.3456	0.3495	0.3534	0.3573	0.3612	0.3651	0.3690	0.3728	0.3767	0.3806	0.3884
0.290	0.3460	0.3500	0.3540	0.3580	0.3620	0.3660	0.3700	0.3740	0.3780	0.3820	0.3860	0.3900	0.3981
0.300	0.3539	0.3581	0.3622	0.3663	0.3704	0.3745	0.3787	0.3828	0.3869	0.3910	0.3951	0.3993	0.4075
0.310	0.3617	0.3659	0.3702	0.3744	0.3786	0.3829	0.3871	0.3913	0.3956	0.3998	0.4040	0.4083	0.4167
0.320	0.3693	0.3736	0.3780	0.3823	0.3867	0.3910	0.3953	0.3997	0.4040	0.4084	0.4127	0.4171	0.4258
0.330	0.3767	0.3811	0.3856	0.3900	0.3945	0.3989	0.4034	0.4078	0.4123	0.4168	0.4212	0.4257	0.4346
0.340	0.3839	0.3884	0.3930	0.3976	0.4021	0.4067	0.4112	0.4158	0.4204	0.4249	0.4295	0.4341	0.4432
0.350	0.3909	0.3955	0.4002	0.4049	0.4095	0.4142	0.4189	0.4236	0.4282	0.4329	0.4376	0.4422	0.4516
0.360	0.3977	0.4025	0.4072	0.4120	0.4168	0.4216	0.4263	0.4311	0.4359	0.4407	0.4454	0.4502	0.4597
0.370	0.4043	0.4092	0.4141	0.4190	0.4238	0.4287	0.4336	0.4385	0.4433	0.4482	0.4531	0.4580	0.4677
0.380	0.4142	0.4193	0.4243	0.4293	0.4343	0.4394	0.4444	0.4494	0.4544	0.4595	0.4645	0.4695	0.4796
0.390	0.4193	0.4244	0.4296	0.4347	0.4398	0.4449	0.4500	0.4551	0.4602	0.4653	0.4704	0.4755	0.4858
0.400	0.4244	0.4296	0.4348	0.4400	0.4452	0.4504	0.4556	0.4608	0.4660	0.4712	0.4764	0.4816	0.4920
0.410	0.4296	0.4348	0.4401	0.4454	0.4507	0.4559	0.4612	0.4665	0.4718	0.4771	0.4823	0.4876	0.4982
0.420	0.4347	0.4400	0.4454	0.4508	0.4561	0.4615	0.4668	0.4722	0.4776	0.4829	0.4883	0.4937	0.5044

β = α	0.390	0.400	0.410	0.420	0.430	0.440	0.450	0.460	0.470	0.480	0.490	0.500	0.520
0.430	0.4398	0.4452	0.4507	0.4561	0.4616	0.4670	0.4725	0.4779	0.4833	0.4888	0.4942	0.4997	0.5106
0.440	0.4449	0.4504	0.4559	0.4615	0.4670	0.4725	0.4781	0.4836	0.4891	0.4947	0.5002	0.5057	0.5168
0.450	0.4500	0.4556	0.4612	0.4668	0.4725	0.4781	0.4837	0.4893	0.4949	0.5005	0.5061	0.5118	0.5230
0.460	0.4551	0.4608	0.4665	0.4722	0.4779	0.4836	0.4893	0.4950	0.5007	0.5064	0.5121	0.5178	0.5292
0.470	0.4602	0.4660	0.4718	0.4776	0.4833	0.4891	0.4949	0.5007	0.5065	0.5123	0.5181	0.5238	0.5354
0.480	0.4653	0.4712	0.4771	0.4829	0.4888	0.4947	0.5005	0.5064	0.5123	0.5181	0.5240	0.5299	0.5416
0.490	0.4704	0.4764	0.4823	0.4883	0.4942	0.5002	0.5061	0.5121	0.5181	0.5240	0.5300	0.5359	0.5478
0.500	0.4755	0.4816	0.4876	0.4937	0.4997	0.5057	0.5118	0.5178	0.5238	0.5299	0.5359	0.5420	0.5540
0.520	0.4850	0.4913	0.4975	0.5037	0.5099	0.5161	0.5223	0.5286	0.5348	0.5410	0.5472	0.5534	0.5658
0.540	0.4931	0.4995	0.5058	0.5122	0.5185	0.5249	0.5312	0.5376	0.5439	0.5503	0.5566	0.5630	0.5757
0.560	0.5008	0.5073	0.5138	0.5203	0.5268	0.5333	0.5397	0.5462	0.5527	0.5592	0.5657	0.5722	0.5851
0.580	0.5082	0.5148	0.5214	0.5281	0.5347	0.5413	0.5479	0.5545	0.5611	0.5677	0.5743	0.5810	0.5942
0.600	0.5153	0.5220	0.5287	0.5355	0.5422	0.5489	0.5557	0.5624	0.5691	0.5759	0.5826	0.5893	0.6028
0.620	0.5220	0.5288	0.5357	0.5425	0.5494	0.5562	0.5631	0.5699	0.5768	0.5836	0.5905	0.5973	0.6110
0.640	0.5283	0.5353	0.5422	0.5492	0.5562	0.5631	0.5701	0.5771	0.5840	0.5910	0.5980	0.6049	0.6189
0.660	0.5343	0.5414	0.5485	0.5556	0.5626	0.5697	0.5768	0.5839	0.5909	0.5980	0.6051	0.6122	0.6263
0.680	0.5400	0.5472	0.5544	0.5615	0.5687	0.5759	0.5831	0.5903	0.5974	0.6046	0.6118	0.6190	0.6333
0.700	0.5453	0.5526	0.5599	0.5672	0.5744	0.5817	0.5890	0.5963	0.6036	0.6108	0.6181	0.6254	0.6399
0.720	0.5503	0.5577	0.5651	0.5724	0.5798	0.5872	0.5946	0.6019	0.6093	0.6167	0.6240	0.6314	0.6462
0.740	0.5550	0.5624	0.5699	0.5774	0.5848	0.5923	0.5997	0.6072	0.6147	0.6221	0.6296	0.6371	0.6520
0.760	0.5593	0.5668	0.5744	0.5819	0.5895	0.5970	0.6046	0.6121	0.6196	0.6272	0.6347	0.6423	0.6574
0.780	0.5637	0.5713	0.5789	0.5865	0.5941	0.6017	0.6093	0.6169	0.6245	0.6322	0.6398	0.6474	0.6626
0.800	0.5676	0.5753	0.5830	0.5907	0.5983	0.6060	0.6137	0.6214	0.6291	0.6368	0.6445	0.6522	0.6675
0.820	0.5713	0.5790	0.5868	0.5946	0.6023	0.6101	0.6178	0.6256	0.6334	0.6411	0.6489	0.6567	0.6722
0.840	0.5747	0.5826	0.5904	0.5982	0.6060	0.6139	0.6217	0.6295	0.6374	0.6452	0.6530	0.6609	0.6765
0.860	0.5780	0.5859	0.5938	0.6017	0.6096	0.6175	0.6254	0.6333	0.6411	0.6490	0.6569	0.6648	0.6806

0.880	0.5810	0.5890	0.5970	0.6049	0.6129	0.6208	0.6288	0.6367	0.6447	0.6526	0.6606	0.6686	0.6845
0.900	0.5839	0.5919	0.5999	0.6080	0.6160	0.6240	0.6320	0.6400	0.6480	0.6560	0.6640	0.6721	0.6881
0.920	0.5866	0.5947	0.6027	0.6108	0.6189	0.6269	0.6350	0.6431	0.6511	0.6592	0.6673	0.6753	0.6915
0.940	0.5891	0.5973	0.6054	0.6135	0.6216	0.6297	0.6378	0.6459	0.6540	0.6622	0.6703	0.6784	0.6946
0.960	0.5915	0.5997	0.6078	0.6160	0.6242	0.6323	0.6405	0.6486	0.6568	0.6649	0.6731	0.6813	0.6976
0.980	0.5938	0.6020	0.6102	0.6184	0.6266	0.6348	0.6430	0.6512	0.6594	0.6676	0.6758	0.6840	0.7004
1.000	0.5958	0.6041	0.6123	0.6206	0.6288	0.6370	0.6453	0.6535	0.6618	0.6700	0.6782	0.6865	0.7029
1.200	0.6106	0.6191	0.6276	0.6362	0.6447	0.6532	0.6617	0.6703	0.6788	0.6873	0.6958	0.7044	0.7214
1.400	0.6166	0.6253	0.6339	0.6425	0.6512	0.6598	0.6685	0.6771	0.6858	0.6944	0.7030	0.7117	0.7290
1.600	0.6205	0.6292	0.6379	0.6466	0.6553	0.6641	0.6728	0.6815	0.6902	0.6989	0.7076	0.7164	0.7338
1.800	0.6226	0.6313	0.6401	0.6489	0.6576	0.6664	0.6751	0.6839	0.6927	0.7014	0.7102	0.7190	0.7365
2.000	0.6234	0.6321	0.6409	0.6497	0.6585	0.6673	0.6761	0.6848	0.6936	0.7024	0.7112	0.7200	0.7375
2.200	0.6234	0.6321	0.6409	0.6497	0.6585	0.6673	0.6761	0.6848	0.6936	0.7024	0.7112	0.7200	0.7375
2.400	0.6234	0.6321	0.6409	0.6497	0.6585	0.6673	0.6761	0.6848	0.6936	0.7024	0.7112	0.7200	0.7375
2.600	0.6234	0.6321	0.6409	0.6497	0.6585	0.6673	0.6761	0.6848	0.6936	0.7024	0.7112	0.7200	0.7375
2.800	0.6234	0.6321	0.6409	0.6497	0.6585	0.6673	0.6761	0.6848	0.6936	0.7024	0.7112	0.7200	0.7375
3.000	0.6234	0.6321	0.6409	0.6497	0.6585	0.6673	0.6761	0.6848	0.6936	0.7024	0.7112	0.7200	0.7375

β =	0.540	0.560	0.580	0.600	0.620	0.640	0.660	0.680	0.700	0.720	0.740	0.760	0.780
α													
0.005	0.0098	0.0098	0.0098	0.0098	0.0099	0.0099	0.0100	0.0100	0.0101	0.0102	0.0103	0.0104	0.0105
0.010	0.0190	0.0191	0.0193	0.0194	0.0196	0.0197	0.0199	0.0200	0.0202	0.0203	0.0205	0.0206	0.0208
0.020	0.0372	0.0376	0.0380	0.0383	0.0387	0.0390	0.0394	0.0397	0.0400	0.0403	0.0406	0.0408	0.0411
0.030	0.0550	0.0557	0.0563	0.0569	0.0574	0.0580	0.0585	0.0589	0.0594	0.0598	0.0602	0.0606	0.0610
0.040	0.0725	0.0734	0.0742	0.0750	0.0758	0.0765	0.0772	0.0778	0.0785	0.0790	0.0796	0.0800	0.0805
0.050	0.0896	0.0908	0.0918	0.0928	0.0938	0.0947	0.0956	0.0964	0.0971	0.0978	0.0985	0.0991	0.0996
0.060	0.1064	0.1078	0.1091	0.1103	0.1114	0.1125	0.1136	0.1145	0.1154	0.1163	0.1171	0.1178	0.1184
0.070	0.1229	0.1244	0.1259	0.1273	0.1287	0.1300	0.1312	0.1323	0.1333	0.1343	0.1352	0.1361	0.1368
0.080	0.1389	0.1407	0.1424	0.1440	0.1456	0.1470	0.1484	0.1497	0.1509	0.1520	0.1531	0.1540	0.1549
0.090	0.1547	0.1567	0.1586	0.1604	0.1621	0.1637	0.1653	0.1667	0.1681	0.1693	0.1705	0.1716	0.1725
0.100	0.1701	0.1723	0.1744	0.1764	0.1783	0.1801	0.1818	0.1834	0.1849	0.1863	0.1876	0.1888	0.1899
0.110	0.1851	0.1875	0.1898	0.1920	0.1940	0.1960	0.1979	0.1996	0.2013	0.2028	0.2042	0.2056	0.2068
0.120	0.1997	0.2023	0.2048	0.2072	0.2094	0.2116	0.2136	0.2155	0.2173	0.2190	0.2206	0.2220	0.2234
0.130	0.2141	0.2169	0.2196	0.2221	0.2246	0.2269	0.2291	0.2312	0.2331	0.2349	0.2366	0.2382	0.2396
0.140	0.2281	0.2312	0.2341	0.2368	0.2394	0.2419	0.2443	0.2465	0.2486	0.2506	0.2524	0.2541	0.2556
0.150	0.2419	0.2451	0.2482	0.2512	0.2540	0.2566	0.2592	0.2615	0.2638	0.2659	0.2678	0.2696	0.2713
0.160	0.2554	0.2588	0.2621	0.2652	0.2682	0.2710	0.2737	0.2763	0.2786	0.2809	0.2829	0.2849	0.2866
0.170	0.2686	0.2722	0.2757	0.2790	0.2822	0.2852	0.2880	0.2907	0.2932	0.2955	0.2977	0.2998	0.3017
0.180	0.2814	0.2853	0.2890	0.2925	0.2958	0.2990	0.3020	0.3048	0.3074	0.3099	0.3123	0.3144	0.3164
0.190	0.2940	0.2981	0.3019	0.3056	0.3091	0.3125	0.3156	0.3186	0.3214	0.3240	0.3264	0.3287	0.3308
0.200	0.3063	0.3105	0.3146	0.3185	0.3221	0.3256	0.3290	0.3321	0.3350	0.3378	0.3403	0.3427	0.3449
0.210	0.3183	0.3227	0.3270	0.3310	0.3349	0.3385	0.3420	0.3453	0.3483	0.3512	0.3539	0.3564	0.3587
0.220	0.3300	0.3346	0.3391	0.3433	0.3473	0.3511	0.3547	0.3582	0.3614	0.3644	0.3672	0.3698	0.3722
0.230	0.3414	0.3462	0.3508	0.3552	0.3594	0.3634	0.3672	0.3707	0.3741	0.3772	0.3801	0.3828	0.3853
0.240	0.3525	0.3575	0.3623	0.3669	0.3712	0.3754	0.3793	0.3830	0.3865	0.3897	0.3928	0.3956	0.3982
0.250	0.3633	0.3685	0.3735	0.3782	0.3828	0.3871	0.3911	0.3950	0.3986	0.4020	0.4051	0.4081	0.4108

0.260	0.3738	0.3792	0.3844	0.3893	0.3940	0.3984	0.4027	0.4066	0.4104	0.4139	0.4172	0.4202	0.4230
0.270	0.3840	0.3896	0.3950	0.4001	0.4050	0.4096	0.4139	0.4180	0.4219	0.4255	0.4289	0.4321	0.4349
0.280	0.3941	0.3999	0.4054	0.4107	0.4157	0.4205	0.4250	0.4292	0.4332	0.4370	0.4405	0.4437	0.4467
0.290	0.4039	0.4099	0.4156	0.4210	0.4262	0.4311	0.4358	0.4402	0.4443	0.4482	0.4518	0.4551	0.4581
0.300	0.4135	0.4197	0.4256	0.4312	0.4365	0.4416	0.4464	0.4509	0.4552	0.4591	0.4628	0.4662	0.4694
0.310	0.4229	0.4292	0.4353	0.4411	0.4466	0.4518	0.4567	0.4614	0.4658	0.4699	0.4737	0.4772	0.4804
0.320	0.4321	0.4386	0.4448	0.4508	0.4564	0.4618	0.4669	0.4717	0.4761	0.4803	0.4843	0.4879	0.4912
0.330	0.4411	0.4478	0.4542	0.4603	0.4661	0.4716	0.4768	0.4817	0.4863	0.4906	0.4946	0.4983	0.5017
0.340	0.4498	0.4567	0.4633	0.4695	0.4755	0.4811	0.4865	0.4915	0.4962	0.5006	0.5048	0.5086	0.5121
0.350	0.4584	0.4654	0.4721	0.4785	0.4846	0.4904	0.4959	0.5011	0.5059	0.5104	0.5146	0.5185	0.5221
0.360	0.4667	0.4739	0.4808	0.4873	0.4936	0.4995	0.5051	0.5104	0.5153	0.5200	0.5243	0.5283	0.5320
0.370	0.4748	0.4822	0.4892	0.4959	0.5023	0.5084	0.5141	0.5195	0.5246	0.5293	0.5337	0.5378	0.5416
0.380	0.4868	0.4943	0.5016	0.5085	0.5151	0.5213	0.5273	0.5328	0.5381	0.5429	0.5475	0.5517	0.5556
0.390	0.4931	0.5008	0.5082	0.5153	0.5220	0.5283	0.5343	0.5400	0.5453	0.5503	0.5550	0.5593	0.5632
0.400	0.4995	0.5073	0.5148	0.5220	0.5288	0.5353	0.5414	0.5472	0.5526	0.5577	0.5624	0.5668	0.5709
0.410	0.5058	0.5138	0.5214	0.5287	0.5357	0.5422	0.5485	0.5544	0.5599	0.5651	0.5699	0.5744	0.5785
0.420	0.5122	0.5203	0.5281	0.5355	0.5425	0.5492	0.5556	0.5615	0.5672	0.5724	0.5774	0.5819	0.5861

β = α	0.540	0.560	0.580	0.600	0.620	0.640	0.660	0.680	0.700	0.720	0.740	0.760	0.780
0.430	0.5185	0.5268	0.5347	0.5422	0.5494	0.5562	0.5626	0.5687	0.5744	0.5798	0.5848	0.5895	0.5937
0.440	0.5249	0.5333	0.5413	0.5489	0.5562	0.5631	0.5697	0.5759	0.5817	0.5872	0.5923	0.5970	0.6014
0.450	0.5312	0.5397	0.5479	0.5557	0.5631	0.5701	0.5768	0.5831	0.5890	0.5946	0.5997	0.6046	0.6090
0.460	0.5376	0.5462	0.5545	0.5624	0.5699	0.5771	0.5839	0.5903	0.5963	0.6019	0.6072	0.6121	0.6166
0.470	0.5439	0.5527	0.5611	0.5691	0.5768	0.5840	0.5909	0.5974	0.6036	0.6093	0.6147	0.6196	0.6243
0.480	0.5503	0.5592	0.5677	0.5759	0.5836	0.5910	0.5980	0.6046	0.6108	0.6167	0.6221	0.6272	0.6319
0.490	0.5566	0.5657	0.5743	0.5826	0.5905	0.5980	0.6051	0.6118	0.6181	0.6240	0.6296	0.6347	0.6395
0.500	0.5630	0.5722	0.5810	0.5893	0.5973	0.6049	0.6122	0.6190	0.6254	0.6314	0.6371	0.6423	0.6471
0.520	0.6051	0.5979	0.5934	0.5915	0.5973	0.6045	0.6122	0.6190	0.6254	0.6348	0.6511	0.6701	0.6916
0.540	0.6085	0.6053	0.6040	0.6046	0.6072	0.6116	0.6181	0.6264	0.6367	0.6490	0.6631	0.6792	0.6972
0.560	0.6123	0.6125	0.6141	0.6170	0.6212	0.6268	0.6337	0.6419	0.6515	0.6624	0.6746	0.6882	0.7031
0.580	0.6165	0.6197	0.6238	0.6286	0.6343	0.6408	0.6482	0.6563	0.6653	0.6750	0.6857	0.6971	0.7093
0.600	0.6210	0.6268	0.6330	0.6396	0.6465	0.6538	0.6616	0.6696	0.6781	0.6870	0.6962	0.7058	0.7158
0.620	0.6259	0.6338	0.6418	0.6498	0.6578	0.6658	0.6739	0.6819	0.6900	0.6981	0.7063	0.7144	0.7226
0.640	0.6312	0.6408	0.6502	0.6593	0.6682	0.6768	0.6851	0.6932	0.7010	0.7086	0.7158	0.7229	0.7297
0.660	0.6368	0.6477	0.6582	0.6681	0.6776	0.6867	0.6953	0.7034	0.7110	0.7182	0.7250	0.7312	0.7370
0.680	0.6428	0.6545	0.6657	0.6762	0.6862	0.6955	0.7043	0.7125	0.7201	0.7271	0.7336	0.7394	0.7447
0.700	0.6492	0.6613	0.6728	0.6836	0.6938	0.7033	0.7123	0.7206	0.7283	0.7353	0.7417	0.7475	0.7527
0.720	0.6560	0.6680	0.6794	0.6902	0.7005	0.7101	0.7192	0.7276	0.7355	0.7427	0.7494	0.7554	0.7609
0.740	0.6631	0.6746	0.6857	0.6962	0.7063	0.7158	0.7250	0.7336	0.7417	0.7494	0.7566	0.7632	0.7695
0.760	0.6706	0.6812	0.6914	0.7014	0.7111	0.7205	0.7297	0.7385	0.7470	0.7553	0.7632	0.7709	0.7783
0.780	0.6738	0.6857	0.6971	0.7080	0.7184	0.7282	0.7376	0.7464	0.7547	0.7624	0.7697	0.7764	0.7826
0.800	0.6789	0.6909	0.7024	0.7134	0.7238	0.7338	0.7432	0.7521	0.7605	0.7684	0.7758	0.7827	0.7890
0.820	0.6837	0.6958	0.7074	0.7184	0.7290	0.7390	0.7486	0.7576	0.7661	0.7741	0.7816	0.7885	0.7950
0.840	0.6882	0.7004	0.7120	0.7232	0.7338	0.7440	0.7536	0.7627	0.7713	0.7794	0.7870	0.7940	0.8006
0.860	0.6924	0.7047	0.7164	0.7277	0.7384	0.7486	0.7583	0.7675	0.7762	0.7844	0.7921	0.7992	0.8058

0.880	0.6963	0.7087	0.7206	0.7319	0.7427	0.7530	0.7628	0.7721	0.7809	0.7891	0.7969	0.8041	0.8108
0.900	0.7000	0.7125	0.7245	0.7359	0.7468	0.7572	0.7671	0.7764	0.7852	0.7936	0.8014	0.8086	0.8154
0.920	0.7035	0.7161	0.7281	0.7396	0.7506	0.7611	0.7710	0.7804	0.7893	0.7977	0.8056	0.8129	0.8197
0.940	0.7067	0.7194	0.7315	0.7431	0.7542	0.7647	0.7748	0.7842	0.7932	0.8016	0.8095	0.8169	0.8238
0.960	0.7098	0.7225	0.7347	0.7464	0.7576	0.7682	0.7783	0.7878	0.7968	0.8053	0.8133	0.8207	0.8276
0.980	0.7126	0.7255	0.7378	0.7495	0.7607	0.7714	0.7816	0.7912	0.8002	0.8088	0.8168	0.8242	0.8311
1.000	0.7153	0.7282	0.7406	0.7524	0.7637	0.7744	0.7846	0.7943	0.8034	0.8120	0.8200	0.8275	0.8345
1.200	0.7342	0.7476	0.7605	0.7728	0.7845	0.7957	0.8064	0.8164	0.8260	0.8349	0.8434	0.8512	0.8585
1.400	0.7419	0.7556	0.7686	0.7812	0.7931	0.8045	0.8153	0.8256	0.8353	0.8445	0.8530	0.8611	0.8685
1.600	0.7469	0.7607	0.7739	0.7866	0.7986	0.8101	0.8211	0.8315	0.8413	0.8505	0.8592	0.8674	0.8749
1.800	0.7498	0.7636	0.7769	0.7895	0.8017	0.8132	0.8243	0.8347	0.8446	0.8539	0.8627	0.8708	0.8785
2.000	0.7509	0.7648	0.7780	0.7907	0.8029	0.8145	0.8255	0.8360	0.8459	0.8552	0.8640	0.8722	0.8799
2.200	0.7510	0.7648	0.7781	0.7907	0.8029	0.8145	0.8255	0.8360	0.8459	0.8552	0.8640	0.8722	0.8799
2.400	0.7510	0.7648	0.7781	0.7907	0.8029	0.8145	0.8255	0.8360	0.8459	0.8552	0.8640	0.8722	0.8799
2.600	0.7510	0.7648	0.7781	0.7907	0.8029	0.8145	0.8255	0.8360	0.8459	0.8552	0.8640	0.8722	0.8799
2.800	0.7510	0.7648	0.7781	0.7907	0.8029	0.8145	0.8255	0.8360	0.8459	0.8552	0.8640	0.8722	0.8799
3.000	0.7510	0.7648	0.7781	0.7907	0.8029	0.8145	0.8255	0.8360	0.8459	0.8552	0.8640	0.8722	0.8799

α \ β =	0.800	0.820	0.840	0.860	0.880	0.900	0.920	0.940	0.960	0.980	1.000	1.200	1.400
0.005	0.0105	0.0105	0.0106	0.0106	0.0107	0.0107	0.0108	0.0108	0.0109	0.0109	0.0109	0.0112	0.0112
0.010	0.0208	0.0209	0.0210	0.0211	0.0212	0.0213	0.0214	0.0215	0.0215	0.0216	0.0217	0.0221	0.0221
0.020	0.0413	0.0415	0.0416	0.0418	0.0420	0.0422	0.0423	0.0425	0.0426	0.0428	0.0429	0.0437	0.0436
0.030	0.0613	0.0616	0.0619	0.0621	0.0624	0.0626	0.0629	0.0631	0.0633	0.0635	0.0637	0.0649	0.0648
0.040	0.0810	0.0814	0.0817	0.0821	0.0824	0.0827	0.0831	0.0834	0.0836	0.0839	0.0842	0.0857	0.0857
0.050	0.1003	0.1007	0.1012	0.1016	0.1021	0.1025	0.1029	0.1032	0.1036	0.1039	0.1042	0.1062	0.1064
0.060	0.1192	0.1198	0.1203	0.1208	0.1214	0.1219	0.1223	0.1228	0.1232	0.1236	0.1239	0.1263	0.1268
0.070	0.1377	0.1384	0.1391	0.1397	0.1403	0.1409	0.1414	0.1419	0.1424	0.1428	0.1432	0.1460	0.1469
0.080	0.1558	0.1567	0.1574	0.1582	0.1588	0.1595	0.1601	0.1607	0.1612	0.1617	0.1622	0.1653	0.1667
0.090	0.1736	0.1745	0.1754	0.1763	0.1770	0.1778	0.1784	0.1791	0.1797	0.1802	0.1807	0.1843	0.1862
0.100	0.1910	0.1921	0.1930	0.1940	0.1948	0.1957	0.1964	0.1971	0.1977	0.1984	0.1989	0.2030	0.2054
0.110	0.2080	0.2092	0.2103	0.2113	0.2123	0.2132	0.2140	0.2148	0.2155	0.2161	0.2167	0.2212	0.2244
0.120	0.2247	0.2260	0.2272	0.2283	0.2293	0.2303	0.2312	0.2320	0.2328	0.2335	0.2342	0.2390	0.2419
0.130	0.2411	0.2425	0.2437	0.2449	0.2461	0.2471	0.2481	0.2490	0.2498	0.2506	0.2513	0.2566	0.2597
0.140	0.2572	0.2586	0.2600	0.2613	0.2625	0.2636	0.2647	0.2656	0.2665	0.2674	0.2682	0.2738	0.2772
0.150	0.2729	0.2745	0.2760	0.2773	0.2786	0.2798	0.2809	0.2820	0.2829	0.2838	0.2847	0.2907	0.2943
0.160	0.2884	0.2901	0.2916	0.2931	0.2944	0.2957	0.2969	0.2980	0.2990	0.3000	0.3009	0.3073	0.3111
0.170	0.3035	0.3053	0.3070	0.3085	0.3099	0.3113	0.3125	0.3137	0.3148	0.3158	0.3168	0.3236	0.3275
0.180	0.3184	0.3203	0.3220	0.3236	0.3251	0.3266	0.3279	0.3291	0.3303	0.3314	0.3324	0.3396	0.3437
0.190	0.3329	0.3349	0.3367	0.3384	0.3400	0.3415	0.3429	0.3442	0.3454	0.3466	0.3476	0.3552	0.3595
0.200	0.3471	0.3492	0.3511	0.3529	0.3546	0.3561	0.3576	0.3590	0.3602	0.3614	0.3626	0.3705	0.3750
0.210	0.3611	0.3632	0.3652	0.3671	0.3688	0.3705	0.3720	0.3734	0.3748	0.3760	0.3772	0.3855	0.3902
0.220	0.3747	0.3769	0.3790	0.3809	0.3828	0.3845	0.3861	0.3876	0.3890	0.3903	0.3915	0.4002	0.4050
0.230	0.3880	0.3903	0.3924	0.3945	0.3964	0.3982	0.3998	0.4014	0.4028	0.4042	0.4055	0.4146	0.4195
0.240	0.4009	0.4033	0.4056	0.4077	0.4097	0.4115	0.4133	0.4149	0.4164	0.4178	0.4192	0.4286	0.4337
0.250	0.4137	0.4162	0.4185	0.4207	0.4227	0.4246	0.4264	0.4281	0.4297	0.4312	0.4326	0.4424	0.4475

0.260	0.4260	0.4286	0.4310	0.4332	0.4354	0.4374	0.4392	0.4410	0.4426	0.4442	0.4456	0.4558	0.4611
0.270	0.4380	0.4407	0.4432	0.4456	0.4478	0.4498	0.4518	0.4536	0.4553	0.4569	0.4584	0.4689	0.4745
0.280	0.4499	0.4526	0.4552	0.4577	0.4599	0.4621	0.4641	0.4660	0.4677	0.4694	0.4709	0.4818	0.4876
0.290	0.4614	0.4643	0.4670	0.4695	0.4719	0.4741	0.4762	0.4781	0.4799	0.4816	0.4832	0.4945	0.5004
0.300	0.4728	0.4758	0.4785	0.4811	0.4836	0.4859	0.4880	0.4900	0.4919	0.4936	0.4953	0.5069	0.5130
0.310	0.4839	0.4870	0.4898	0.4925	0.4950	0.4974	0.4996	0.5016	0.5036	0.5054	0.5071	0.5190	0.5254
0.320	0.4948	0.4979	0.5009	0.5036	0.5062	0.5086	0.5109	0.5130	0.5150	0.5169	0.5186	0.5309	0.5375
0.330	0.5054	0.5087	0.5117	0.5145	0.5172	0.5197	0.5220	0.5242	0.5262	0.5281	0.5299	0.5426	0.5493
0.340	0.5158	0.5192	0.5223	0.5252	0.5279	0.5305	0.5328	0.5351	0.5372	0.5392	0.5410	0.5540	0.5609
0.350	0.5260	0.5294	0.5326	0.5356	0.5384	0.5410	0.5434	0.5457	0.5479	0.5499	0.5518	0.5651	0.5722
0.360	0.5360	0.5394	0.5427	0.5457	0.5486	0.5513	0.5538	0.5562	0.5584	0.5604	0.5624	0.5760	0.5832
0.370	0.5457	0.5492	0.5525	0.5557	0.5586	0.5613	0.5639	0.5663	0.5686	0.5707	0.5727	0.5867	0.5941
0.380	0.5599	0.5635	0.5669	0.5701	0.5731	0.5759	0.5785	0.5810	0.5834	0.5856	0.5876	0.6021	0.6094
0.390	0.5676	0.5713	0.5747	0.5780	0.5810	0.5839	0.5866	0.5891	0.5915	0.5938	0.5958	0.6106	0.6182
0.400	0.5753	0.5790	0.5826	0.5859	0.5890	0.5919	0.5947	0.5973	0.5997	0.6020	0.6041	0.6191	0.6269
0.410	0.5830	0.5868	0.5904	0.5938	0.5970	0.5999	0.6027	0.6054	0.6078	0.6102	0.6123	0.6276	0.6356
0.420	0.5907	0.5946	0.5982	0.6017	0.6049	0.6080	0.6108	0.6135	0.6160	0.6184	0.6206	0.6362	0.6443

β = α	0.800	0.820	0.840	0.860	0.880	0.900	0.920	0.940	0.960	0.980	1.000	1.200	1.400
0.430	0.5983	0.6023	0.6060	0.6096	0.6129	0.6160	0.6189	0.6216	0.6242	0.6266	0.6288	0.6447	0.6530
0.440	0.6060	0.6101	0.6139	0.6175	0.6208	0.6240	0.6269	0.6297	0.6323	0.6348	0.6370	0.6532	0.6617
0.450	0.6137	0.6176	0.6217	0.6254	0.6288	0.6320	0.6350	0.6378	0.6405	0.6430	0.6453	0.6617	0.6704
0.460	0.6214	0.6256	0.6295	0.6333	0.6367	0.6400	0.6431	0.6459	0.6486	0.6512	0.6535	0.6703	0.6791
0.470	0.6291	0.6334	0.6374	0.6411	0.6447	0.6480	0.6511	0.6540	0.6568	0.6594	0.6618	0.6788	0.6878
0.480	0.6368	0.6411	0.6452	0.6490	0.6526	0.6560	0.6592	0.6622	0.6649	0.6676	0.6700	0.6873	0.6966
0.490	0.6445	0.6489	0.6530	0.6569	0.6606	0.6640	0.6673	0.6703	0.6731	0.6758	0.6782	0.6958	0.7053
0.500	0.6522	0.6567	0.6609	0.6648	0.6686	0.6721	0.6753	0.6784	0.6813	0.6840	0.6865	0.7044	0.7140
0.520	0.6664	0.6711	0.6755	0.6796	0.6835	0.6871	0.6904	0.6936	0.6965	0.6993	0.7018	0.7202	0.7308
0.540	0.6789	0.6837	0.6882	0.6924	0.6963	0.7000	0.7035	0.7067	0.7098	0.7126	0.7153	0.7342	0.7445
0.560	0.6909	0.6958	0.7004	0.7047	0.7087	0.7125	0.7161	0.7194	0.7225	0.7255	0.7282	0.7476	0.7578
0.580	0.7024	0.7074	0.7120	0.7164	0.7206	0.7245	0.7281	0.7315	0.7347	0.7378	0.7406	0.7605	0.7706
0.600	0.7134	0.7184	0.7232	0.7277	0.7319	0.7359	0.7396	0.7431	0.7464	0.7495	0.7524	0.7728	0.7829
0.620	0.7238	0.7290	0.7338	0.7384	0.7427	0.7468	0.7506	0.7542	0.7576	0.7607	0.7637	0.7845	0.7948
0.640	0.7338	0.7390	0.7440	0.7486	0.7530	0.7572	0.7611	0.7647	0.7682	0.7714	0.7744	0.7957	0.8061
0.660	0.7432	0.7486	0.7536	0.7583	0.7628	0.7671	0.7710	0.7748	0.7783	0.7816	0.7846	0.8064	0.8170
0.680	0.7521	0.7576	0.7627	0.7675	0.7721	0.7764	0.7804	0.7842	0.7878	0.7912	0.7943	0.8164	0.8274
0.700	0.7605	0.7661	0.7713	0.7762	0.7809	0.7852	0.7893	0.7932	0.7968	0.8002	0.8034	0.8260	0.8374
0.720	0.7684	0.7741	0.7794	0.7844	0.7891	0.7936	0.7977	0.8016	0.8053	0.8088	0.8120	0.8349	0.8468
0.740	0.7758	0.7816	0.7870	0.7921	0.7969	0.8014	0.8056	0.8095	0.8133	0.8168	0.8200	0.8434	0.8558
0.760	0.7827	0.7885	0.7940	0.7992	0.8041	0.8086	0.8129	0.8169	0.8207	0.8242	0.8275	0.8512	0.8643
0.780	0.7894	0.7955	0.8011	0.8064	0.8113	0.8160	0.8203	0.8243	0.8281	0.8316	0.8349	0.8591	0.8744
0.800	0.7959	0.8019	0.8076	0.8129	0.8179	0.8225	0.8269	0.8310	0.8349	0.8385	0.8418	0.8663	0.8804
0.820	0.8019	0.8080	0.8136	0.8190	0.8240	0.8288	0.8332	0.8374	0.8413	0.8450	0.8484	0.8731	0.8864
0.840	0.8076	0.8136	0.8194	0.8248	0.8298	0.8346	0.8391	0.8434	0.8473	0.8511	0.8546	0.8794	0.8922
0.860	0.8129	0.8190	0.8248	0.8302	0.8353	0.8402	0.8447	0.8490	0.8530	0.8568	0.8604	0.8855	0.8980

0.880	0.8179	0.8240	0.8298	0.8353	0.8405	0.8454	0.8500	0.8544	0.8584	0.8622	0.8658	0.8912	0.9036
0.900	0.8225	0.8288	0.8346	0.8402	0.8454	0.8504	0.8550	0.8594	0.8635	0.8674	0.8710	0.8966	0.9091
0.920	0.8269	0.8332	0.8391	0.8447	0.8500	0.8550	0.8597	0.8641	0.8682	0.8721	0.8758	0.9016	0.9143
0.940	0.8310	0.8374	0.8434	0.8490	0.8544	0.8594	0.8641	0.8685	0.8727	0.8766	0.8803	0.9064	0.9194
0.960	0.8349	0.8413	0.8473	0.8530	0.8584	0.8635	0.8682	0.8727	0.8769	0.8809	0.8846	0.9108	0.9242
0.980	0.8385	0.8450	0.8511	0.8568	0.8622	0.8674	0.8721	0.8766	0.8809	0.8849	0.8886	0.9151	0.9288
1.000	0.8418	0.8484	0.8546	0.8604	0.8658	0.8710	0.8758	0.8803	0.8846	0.8886	0.8923	0.9190	0.9332
1.200	0.8663	0.8731	0.8794	0.8855	0.8912	0.8966	0.9016	0.9064	0.9108	0.9151	0.9190	0.9472	0.9615
1.400	0.8764	0.8833	0.8898	0.8960	0.9018	0.9073	0.9124	0.9173	0.9219	0.9262	0.9302	0.9590	0.9739
1.600	0.8828	0.8898	0.8965	0.9027	0.9086	0.9141	0.9194	0.9243	0.9289	0.9333	0.9374	0.9666	0.9817
1.800	0.8864	0.8935	0.9001	0.9064	0.9124	0.9180	0.9232	0.9282	0.9329	0.9373	0.9414	0.9709	0.9860
2.000	0.8878	0.8949	0.9016	0.9079	0.9138	0.9195	0.9248	0.9297	0.9344	0.9388	0.9430	0.9725	0.9877
2.200	0.8878	0.8949	0.9016	0.9079	0.9138	0.9195	0.9248	0.9297	0.9344	0.9389	0.9430	0.9726	0.9877
2.400	0.8878	0.8949	0.9016	0.9079	0.9138	0.9195	0.9248	0.9297	0.9344	0.9389	0.9430	0.9726	0.9877
2.600	0.8878	0.8949	0.9016	0.9079	0.9138	0.9195	0.9248	0.9297	0.9344	0.9389	0.9430	0.9726	0.9877
2.800	0.8878	0.8949	0.9016	0.9079	0.9138	0.9195	0.9248	0.9297	0.9344	0.9389	0.9430	0.9726	0.9877
3.000	0.8878	0.8949	0.9016	0.9079	0.9138	0.9195	0.9248	0.9297	0.9344	0.9389	0.9430	0.9726	0.9877

β =	1.600	1.800	2.000	2.200	2.400	2.600	2.800	3.000
α								
0.005	0.0113	0.0113	0.0113	0.0114	0.0114	0.0114	0.0114	0.0114
0.010	0.0224	0.0224	0.0224	0.0225	0.0225	0.0225	0.0225	0.0225
0.020	0.0443	0.0444	0.0444	0.0444	0.0444	0.0444	0.0444	0.0444
0.030	0.0657	0.0659	0.0659	0.0659	0.0659	0.0659	0.0659	0.0659
0.040	0.0868	0.0871	0.0871	0.0871	0.0871	0.0871	0.0871	0.0871
0.050	0.1076	0.1078	0.1079	0.1079	0.1079	0.1079	0.1079	0.1079
0.060	0.1279	0.1283	0.1284	0.1284	0.1284	0.1284	0.1284	0.1284
0.070	0.1479	0.1483	0.1484	0.1484	0.1484	0.1484	0.1484	0.1484
0.080	0.1675	0.1680	0.1681	0.1682	0.1682	0.1682	0.1682	0.1682
0.090	0.1867	0.1872	0.1875	0.1875	0.1875	0.1875	0.1875	0.1875
0.100	0.2056	0.2062	0.2064	0.2065	0.2065	0.2065	0.2065	0.2065
0.110	0.2240	0.2247	0.2250	0.2250	0.2250	0.2250	0.2250	0.2250
0.120	0.2422	0.2429	0.2432	0.2432	0.2432	0.2432	0.2432	0.2432
0.130	0.2600	0.2608	0.2611	0.2611	0.2611	0.2611	0.2611	0.2611
0.140	0.2775	0.2783	0.2786	0.2787	0.2787	0.2787	0.2787	0.2787
0.150	0.2947	0.2956	0.2959	0.2959	0.2959	0.2959	0.2959	0.2959
0.160	0.3115	0.3125	0.3128	0.3128	0.3128	0.3128	0.3128	0.3128
0.170	0.3281	0.3291	0.3295	0.3295	0.3295	0.3295	0.3295	0.3295
0.180	0.3443	0.3454	0.3458	0.3458	0.3458	0.3458	0.3458	0.3458
0.190	0.3602	0.3613	0.3617	0.3617	0.3617	0.3617	0.3617	0.3617
0.200	0.3758	0.3769	0.3774	0.3774	0.3774	0.3774	0.3774	0.3774
0.210	0.3911	0.3923	0.3927	0.3927	0.3927	0.3927	0.3927	0.3927
0.220	0.4060	0.4073	0.4077	0.4077	0.4077	0.4077	0.4077	0.4078
0.230	0.4206	0.4219	0.4224	0.4224	0.4224	0.4224	0.4224	0.4225
0.240	0.4350	0.4363	0.4368	0.4368	0.4368	0.4368	0.4368	0.4368
0.250	0.4490	0.4504	0.4509	0.4509	0.4509	0.4509	0.4509	0.4509

0.260	0.4626	0.4641	0.4646	0.4646	0.4646	0.4646	0.4646	0.4647
0.270	0.4760	0.4775	0.4781	0.4781	0.4781	0.4781	0.4781	0.4781
0.280	0.4891	0.4907	0.4913	0.4913	0.4913	0.4913	0.4913	0.4913
0.290	0.5019	0.5036	0.5042	0.5042	0.5042	0.5042	0.5042	0.5043
0.300	0.5146	0.5163	0.5169	0.5169	0.5169	0.5169	0.5169	0.5170
0.310	0.5269	0.5287	0.5294	0.5294	0.5294	0.5294	0.5294	0.5295
0.320	0.5391	0.5409	0.5416	0.5416	0.5416	0.5416	0.5416	0.5417
0.330	0.5509	0.5528	0.5535	0.5536	0.5536	0.5536	0.5536	0.5536
0.340	0.5626	0.5645	0.5652	0.5653	0.5653	0.5653	0.5653	0.5654
0.350	0.5739	0.5759	0.5767	0.5767	0.5767	0.5767	0.5767	0.5768
0.360	0.5851	0.5871	0.5879	0.5879	0.5879	0.5879	0.5879	0.5880
0.370	0.5959	0.5980	0.5988	0.5988	0.5988	0.5988	0.5988	0.5989
0.380	0.6118	0.6138	0.6146	0.6146	0.6146	0.6146	0.6146	0.6147
0.390	0.6205	0.6226	0.6234	0.6234	0.6234	0.6234	0.6234	0.6235
0.400	0.6292	0.6313	0.6321	0.6321	0.6321	0.6321	0.6321	0.6323
0.410	0.6379	0.6401	0.6409	0.6409	0.6409	0.6409	0.6409	0.6411
0.420	0.6466	0.6489	0.6497	0.6497	0.6497	0.6497	0.6497	0.6499

β = α	1.600	1.800	2.000	2.200	2.400	2.600	2.800	3.000
0.430	0.6553	0.6576	0.6585	0.6585	0.6585	0.6585	0.6585	0.6586
0.440	0.6641	0.6664	0.6673	0.6673	0.6673	0.6673	0.6673	0.6674
0.450	0.6728	0.6751	0.6761	0.6761	0.6761	0.6761	0.6761	0.6762
0.460	0.6815	0.6839	0.6848	0.6848	0.6848	0.6848	0.6848	0.6850
0.470	0.6902	0.6927	0.6936	0.6936	0.6936	0.6936	0.6936	0.6938
0.480	0.6989	0.7014	0.7024	0.7024	0.7024	0.7024	0.7024	0.7026
0.490	0.7076	0.7102	0.7112	0.7112	0.7112	0.7112	0.7112	0.7114
0.500	0.7164	0.7190	0.7200	0.7200	0.7200	0.7200	0.7200	0.7202
0.520	0.7326	0.7354	0.7366	0.7366	0.7366	0.7366	0.7366	0.7366
0.540	0.7469	0.7498	0.7509	0.7510	0.7510	0.7510	0.7510	0.7510
0.560	0.7607	0.7636	0.7648	0.7648	0.7648	0.7648	0.7648	0.7649
0.580	0.7739	0.7769	0.7780	0.7781	0.7781	0.7781	0.7731	0.7782
0.600	0.7866	0.7895	0.7907	0.7907	0.7907	0.7907	0.7907	0.7909
0.620	0.7986	0.8017	0.8029	0.8029	0.8029	0.8029	0.8029	0.8030
0.640	0.8101	0.8132	0.8145	0.8145	0.8145	0.8145	0.8145	0.8146
0.660	0.8211	0.8243	0.8255	0.8255	0.8255	0.8255	0.8255	0.8257
0.680	0.8315	0.8347	0.8360	0.8359	0.8359	0.8359	0.8359	0.8361
0.700	0.8413	0.8446	0.8459	0.8459	0.8459	0.8459	0.8459	0.8460
0.720	0.8505	0.8539	0.8552	0.8552	0.8552	0.8552	0.8552	0.8554
0.740	0.8592	0.8627	0.8640	0.8640	0.8640	0.8640	0.8640	0.8642
0.760	0.8674	0.8708	0.8722	0.8722	0.8722	0.8722	0.8722	0.8724
0.780	0.8754	0.8789	0.8803	0.8803	0.8803	0.8803	0.8803	0.8805
0.800	0.8828	0.8864	0.8878	0.8878	0.8878	0.8878	0.8878	0.8880
0.820	0.8898	0.8935	0.8949	0.8949	0.8949	0.8949	0.8949	0.8951
0.840	0.8965	0.9001	0.9016	0.9016	0.9016	0.9016	0.9016	0.9017
0.860	0.9027	0.9064	0.9079	0.9079	0.9079	0.9079	0.9079	0.9081

0.880	0.9086	0.9124	0.9138	0.9138	0.9138	0.9138	0.9138	0.9140
0.900	0.9141	0.9180	0.9195	0.9195	0.9195	0.9195	0.9195	0.9197
0.920	0.9194	0.9232	0.9248	0.9248	0.9248	0.9248	0.9248	0.9249
0.940	0.9243	0.9282	0.9297	0.9297	0.9297	0.9297	0.9297	0.9299
0.960	0.9289	0.9329	0.9344	0.9344	0.9344	0.9344	0.9344	0.9346
0.980	0.9333	0.9373	0.9388	0.9389	0.9389	0.9389	0.9389	0.9391
1.000	0.9374	0.9414	0.9430	0.9430	0.9430	0.9430	0.9430	0.9432
1.200	0.9666	0.9709	0.9725	0.9726	0.9726	0.9726	0.9726	0.9729
1.400	0.9792	0.9834	0.9849	0.9849	0.9849	0.9849	0.9849	0.9854
1.600	0.9872	0.9914	0.9929	0.9929	0.9929	0.9929	0.9929	0.9935
1.800	0.9916	0.9959	0.9974	0.9974	0.9974	0.9974	0.9974	0.9980
2.000	0.9935	0.9978	0.9995	0.9995	0.9995	0.9995	0.9995	0.9999
2.200	0.9935	0.9979	0.9996	0.9997	0.9997	0.9997	0.9997	1.0000
2.400	0.9935	0.9979	0.9996	0.9997	0.9997	0.9997	0.9997	1.0000
2.600	0.9935	0.9979	0.9996	0.9997	0.9997	0.9997	1.0000	1.0000
2.800	0.9935	0.9979	0.9996	0.9997	0.9997	1.0000	1.0000	1.0000
3.000	0.9935	0.9979	0.9996	0.9997	0.9997	1.0000	1.0000	1.0000

APPENDIX E

KONIKOW-BREDEHOEFT SOLUTE TRANSPORT MODEL

FORTRAN IV Program Listing

```
C     **************************************************************   A  10
C     *                                                            *   A  20
C     *      SOLUTE TRANSPORT AND DISPERSION IN A POROUS MEDIUM    *   A  30
C     *       NUMERICAL SOLUTION --- METHOD OF CHARACTERISTICS     *   A  40
C     *       PROGRAMMED BY J. D. BREDEHOEFT AND L. F. KONIKOW     *   A  50
C     *                                                            *   A  60
C     **************************************************************   A  70
      DOUBLE PRECISION DMIN1,DEXP,DLOG,DABS                            A  80
      REAL *8TMRX,VPRM,HI,HR,HC,HK,WT,REC,RECH,TIM,AOPT,TITLE          A  90
      REAL *8XDEL,YDEL,S,AREA,SUMT,RHO,PARAM,TEST,TOL,PINT,HMIN,PYR    A 100
      REAL *8TINT,ALPHA1,ANITP                                         A 110
      COMMON /PRMI/ NTIM,NPMP,NPNT,NITP,N,NX,NY,NP,NREC,INT,NNX,NNY,NUMO A 120
     1BS,NMOV,IMOV,NPMAX,ITMAX,NZCRIT,IPRNT,NPTPND,NPNTMV,NPNTVL,NPNTD,N A 130
     2PNCHV,NPDELC                                                     A 140
      COMMON /PRMK/ NODEID(20,20),NPCELL(20,20),LIMBO(500),IXOBS(5),IYOB A 150
     1S(5)                                                             A 160
      COMMON /HEDA/ THCK(20,20),PERM(20,20),TMWL(5,50),TMOBS(50),ANFCTR A 170
      COMMON /HEDB/ TMRX(20,20,2),VPRM(20,20),HI(20,20),HR(20,20),HC(20, A 180
     120),HK(20,20),WT(20,20),REC(20,20),RECH(20,20),TIM(100),AOPT(20),T A 190
     2ITLE(10),XDEL,YDEL,S,AREA,SUMT,RHO,PARAM,TEST,TOL,PINT,HMIN,PYR  A 200
      COMMON /CHMA/ PART(3,3200),CONC(20,20),TMCN(5,50),VX(20,20),VY(20, A 210
     120),CONINT(20,20),CNRECH(20,20),POROS,SUMTCH,BETA,TIMV,STORM,STORM A 220
     2I,CMSIN,CMSOUT,FLMIN,FLMOT,SUMIO,CELDIS,DLTRAT,CSTORM            A 230
C     *************************************************************    A 240
C      ---LOAD DATA---                                                 A 250
      INT=0                                                            A 260
      CALL PARLOD                                                      A 270
      CALL GENPT                                                       A 280
C     *************************************************************    A 290
C     ---START COMPUTATIONS---                                         A 300
C        ---COMPUTE ONE PUMPING PERIOD---                              A 310
      DO 150 INT=1,NPMP                                                A 320
      IF (INT.GT.1) CALL PARLOD                                        A 330
```

```
C          ---COMPUTE ONE TIME STEP---                                  A 340
      DO 130 N=1,NTIM                                                   A 350
      IPRNT=0                                                           A 360
C          ---LOAD NEW DELTA T---                                       A 370
      TINT=SUMT-PYR*(INT-1)                                             A 380
      TDEL=DMIN1(TIM(N),PYR-TINT)                                       A 390
      SUMT=SUMT+TDEL                                                    A 400
      TIM(N)=TDEL                                                       A 410
      REMN=MOD(N,NPNT)                                                  A 420
C     ******************************************************************* A 430
      CALL ITERAT                                                       A 440
      IF (REMN.EQ.0.0.OR.N.EQ.NTIM) CALL OUTPT                          A 450
      CALL VELO                                                         A 460
      CALL MOVE                                                         A 470
C     ******************************************************************* A 480
C     ---STORE OBS. WELL DATA FOR TRANSIENT FLOW PROBLEMS---            A 490
      IF (S.EQ.0.0) GO TO 120                                           A 500
      IF (NUMOBS.LE.0) GO TO 120                                        A 510
      J=MOD(N,50)                                                       A 520
      IF (J.EQ.0) J=50                                                  A 530
      TMOBS(J)=SUMT                                                     A 540
      DO 110 I=1,NUMOBS                                                 A 550
      TMWL(I,J)=HK(IXOBS(I),IYOBS(I))                                   A 560
      TMCN(I,J)=CONC(IXOBS(I),IYOBS(I))                                 A 570
  110 CONTINUE                                                          A 580
```

FORTRAN IV program listing—Continued

```
C       ****************************************************************    A 590
C       ---OUTPUT ROUTINES---                                               A 600
  120 IF (REMN.EQ.0.0.OR.N.EQ.NTIM.OR.MOD(N,50).EQ.0) CALL CHMOT            A 610
      IF (SUMT.GE.(PYR*INT)) GO TO 140                                      A 620
  130 CONTINUE                                                              A 630
C       ****************************************************************    A 640
C       ---SUMMARY OUTPUT---                                                A 650
  140 CONTINUE                                                              A 660
      IPRNT=1                                                               A 670
      CALL CHMOT                                                            A 680
  150 CONTINUE                                                              A 690
C       ****************************************************************    A 700
      STOP                                                                  A 710
C       ****************************************************************    A 720
      END                                                                   A 730-
      SUBROUTINE PARLOD                                                     B  10
      DOUBLE PRECISION DMIN1,DEXP,DLOG,DABS                                 B  20
      REAL *8TMRX,VPRM,HI,HR,HC,HK,WT,REC,RECH,TIM,AOPT,TITLE               B  30
      REAL *8XDEL,YDEL,S,AREA,SUMT,RHO,PARAM,TEST,TOL,PINT,HMIN,PYR         B  40
      REAL *8FCTR,TIMX,TINIT,PIES,YNS,XNS,RAT,HMX,HMY                       B  50
      REAL *8TINT,ALPHA1,ANITP                                              B  60
      COMMON /PRMI/ NTIM,NPMP,NPNT,NITP,N,WX,NY,NP,NREC,INT,NNX,NNY,NUMO    B  70
     1BS,NMOV,IMOV,NPMAX,ITMAX,NZCRIT,IPRNT,NPTPND,NPNTMV,NPNTVL,NPNTD,N    B  80
     2PNCHV,NPDELC                                                          B  90
      COMMON /PRMK/ NODEID(20,20),NPCELL(20,20),LIMBO(500),IXOBS(5),IYOB    B 100
     1S(5)                                                                  B 110
      COMMON /HEDA/ THCK(20,20),PERM(20,20),TMWL(5,50),TMOBS(50),ANFCTR     B 120
      COMMON /HEDB/ TMRX(20,20,2),VPRM(20,20),HI(20,20),HR(20,20),HC(20,    B 130
     120),HK(20,20),WT(20,20),REC(20,20),RECH(20,20),TIM(100),AOPT(20),T    B 140
     2ITLE(10),XDEL,YDEL,S,AREA,SUMT,RHO,PARAM,TEST,TOL,PINT,HMIN,PYR       B 150
      COMMON /CHMA/ PART(3,3200),CONC(20,20),TMCN(5,50),VX(20,20),VY(20,    B 160
     120),CONINT(20,20),CNRECH(20,20),POROS,SUMTCH,BETA,TIMV,STORM,STORM    B 170
     2I,CMSIN,CMSOUT,FLMIN,FLMOT,SUMIO,CELDIS,DLTRAT,CSTORM                 B 180
      COMMON /BALM/ TOTLQ                                                   B 190
      COMMON /XINV/ DXINV,DYINV,ARINV,PORINV                                B 200
      COMMON /CHMC/ SUMC(20,20),VXBDY(20,20),VYBDY(20,20)                   B 210
```

```
C     ****************************************************************     B 220
      IF (INT.GT.1) GO TO 10                                               B 230
      WRITE (6,750)                                                        B 240
      READ (5,720) TITLE                                                   B 250
      WRITE (6,730) TITLE                                                  B 260
C     ****************************************************************     B 270
C     ---INITIALIZE TEST AND CONTROL VARIABLES---                          B 280
      STORMI=0.0                                                           B 290
      TEST=0.0                                                             B 300
      TOTLQ=0.0                                                            B 310
      SUMT=0.0                                                             B 320
      SUMTCH=0.0                                                           B 330
      INT=0                                                                B 340
      IPRNT=0                                                              B 350
      NCA=0                                                                B 360
      N=0                                                                  B 370
      IMOV=0                                                               B 380
      NMOV=0                                                               B 390
C     ****************************************************************     B 400
C     ---LOAD CONTROL PARAMETERS---                                        B 410
      READ (5,740) NTIM,NPMP,NX,NY,NPMAX,NPNT,NITP,NUMOBS,ITMAX,NREC,NPT   B 420
     1PND,NCODES,NPNTMV,NPNTVL,NPNTD,NPDELC,NPNCHV                         B 430
      READ (5,800) PINT,TOL,POROS,BETA,S,TIMX,TINIT,XDEL,YDEL,DLTRAT,CEL   B 440
     1DIS,ANFCTR                                                           B 450
      PYR=PINT*86400.0*365.25                                              B 460
      NNX=NX-1                                                             B 470
```

FORTRAN IV program listing—Continued

```
      NNY=NY-1                                                          B 480
      NP=NPMAX                                                          B 490
      DXINV=1.0/XDEL                                                    B 500
      DYINV=1.0/YDEL                                                    B 510
      ARINV=DXINV*DYINV                                                 B 520
      PORINV=1.0/POROS                                                  B 530
C     ---PRINT CONTROL PARAMETERS---                                    B 540
      WRITE (6,760)                                                     B 550
      WRITE (6,770) NX,NY,XDEL,YDEL                                     B 560
      WRITE (6,780) NTIM,NPMP,PINT,TIMX,TINIT                           B 570
      WRITE (6,790) S,POROS,BETA,DLTRAT,ANFCTR                          B 580
      WRITE (6,870) NITP,TOL,ITMAX,CELDIS,NPMAX,NPTPND                  B 590
      IF (NPTPND.LT.4.OR.NPTPND.GT.9.OR.NPTPND.EQ.6.OR.NPTPND.EQ.7) WRIT B 600
     1E (6,880)                                                         B 610
      WRITE (6,890) NPNT,NPNTMV,NPNTVL,NPNTD,NUMOBS,NREC,NCODES,NPNCHV,N B 620
     1PDELC                                                             B 630
      IF (NPNTMV.EQ.0) NPNTMV=999                                       B 640
      GO TO 20                                                          B 650
C     ****************************************************************  B 660
C     ---READ DATA TO REVISE TIME STEPS AND STRESSES FOR SUBSEQUENT     B 670
C        PUMPING PERIODS---                                             B 680
   10 READ (5,1060) ICHK                                                B 690
      IF (ICHK.LE.0) RETURN                                             B 700
      READ (5,1070) NTIM,NPNT,NITP,ITMAX,NREC,NPNTMV,NPNTVL,NPNTD,NPDELC B 710
     1,NPNCHV,PINT,TIMX,TINIT                                           B 720
      WRITE (6,1080) INT                                                B 730
      WRITE (6,1090) NTIM,NPNT,NITP,ITMAX,NREC,NPNTMV,NPNTVL,NPNTD,NPDEL B 740
     1C,NPNCHV,PINT,TIMX,TINIT                                          B 750
C     ****************************************************************  B 760
C     ---LIST TIME INCREMENTS---                                        B 770
   20 DO 30 J=1,100                                                     B 780
      TIM(J)=0.0                                                        B 790
```

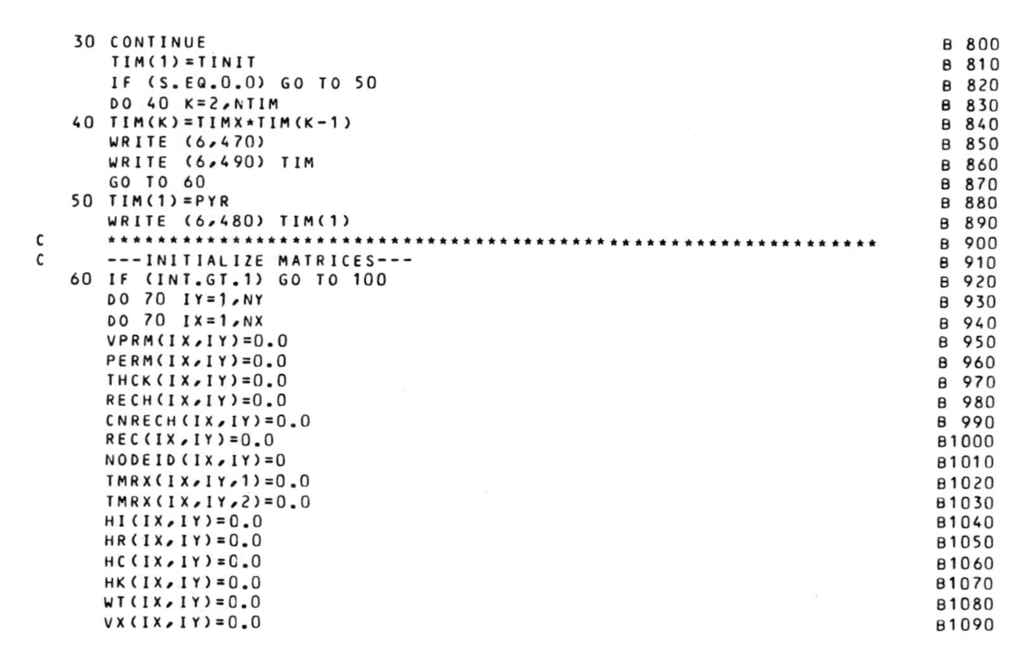

```
   30 CONTINUE                                                                    B 800
      TIM(1)=TINIT                                                                B 810
      IF (S.EQ.0.0) GO TO 50                                                      B 820
      DO 40 K=2,NTIM                                                              B 830
   40 TIM(K)=TIMX*TIM(K-1)                                                        B 840
      WRITE (6,470)                                                               B 850
      WRITE (6,490) TIM                                                           B 860
      GO TO 60                                                                    B 870
   50 TIM(1)=PYR                                                                  B 880
      WRITE (6,480) TIM(1)                                                        B 890
C     ***************************************************************             B 900
C     ---INITIALIZE MATRICES---                                                   B 910
   60 IF (INT.GT.1) GO TO 100                                                     B 920
      DO 70 IY=1,NY                                                               B 930
      DO 70 IX=1,NX                                                               B 940
      VPRM(IX,IY)=0.0                                                             B 950
      PERM(IX,IY)=0.0                                                             B 960
      THCK(IX,IY)=0.0                                                             B 970
      RECH(IX,IY)=0.0                                                             B 980
      CNRECH(IX,IY)=0.0                                                           B 990
      REC(IX,IY)=0.0                                                              B1000
      NODEID(IX,IY)=0                                                             B1010
      TMRX(IX,IY,1)=0.0                                                           B1020
      TMRX(IX,IY,2)=0.0                                                           B1030
      HI(IX,IY)=0.0                                                               B1040
      HR(IX,IY)=0.0                                                               B1050
      HC(IX,IY)=0.0                                                               B1060
      HK(IX,IY)=0.0                                                               B1070
      WT(IX,IY)=0.0                                                               B1080
      VX(IX,IY)=0.0                                                               B1090
```

FORTRAN IV program listing—Continued

```
      VY(IX,IY)=0.0                                                     B1100
      VXBDY(IX,IY)=0.0                                                  B1110
      VYBDY(IX,IY)=0.0                                                  B1120
      CONC(IX,IY)=0.0                                                   B1130
      CONINT(IX,IY)=0.0                                                 B1140
      SUMC(IX,IY)=0.0                                                   B1150
   70 CONTINUE                                                          B1160
C     ******************************************************************B1170
C     ---READ OBSERVATION WELL LOCATIONS---                             B1180
      IF (NUMOBS.LE.0) GO TO 100                                        B1190
      WRITE (6,900)                                                     B1200
      DO 80 J=1,NUMOBS                                                  B1210
      READ (5,700) IX,IY                                                B1220
      WRITE (6,810) J,IX,IY                                             B1230
      IXOBS(J)=IX                                                       B1240
   80 IYOBS(J)=IY                                                       B1250
      DO 90 I=1,NUMOBS                                                  B1260
      DO 90 J=1,50                                                      B1270
      TMWL(I,J)=0.0                                                     B1280
   90 TMCN(I,J)=0.0                                                     B1290
C     ******************************************************************B1300
C     ---READ PUMPAGE DATA -- (X-Y COORDINATES AND RATE IN CFS)---      B1310
C         ---SIGNS : WITHDRAWAL = POS.;  INJECTION = NEG.---            B1320
C         ---IF INJ. WELL, ALSO READ CONCENTRATION OF INJECTED WATER--- B1330
  100 IF (NREC.LE.0) GO TO 120                                          B1340
      WRITE (6,910)                                                     B1350
      DO 110 I=1,NREC                                                   B1360
      READ (5,710) IX,IY,FCTR,CNREC                                     B1370
      IF (FCTR.LT.0.0) CNRECH(IX,IY)=CNREC                              B1380
      REC(IX,IY)=FCTR                                                   B1390
  110 WRITE (6,820) IX,IY,REC(IX,IY),CNRECH(IX,IY)                      B1400
C     ******************************************************************B1410
```

```
  120 IF (INT.GT.1) RETURN                                              B1420
      AREA=XDEL*YDEL                                                    B1430
      WRITE (6,690) AREA                                                B1440
      WRITE (6,600)                                                     B1450
      WRITE (6,610) XDEL                                                B1460
      WRITE (6,610) YDEL                                                B1470
C     ****************************************************************  B1480
C     ---READ TRANSMISSIVITY IN FT**2/SEC INTO VPRM ARRAY---            B1490
C        ---FCTR = TRANSMISSIVITY MULTIPLIER  --->  FT**2/SEC---        B1500
      WRITE (6,530)                                                     B1510
      READ (5,550) INPUT,FCTR                                           B1520
      DO 160 IY=1,NY                                                    B1530
      IF (INPUT.EQ.1) READ (5,560) (VPRM(IX,IY),IX=1,NX)                B1540
      DO 150 IX=1,NX                                                    B1550
      IF (INPUT.NE.1) GO TO 130                                         B1560
      VPRM(IX,IY)=VPRM(IX,IY)*FCTR                                      B1570
      GO TO 140                                                         B1580
  130 VPRM(IX,IY)=FCTR                                                  B1590
  140 IF (IX.EQ.1.OR.IX.EQ.NX) VPRM(IX,IY)=0.0                          B1600
      IF (IY.EQ.1.OR.IY.EQ.NY) VPRM(IX,IY)=0.0                          B1610
  150 CONTINUE                                                          B1620
  160 WRITE (6,520) (VPRM(IX,IY),IX=1,NX)                               B1630
C     ****************************************************************  B1640
C     ---SET UP COEFFICIENT MATRIX --- BLOCK-CENTERED GRID---           B1650
C     ---AVERAGE TRANSMISSIVITY --- HARMONIC MEAN---                    B1660
      IF (ANFCTR.NE.0.0) GO TO 170                                      B1670
      WRITE (6,1050)                                                    B1680
      ANFCTR=1.0                                                        B1690
  170 PIES=3.1415927*3.1415927/2.0                                      B1700
      YNS=NY*NY                                                         B1710
```

FORTRAN IV program listing—Continued

```
      XNS=NX*NX                                                         B1720
      HMIN=2.0                                                          B1730
      DO 180 IY=2,NNY                                                   B1740
      DO 180 IX=2,NNX                                                   B1750
      IF (VPRM(IX,IY).EQ.0.0) GO TO 180                                 B1760
      TMRX(IX,IY,1)=2.0*VPRM(IX,IY)*VPRM(IX+1,IY)/(VPRM(IX,IY)*XDEL+VPRM B1770
     1(IX+1,IY)*XDEL)                                                   B1780
      TMRX(IX,IY,2)=2.0*VPRM(IX,IY)*VPRM(IX,IY+1)/(VPRM(IX,IY)*YDEL+VPRM B1790
     1(IX,IY+1)*YDEL)                                                   B1800
C        ---ADJUST COEFFICIENT FOR ANISOTROPY---                        B1810
      TMRX(IX,IY,2)=TMRX(IX,IY,2)*ANFCTR                                B1820
C        ---COMPUTE MINIMUM ITERATION PARAMETER (HMIN)---               B1830
      IF (TMRX(IX,IY,1).EQ.0.0) GO TO 180                               B1840
      IF (TMRX(IX,IY,2).EQ.0.0) GO TO 180                               B1850
      RAT=TMRX(IX,IY,1)*YDEL/(TMRX(IX,IY,2)*XDEL)                       B1860
      HMX=PIES/(XNS*(1.0+RAT))                                          B1870
      HMY=PIES/(YNS*(1.0+(1.0/RAT)))                                    B1880
      IF (HMX.LT.HMIN) HMIN=HMX                                         B1890
      IF (HMY.LT.HMIN) HMIN=HMY                                         B1900
  180 CONTINUE                                                          B1910
C     ****************************************************************  B1920
C     ---READ AQUIFER THICKNESS---                                      B1930
      WRITE (6,510)                                                     B1940
      READ (5,550) INPUT,FCTR                                           B1950
      DO 210 IY=1,NY                                                    B1960
      IF (INPUT.EQ.1) READ (5,540) (THCK(IX,IY),IX=1,NX)                B1970
      DO 200 IX=1,NX                                                    B1980
      IF (INPUT.NE.1) GO TO 190                                         B1990
      THCK(IX,IY)=THCK(IX,IY)*FCTR                                      B2000
      GO TO 200                                                         B2010
  190 IF (VPRM(IX,IY).NE.0.0) THCK(IX,IY)=FCTR                          B2020
  200 CONTINUE                                                          B2030
  210 WRITE (6,500) (THCK(IX,IY),IX=1,NX)                               B2040
C     ****************************************************************  B2050
```

```
C      ---READ DIFFUSE RECHARGE AND DISCHARGE---                             B2060
       WRITE (6,830)                                                         B2070
       READ (5,550) INPUT,FCTR                                               B2080
       DO 240 IY=1,NY                                                        B2090
       IF (INPUT.EQ.1) READ (5,560) (RECH(IX,IY),IX=1,NX)                    B2100
       DO 230 IX=1,NX                                                        B2110
       IF (INPUT.NE.1) GO TO 220                                             B2120
       RECH(IX,IY)=RECH(IX,IY)*FCTR                                          B2130
       GO TO 230                                                             B2140
  220  IF (THCK(IX,IY).NE.0.0) RECH(IX,IY)=FCTR                              B2150
  230  CONTINUE                                                              B2160
  240  WRITE (6,840) (RECH(IX,IY),IX=1,NX)                                   B2170
C      ******************************************************************   B2180
C      ---COMPUTE PERMEABILITY FROM TRANSMISSIVITY---                        B2190
C      ---COUNT NO. OF CELLS IN AQUIFER---                                   B2200
C      ---SET NZCRIT = 2% OF THE NO. OF CELLS IN THE AQUIFER---              B2210
       DO 250 IX=1,NX                                                        B2220
       DO 250 IY=1,NY                                                        B2230
       IF (THCK(IX,IY).EQ.0.0) GO TO 250                                     B2240
       PERM(IX,IY)=VPRM(IX,IY)/THCK(IX,IY)                                   B2250
       NCA=NCA+1                                                             B2260
  250  VPRM(IX,IY)=0.0                                                       B2270
C                                                                            B2280
       AAQ=NCA*AREA                                                          B2290
       NZCRIT=(NCA+25)/50                                                    B2300
       WRITE (6,620)                                                         B2310
       DO 260 IY=1,NY                                                        B2320
  260  WRITE (6,650) (PERM(IX,IY),IX=1,NX)                                   B2330
```

FORTRAN IV program listing—Continued

```
      WRITE (6,630) NCA,AAQ,NZCRIT                                      B2340
C     ****************************************************************  B2350
C     ---READ NODE IDENTIFICATION CARDS---                              B2360
C        ---SET VERT. PERM., SOURCE CONC., AND DIFFUSE RECHARGE---      B2370
C           ---SPECIFY CODES TO FIT YOUR NEEDS---                       B2380
      WRITE (6,570)                                                     B2390
      READ (5,550) INPUT,FCTR                                           B2400
      DO 280 IY=1,NY                                                    B2410
      IF (INPUT.EQ.1) READ (5,640) (NODEID(IX,IY),IX=1,NX)              B2420
      DO 270 IX=1,NX                                                    B2430
  270 IF (INPUT.NE.1.AND.THCK(IX,IY).NE.0.0) NODEID(IX,IY)=FCTR         B2440
  280 WRITE (6,580) (NODEID(IX,IY),IX=1,NX)                             B2450
      WRITE (6,920) NCODES                                              B2460
      IF (NCODES.LE.0) GO TO 310                                        B2470
      WRITE (6,930)                                                     B2480
      DO 300 IJ=1,NCODES                                                B2490
      READ (5,850) ICODE,FCTR1,FCTR2,FCTR3,OVERRD                       B2500
      DO 290 IX=1,NX                                                    B2510
      DO 290 IY=1,NY                                                    B2520
      IF (NODEID(IX,IY).NE.ICODE) GO TO 290                             B2530
      VPRM(IX,IY)=FCTR1                                                 B2540
      CNRECH(IX,IY)=FCTR2                                               B2550
      IF (OVERRD.NE.0) RECH(IX,IY)=FCTR3                                B2560
  290 CONTINUE                                                          B2570
      WRITE (6,860) ICODE,FCTR1,FCTR2                                   B2580
  300 IF (OVERRD.NE.0) WRITE (6,1100) FCTR3                             B2590
  310 WRITE (6,590)                                                     B2600
      DO 320 IY=1,NY                                                    B2610
  320 WRITE (6,520) (VPRM(IX,IY),IX=1,NX)                               B2620
C     ****************************************************************  B2630
```

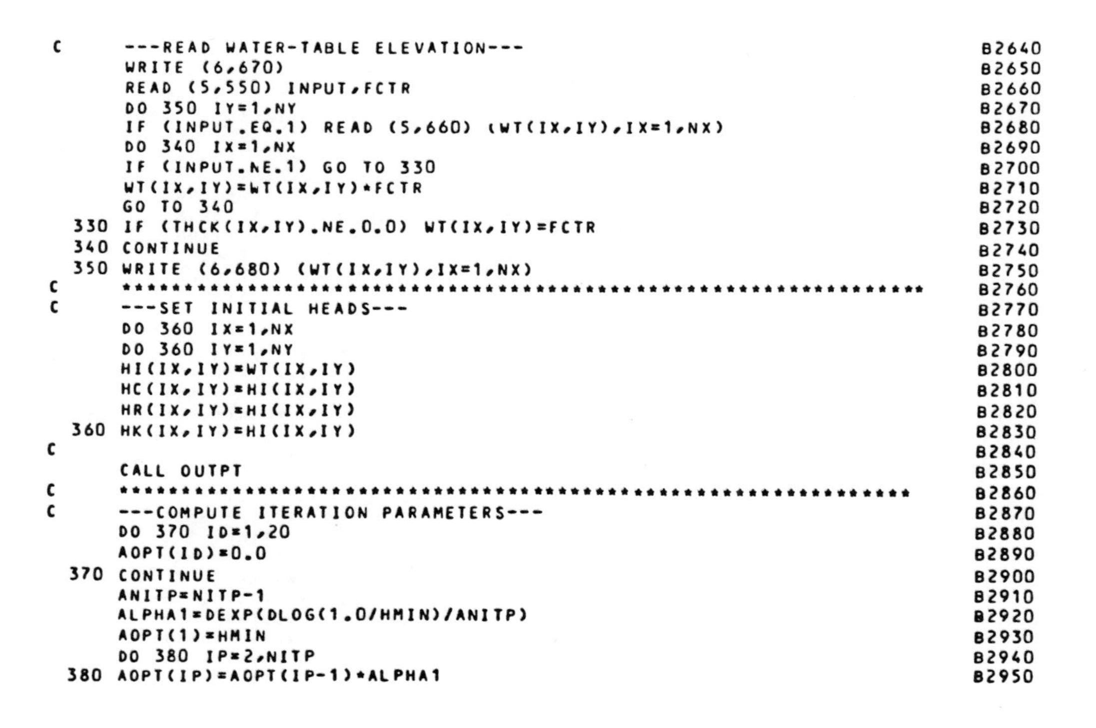

```
C       ---READ WATER-TABLE ELEVATION---                                   B2640
        WRITE (6,670)                                                      B2650
        READ (5,550) INPUT,FCTR                                            B2660
        DO 350 IY=1,NY                                                     B2670
        IF (INPUT.EQ.1) READ (5,660) (WT(IX,IY),IX=1,NX)                   B2680
        DO 340 IX=1,NX                                                     B2690
        IF (INPUT.NE.1) GO TO 330                                          B2700
        WT(IX,IY)=WT(IX,IY)*FCTR                                           B2710
        GO TO 340                                                          B2720
  330   IF (THCK(IX,IY).NE.0.0) WT(IX,IY)=FCTR                             B2730
  340   CONTINUE                                                           B2740
  350   WRITE (6,680) (WT(IX,IY),IX=1,NX)                                  B2750
C       ************************************************************       B2760
C       ---SET INITIAL HEADS---                                            B2770
        DO 360 IX=1,NX                                                     B2780
        DO 360 IY=1,NY                                                     B2790
        HI(IX,IY)=WT(IX,IY)                                                B2800
        HC(IX,IY)=HI(IX,IY)                                                B2810
        HR(IX,IY)=HI(IX,IY)                                                B2820
  360   HK(IX,IY)=HI(IX,IY)                                                B2830
C                                                                          B2840
        CALL OUTPT                                                         B2850
C       ************************************************************       B2860
C       ---COMPUTE ITERATION PARAMETERS---                                 B2870
        DO 370 ID=1,20                                                     B2880
        AOPT(ID)=0.0                                                       B2890
  370   CONTINUE                                                           B2900
        ANITP=NITP-1                                                       B2910
        ALPHA1=DEXP(DLOG(1.0/HMIN)/ANITP)                                  B2920
        AOPT(1)=HMIN                                                       B2930
        DO 380 IP=2,NITP                                                   B2940
  380   AOPT(IP)=AOPT(IP-1)*ALPHA1                                         B2950
```

FORTRAN IV program listing—Continued

```
C                                                                         B2960
      WRITE (6,450)                                                       B2970
      WRITE (6,460) AOPT                                                  B2980
C     ****************************************************************    B2990
C     ---READ INITIAL CONCENTRATIONS AND COMPUTE INITIAL MASS STORED---   B3000
      READ (5,550) INPUT,FCTR                                             B3010
      DO 420 IY=1,NY                                                      B3020
      IF (INPUT.EQ.1) READ (5,660) (CONC(IX,IY),IX=1,NX)                  B3030
      DO 410 IX=1,NX                                                      B3040
      IF (INPUT.NE.1) GO TO 390                                           B3050
      CONC(IX,IY)=CONC(IX,IY)*FCTR                                        B3060
      GO TO 400                                                           B3070
  390 IF (THCK(IX,IY).NE.0.0) CONC(IX,IY)=FCTR                            B3080
  400 CONINT(IX,IY)=CONC(IX,IY)                                           B3090
  410 STORMI=STORMI+CONINT(IX,IY)*THCK(IX,IY)*AREA*POROS                  B3100
  420 CONTINUE                                                            B3110
C     ****************************************************************    B3120
C     ---CHECK DATA SETS FOR INTERNAL CONSISTENCY---                      B3130
      DO 440 IX=1,NX                                                      B3140
      DO 440 IY=1,NY                                                      B3150
      IF (THCK(IX,IY).GT.0.0) GO TO 430                                   B3160
      IF (TMRX(IX,IY,1).GT.0.0) WRITE (6,940) IX,IY                       B3170
      IF (TMRX(IX,IY,2).GT.0.0) WRITE (6,950) IX,IY                       B3180
      IF (NODEID(IX,IY).GT.0) WRITE (6,960) IX,IY                         B3190
      IF (WT(IX,IY).NE.0.0) WRITE (6,970) IX,IY                           B3200
      IF (RECH(IX,IY).NE.0.0) WRITE (6,980) IX,IY                         B3210
      IF (REC(IX,IY).NE.0.0) WRITE (6,990) IX,IY                          B3220
  430 IF (PERM(IX,IY).GT.0.0) GO TO 440                                   B3230
      IF (NODEID(IX,IY).GT.0.0) WRITE (6,1000) IX,IY                      B3240
      IF (WT(IX,IY).NE.0.0) WRITE (6,1010) IX,IY                          B3250
      IF (RECH(IX,IY).NE.0.0) WRITE (6,1020) IX,IY                        B3260
      IF (REC(IX,IY).NE.0.0) WRITE (6,1030) IX,IY                         B3270
      IF (THCK(IX,IY).GT.0.0) WRITE (6,1040) IX,IY                        B3280
  440 CONTINUE                                                            B3290
```

```
C     ************************************************************     B3300
      RETURN                                                           B3310
C     ************************************************************     B3320
C                                                                      B3330
C                                                                      B3340
C                                                                      B3350
  450 FORMAT (1H1,20HITERATION PARAMETERS)                             B3360
  460 FORMAT (3H    ,1G20.6)                                           B3370
  470 FORMAT (1H1,27HTIME INTERVALS (IN SECONDS))                      B3380
  480 FORMAT (1H1,15X,17HSTEADY-STATE FLOW//5X,57HTIME INTERVAL (IN SEC) B3390
     1 FOR SOLUTE-TRANSPORT SIMULATION = ,G12.5)                       B3400
  490 FORMAT (3H    ,10G12.5)                                          B3410
  500 FORMAT (3H    ,20F5.1)                                           B3420
  510 FORMAT (1H1,22HAQUIFER THICKNESS (FT))                           B3430
  520 FORMAT (3H    ,20F5.2)                                           B3440
  530 FORMAT (1H1,30HTRANSMISSIVITY MAP (FT*FT/SEC))                   B3450
  540 FORMAT (20G3.0)                                                  B3460
  550 FORMAT (I1,G10.0)                                                B3470
  560 FORMAT (20G4.1)                                                  B3480
  570 FORMAT (1H1,23HNODE IDENTIFICATION MAP//)                        B3490
  580 FORMAT (1H ,20I5)                                                B3500
  590 FORMAT (1H1,45HVERTICAL PERMEABILITY/THICKNESS (FT/(FT*SEC)))    B3510
  600 FORMAT (1H0,10X,12HX-Y SPACING:)                                 B3520
  610 FORMAT (1H ,12X,10G12.5)                                         B3530
  620 FORMAT (1H1,24HPERMEABILTY MAP (FT/SEC))                         B3540
  630 FORMAT (1H0,////10X,44HNO. OF FINITE-DIFFERENCE CELLS IN AQUIFER = B3550
     1 ,I4//10X,28HAREA OF AQUIFER IN MODEL  = ,G12.5,10H   SQ. FT.////1 B3560
     20X,47HNZCRIT   (MAX. NO. OF CELLS THAT CAN BE VOID OF/20X,56HPARTI B3570
```

FORTRAN IV program listing—Continued

```
    3CLES;  IF EXCEEDED, PARTICLES ARE REGENERATED)    = ,I4/)              B3580
640 FORMAT (20I1)                                                           B3590
650 FORMAT (3H   ,20F5.3)                                                   B3600
660 FORMAT (20G4.0)                                                         B3610
670 FORMAT (1H1,11HWATER TABLE)                                             B3620
680 FORMAT (1H ,20F5.0)                                                     B3630
690 FORMAT (1H0,10X,19HAREA OF ONE CELL = ,G12.4)                           B3640
700 FORMAT (2I2)                                                            B3650
710 FORMAT (2I2,2G8.2)                                                      B3660
720 FORMAT (10A8)                                                           B3670
730 FORMAT (1H0,10A8)                                                       B3680
740 FORMAT (17I4)                                                           B3690
750 FORMAT (1H1,77HU.S.G.S. METHOD-OF-CHARACTERISTICS MODEL FOR SOLUTE      B3700
   1 TRANSPORT IN GROUND WATER)                                             B3710
760 FORMAT (1H0,21X,21HI N P U T      D A T A)                              B3720
770 FORMAT (1H0,23X,16HGRID DESCRIPTORS//13X,30HNX     (NUMBER OF COLUM     B3730
   1NS)  =  ,I4/13X,28HNY     (NUMBER OF ROWS)      =,I6/13X,29HXDEL  (X    B3740
   2-DISTANCE IN FEET) = ,F7.1/13X,29HYDEL  (Y-DISTANCE IN FEET) = ,F7      B3750
   3.1)                                                                     B3760
780 FORMAT (1H0,23X,16HTIME  PARAMETERS//13X,40HNTIM   (MAX. NO. OF TI      B3770
   1ME STEPS)        = ,I6/13X,40HNPMP   (NO. OF PUMPING PERIODS)           B3780
   2 = ,I6/13X,39HPINT   (PUMPING PERIOD IN YEARS)       =,F10.2/13X,39     B3790
   3HTIMX    (TIME INCREMENT MULTIPLIER)    =,F10.2/13X,39HTINIT  (INIT     B3800
   4IAL TIME STEP IN SEC.)     =,F8.0)                                      B3810
790 FORMAT (1H0,14X,34HHYDROLOGIC AND CHEMICAL PARAMETERS//13X,1HS,7X,      B3820
   129H(STORAGE COEFFICIENT)        =,5X,F9.6/13X,28HPOROS   (EFFECTIVE     B3830
   2 POROSITY),8X,3H=   ,F8.2/13X,39HBETA    (CHARACTERISTIC LENGTH)        B3840
   3  =   ,F7.1/13X,31HDLTRAT  (RATIO OF TRANSVERSE TO/21X,30HLONGITUDI     B3850
   4NAL DISPERSIVITY)  = ,F9.2/13X,39HANFCTR  (RATIO OF T-YY TO T-XX)       B3860
   5    =   ,F12.6)                                                         B3870
800 FORMAT (12G5.0)                                                         B3880
810 FORMAT (1H ,16X,I2,5X,I2,4X,I2)                                         B3890
820 FORMAT (1H ,7X,2I4,3X,F7.2,5X,F7.1)                                     B3900
830 FORMAT (1H1,39HDIFFUSE RECHARGE AND DISCHARGE (FT/SEC))                 B3910
```

```
840 FORMAT (1H ,1P10E10.2)                                                   B3920
850 FORMAT (I2,3G10.2,I2)                                                    B3930
860 FORMAT (1H0,7X,I2,7X,E10.3,5X,F9.2)                                      B3940
870 FORMAT (1H0,21X,20HEXECUTION PARAMETERS//13X,39HNITP   (NO. OF ITE       B3950
   1RATION PARAMETERS) = ,I4/13X,39HTOL     (CONVERGENCE CRITERIA - ADI      B3960
   2P) = ,F9.4/13X,39HITMAX  (MAX.NO.OF ITERATIONS - ADIP) = ,I4/13X,3       B3970
   34HCELDIS (MAX.CELL DISTANCE PER MOVE/24X,28HOF PARTICLES - M.O.C.)       B3980
   4    = ,F8.3/13X,30HNPMAX  (MAX. NO. OF PARTICLES),7X,2H= ,I4/12X,3       B3990
   52H NPTPND (NO. PARTICLES PER NODE),6X,3H=  ,I4)                          B4000
880 FORMAT (1H0,5X,47H*** WARNING ***  NPTPND MUST EQUAL 4,5,8, OR 9.)       B4010
890 FORMAT (1H0,23X,15HPROGRAM OPTIONS//13X,30HNPNT   (TIME STEP INTER       B4020
   1VAL FOR/21X,18HCOMPLETE PRINTOUT),7X,3H=  ,I4/13X,31HNPNTMV (MOVE        B4030
   2INTERVAL FOR CHEM./21X,28HCONCENTRATION PRINTOUT)  =  ,I4/13X,29HN       B4040
   3PNTVL (PRINT OPTION-VELOCITY/21X,24H0=NO; 1=FIRST TIME STEP;/21X,1       B4050
   47H2=ALL TIME STEPS),8X,3H=  ,I4/13X,31HNPNTD  (PRINT OPTION-DISP.C       B4060
   5OEF./21X,24H0=NO; 1=FIRST TIME STEP;/21X,17H2=ALL TIME STEPS),8X,3       B4070
   6H=  ,I4/13X,32HNUMOBS (NO. OF OBSERVATION WELLS/21X,28HFOR HYDROGR       B4080
   7APH PRINTOUT) =  ,I4/13X,35HNREC   (NO. OF PUMPING WELLS)    = ,I5       B4090
   8/13X,24HNCODES (FOR NODE IDENT.),9X,2H= ,I5/13X,25HNPNCHV (PUNCH V       B4100
   9ELOCITIES),8X,2H= ,I5/13X,36HNPDELC (PRINT OPT.-CONC. CHANGE) =          B4110
   $I4)                                                                      B4120
900 FORMAT (1H0,10X,29HLOCATION OF OBSERVATION WELLS//17X,3HNO.,5X,1HX       B4130
   1,5X,1HY/)                                                                B4140
910 FORMAT (1H0,10X,28HLOCATION  OF  PUMPING  WELLS//11X,28HX   Y   RA       B4150
   1TE(IN CFS)   CONC./)                                                     B4160
920 FORMAT (1H0,5X,37HNO. OF NODE IDENT. CODES SPECIFIED = ,I2)              B4170
930 FORMAT (1H0,10X,41HTHE FOLLOWING ASSIGNMENTS HAVE BEEN MADE:/5X,51       B4180
   1HCODE NO.     LEAKANCE     SOURCE CONC.     RECHARGE)                    B4190
```

FORTRAN IV program listing—Continued

```
 940 FORMAT (1H ,5X,61H*** WARNING ***   THCK.EQ.0.0 AND TMRX(X).GT.0.0  B4200
    1 AT NODE IX =,I4,6H, IY =,I4)                                       B4210
 950 FORMAT (1H ,5X,61H*** WARNING ***   THCK.EQ.0.0 AND TMRX(Y).GT.0.0  B4220
    1 AT NODE IX =,I4,6H, IY =,I4)                                       B4230
 960 FORMAT (1H ,5X,61H*** WARNING ***    THCK.EQ.0.0 AND NODEID.GT.0.0  B4240
    1 AT NODE IX =,I4,6H, IY =,I4)                                       B4250
 970 FORMAT (1H ,5X,56H*** WARNING ***   THCK.EQ.0.0 AND WT.NE.0.0 AT N  B4260
    1ODE IX =,I4,6H, IY =,I4)                                            B4270
 980 FORMAT (1H ,5X,58H*** WARNING ***   THCK.EQ.0.0 AND RECH.NE.0.0 AT  B4280
    1 NODE IX =,I4,6H, IY =,I4)                                          B4290
 990 FORMAT (1H ,5X,58H*** WARNING ***    THCK.EQ.0.0 AND REC.NE.0.0 AT  B4300
    1 NODE IX =,I4,6H, IY =,I4)                                          B4310
1000 FORMAT (1H ,5X,61H*** WARNING ***    PERM.EQ.0.0 AND NODEID.GT.0.0  B4320
    1 AT NODE IX =,I4,6H, IY =,I4)                                       B4330
1010 FORMAT (1H ,5X,56H*** WARNING ***   PERM.EQ.0.0 AND WT.NE.0.0 AT N  B4340
    1ODE IX =,I4,6H, IY =,I4)                                            B4350
1020 FORMAT (1H ,5X,58H*** WARNING ***   PERM.EQ.0.0 AND RECH.NE.0.0 AT  B4360
    1 NODE IX =,I4,6H, IY =,I4)                                          B4370
1030 FORMAT (1H ,5X,58H*** WARNING ***    PERM.EQ.0.0 AND REC.NE.0.0 AT  B4380
    1 NODE IX =,I4,6H, IY =,I4)                                          B4390
1040 FORMAT (1H ,5X,58H*** WARNING ***   PERM.EQ.0.0 AND THCK.GT.0.0 AT  B4400
    1 NODE IX =,I4,6H, IY =,I4)                                          B4410
1050 FORMAT (1H0,5X,45H*** WARNING ***   ANFCTR WAS SPECIFIED AS 0.0/23  B4420
    1X,34HDEFAULT ACTION: RESET ANFCTR = 1.0)                            B4430
1060 FORMAT (I1)                                                         B4440
1070 FORMAT (10I4,3G5.0)                                                 B4450
1080 FORMAT (1H1,5X,25HSTART PUMPING PERIOD NO. ,I2//2X,75HTHE FOLLOWIN  B4460
    1G TIME STEP, PUMPAGE, AND PRINT PARAMETERS HAVE BEEN REDEFINED:/)   B4470
1090 FORMAT (1HC,14X,9HNTIM    = ,I4/15X,9HNPNT    = ,I4/15X,9HNITP    = ,B4480
    1I4/15X,9HITMAX  = ,I4/15X,9HNREC    = ,I4/15X,9HNPNTMV = ,I4/15X,9H B4490
    2NPNTVL = ,I4/15X,9HNPNTD  = ,I4/15X,9HNPDELC = ,I4/15X,9HNPNCHV =   B4500
    3,I4/15X,9HPINT    = ,F10.3/15X,9HTIMX    = ,F10.3/15X,9HTINIT  = ,F1 B4510
    40.3/)                                                               B4520
```

```
1100 FORMAT (1H ,46X,E10.3)                                              B4530
     END                                                                 B4540-
     SUBROUTINE ITERAT                                                   C  10
     DOUBLE PRECISION DMIN1,DEXP,DLOG,DABS                               C  20
     REAL *8TMRX,VPRM,HI,HR,HC,HK,WT,REC,RECH,TIM,AOPT,TITLE             C  30
     REAL *8XDEL,YDEL,S,AREA,SUMT,RHO,PARAM,TEST,TOL,PINT,HMIN,PYR       C  40
     REAL *8B,G,W,A,C,E,F,DR,DC,TBAR,TMK,COEF,BLH,BRK,CHK,QL,BRH         C  50
     COMMON /PRMI/ NTIM,NPMP,NPNT,NITP,N,NX,NY,NP,NREC,INT,NNX,NNY,NUMO  C  60
    1BS,NMOV,IMCV,NPMAX,ITMAX,NZCRIT,IPRNT,NPTPND,NPNTMV,NPNTVL,NPNTD,N  C  70
    2PNCHV,NPDELC                                                        C  80
     COMMON /PRMK/ NODEID(20,20),NPCELL(20,20),LIMBO(500),IXOBS(5),IYOB  C  90
    1S(5)                                                                C 100
     COMMON /HEDA/ THCK(20,20),PERM(20,20),TMWL(5,50),TMOBS(50),ANFCTR   C 110
     COMMON /HEDB/ TMRX(20,20,2),VPRM(20,20),HI(20,20),HR(20,20),HC(20,  C 120
    120),HK(20,20),WT(20,20),REC(20,20),RECH(20,20),TIM(100),AOPT(20),T  C 130
    2ITLE(10),XDEL,YDEL,S,AREA,SUMT,RHO,PARAM,TEST,TOL,PINT,HMIN,PYR     C 140
     COMMON /BALM/ TOTLQ                                                 C 150
     COMMON /XINV/ DXINV,DYINV,ARINV,PORINV                              C 160
     DIMENSION W(20), B(20), G(20)                                       C 170
     **************************************************************     C 180
     KOUNT=0                                                             C 190
     ---COMPUTE ROW AND COLUMN---                                        C 200
        ---CALL NEW ITERATION PARAMETER---                               C 210
  10 REMN=MOD(KOUNT,NITP)                                                C 220
     IF (REMN.EQ.0) NTH=0                                                C 230
     NTH=NTH+1                                                           C 240
     PARAM=AOPT(NTH)                                                     C 250
     **************************************************************     C 260
     ---ROW COMPUTATIONS---                                              C 270
```

FORTRAN IV program listing—Continued

```
      TEST=0.0                                                               C 280
      RHO=S/TIM(N)                                                           C 290
      BRK=-RHO                                                               C 300
      DO 50 IY=1,NY                                                          C 310
      DO 20 M=1,NX                                                           C 320
      W(M)=0.0                                                               C 330
      B(M)=0.0                                                               C 340
      G(M)=0.0                                                               C 350
   20 CONTINUE                                                               C 360
      DO 30 IX=1,NX                                                          C 370
      IF (THCK(IX,IY).EQ.0.0) GO TO 30                                       C 380
      COEF=VPRM(IX,IY)                                                       C 390
      QL=-COEF*WT(IX,IY)                                                     C 400
      A=TMRX(IX-1,IY,1)*DXINV                                                C 410
      C=TMRX(IX,IY,1)*DXINV                                                  C 420
      E=TMRX(IX,IY-1,2)*DYINV                                                C 430
      F=TMRX(IX,IY,2)*DYINV                                                  C 440
      TBAR=A+C+E+F                                                           C 450
      TMK=TBAR*PARAM                                                         C 460
      BLH=-A-C-RHO-COEF-TMK                                                  C 470
      BRH=E+F-TMK                                                            C 480
      DR=BRH*HC(IX,IY)+BRK*HK(IX,IY)-E*HC(IX,IY-1)-F*HC(IX,IY+1)+QL+RECH     C 490
     1(IX,IY)+REC(IX,IY)*ARINV                                               C 500
      W(IX)=BLH-A*B(IX-1)                                                    C 510
      B(IX)=C/W(IX)                                                          C 520
      G(IX)=(DR-A*G(IX-1))/W(IX)                                             C 530
   30 CONTINUE                                                               C 540
                                                                             C 550
         ---BACK SUBSTITUTION---                                             C 560
      DO 40 J=2,NX                                                           C 570
      IJ=J-1                                                                 C 580
      IS=NX-IJ                                                               C 590
   40 HR(IS,IY)=G(IS)-B(IS)*HR(IS+1,IY)                                      C 600
   50 CONTINUE                                                               C 610
```

```
      *****************************************************************          C 620
      ---COLUMN COMPUTATIONS---                                                  C 630
      DO 90 IX=1,NX                                                              C 640
      DO 60 M=1,NY                                                               C 650
      W(M)=0.0                                                                   C 660
      B(M)=0.0                                                                   C 670
   60 G(M)=0.0                                                                   C 680
      DO 70 IY=1,NY                                                              C 690
      IF (THCK(IX,IY).EQ.0.0) GO TO 70                                           C 700
      COEF=VPRM(IX,IY)                                                           C 710
      QL=-COEF*WT(IX,IY)                                                         C 720
      A=TMRX(IX,IY-1,2)*DYINV                                                    C 730
      C=TMRX(IX,IY,2)*DYINV                                                      C 740
      E=TMRX(IX-1,IY,1)*DXINV                                                    C 750
      F=TMRX(IX,IY,1)*DXINV                                                      C 760
      TBAR=A+C+E+F                                                               C 770
      TMK=TBAR*PARAM                                                             C 780
      BLH=-A-C-RHO-COEF-TMK                                                      C 790
      BRH=E+F-TMK                                                                C 800
      DC=BRH*HR(IX,IY)+BRK*HK(IX,IY)-E*HR(IX-1,IY)-F*HR(IX+1,IY)+QL+RECH         C 810
     1(IX,IY)+REC(IX,IY)*ARINV                                                   C 820
      W(IY)=BLH-A*B(IY-1)                                                        C 830
      B(IY)=C/W(IY)                                                              C 840
      G(IY)=(DC-A*G(IY-1))/W(IY)                                                 C 850
   70 CONTINUE                                                                   C 860
                                                                                 C 870
         ---BACK SUBSTITUTION---                                                 C 880
      DO 80 J=2,NY                                                               C 890
```

FORTRAN IV program listing—Continued

```
      IJ=J-1                                                            C 900
      IB=NY-IJ                                                          C 910
      HC(IX,IB)=G(IB)-B(IB)*HC(IX,IB+1)                                 C 920
      IF (THCK(IX,IB).EQ.0.0) GO TO 80                                  C 930
      CHK=DABS(HC(IX,IB)-HR(IX,IB))                                     C 940
      IF (CHK.GT.TOL) TEST=1.0                                          C 950
   80 CONTINUE                                                          C 960
   90 CONTINUE                                                          C 970
      ***************************************************************   C 980
      KOUNT=KOUNT+1                                                     C 990
      IF (TEST.EQ.0.0) GO TO 120                                        C1000
      IF (KOUNT.GE.ITMAX) GO TO 100                                     C1010
      GO TO 10                                                          C1020
      ***************************************************************   C1030
      ---TERMINATE PROGRAM -- ITMAX EXCEEDED---                         C1040
  100 WRITE (6,160)                                                     C1050
      DO 110 IX=1,NX                                                    C1060
      DO 110 IY=1,NY                                                    C1070
  110 HK(IX,IY)=HC(IX,IY)                                               C1080
      CALL OUTPT                                                        C1090
      STOP                                                              C1100
      ***************************************************************   C1110
      ---SET NEW HEAD (HK)---                                           C1120
  120 DO 130 IY=1,NY                                                    C1130
      DO 130 IX=1,NX                                                    C1140
      IF (THCK(IX,IY).EQ.0.0) GO TO 130                                 C1150
      HR(IX,IY)=HK(IX,IY)                                               C1160
      HK(IX,IY)=HC(IX,IY)                                               C1170
                                                                        C1180
      ---COMPUTE LEAKAGE FOR MASS BALANCE---                            C1190
      IF (VPRM(IX,IY).EQ.0.0) GO TO 130                                 C1200
      DELQ=VPRM(IX,IY)*AREA*(WT(IX,IY)-HK(IX,IY))                       C1210
      TOTLQ=TOTLQ+DELQ*TIM(N)                                           C1220
```

```
  130 CONTINUE                                                              C1230
                                                                            C1240
      WRITE (6,140) N                                                       C1250
      WRITE (6,150) KOUNT                                                   C1260
      ****************************************************************     C1270
      RETURN                                                                C1280
      ****************************************************************     C1290
                                                                            C1300
                                                                            C1310
                                                                            C1320
  140 FORMAT (1HC//3X,4HN = ,1I4)                                           C1330
  150 FORMAT (1H ,2X,23HNUMBER OF ITERATIONS = ,1I4)                        C1340
  160 FORMAT (1H0,5X,64H***    EXECUTION TERMINATED -- MAX. NO. ITERATION   C1350
     1S EXCEEDED   ***/26X,21HFINAL OUTPUT FOLLOWS:)                        C1360
      END                                                                   C1370-
      SUBROUTINE GENPT                                                      D  10
      REAL *8TMRX,VPRM,HI,HR,HC,HK,WT,REC,RECH,TIM,AOPT,TITLE               D  20
      REAL *8XDEL,YDEL,S,AREA,SUMT,RHO,PARAM,TEST,TOL,PINT,HMIN,PYR         D  30
      COMMON /PRMI/ NTIM,NPMP,NPNT,NITP,N,NX,NY,NP,NREC,INT,NNX,NNY,NUMO    D  40
     1BS,NMOV,IMOV,NPMAX,ITMAX,NZCRIT,IPRNT,NPTPND,NPNTMV,NPNTVL,NPNTD,N    D  50
     2PNCHV,NPDELC                                                          D  60
      COMMON /PRMK/ NODEID(20,20),NPCELL(20,20),LIMBO(500),IXOBS(5),IYOB    D  70
     1S(5)                                                                  D  80
      COMMON /HEDA/ THCK(20,20),PERM(20,20),TMWL(5,50),TMOBS(50),ANFCTR     D  90
      COMMON /HEDB/ TMRX(20,20,2),VPRM(20,20),HI(20,20),HR(20,20),HC(20,    D 100
     120),HK(20,20),WT(20,20),REC(20,20),RECH(20,20),TIM(100),AOPT(20),T    D 110
     2ITLE(10),XDEL,YDEL,S,AREA,SUMT,RHO,PARAM,TEST,TOL,PINT,HMIN,PYR       D 120
      COMMON /CHMA/ PART(3,3200),CONC(20,20),TMCN(5,50),VX(20,20),VY(20,    D 130
     120),CONINT(20,20),CNRECH(20,20),POROS,SUMTCH,BETA,TIMV,STORM,STORM    D 140
```

FORTRAN IV program listing—Continued

```
     2I,CMSIN,CMSOUT,FLMIN,FLMOT,SUMIO,CELDIS,DLTRAT,CSTORM              D 150
      DIMENSION RP(8), RN(8), IPT(8)                                     D 160
      ****************************************************************   D 170
      F1=0.30                                                            D 180
      F2=1.0/3.0                                                         D 190
      IF (NPTPND.EQ.4) F1=0.25                                           D 200
      IF (NPTPND.EQ.9) F1=1.0/3.0                                        D 210
      IF (NPTPND.EQ.8) F2=0.25                                           D 220
      NCHK=NPTPND                                                        D 230
      IF (NPTPND.EQ.5.OR.NPTPND.EQ.9) NCHK=NPTPND-1                      D 240
      IF (TEST.GT.98.) GO TO 10                                          D 250
      ****************************************************************   D 260
      ---INITIALIZE VALUES---                                            D 270
      STORM=0.0                                                          D 280
      CMSIN=0.0                                                          D 290
      CMSOUT=0.0                                                         D 300
      FLMIN=0.0                                                          D 310
      FLMOT=0.0                                                          D 320
      SUMIO=0.0                                                          D 330
      ****************************************************************   D 340
   10 DO 20 ID=1,3                                                       D 350
      DO 20 IN=1,NPMAX                                                   D 360
   20 PART(ID,IN)=0.0                                                    D 370
      DO 30 IA=1,8                                                       D 380
      RP(IA)=0.0                                                         D 390
      RN(IA)=0.0                                                         D 400
   30 IPT(IA)=0                                                          D 410
      ---SET UP LIMBO ARRAY---                                           D 420
      DO 40 IN=1,500                                                     D 430
   40 LIMBO(IN)=0.0                                                      D 440
      IND=1                                                              D 450
      DO 50 IL=1,500,2                                                   D 460
      LIMBO(IL)=IND                                                      D 470
   50 IND=IND+1                                                          D 480
      ****************************************************************   D 490
```

```
      ---INSERT PARTICLES---                                            D 500
      DO 410 IX=1,NX                                                    D 510
      DO 410 IY=1,NY                                                    D 520
      IF (THCK(IX,IY).EQ.0.0) GO TO 410                                 D 530
      KR=0                                                              D 540
      TEST2=0.0                                                         D 550
      METH=1                                                            D 560
      NPCELL(IX,IY)=0                                                   D 570
      C1=CONC(IX,IY)                                                    D 580
      IF (C1.LE.1.0E-05) TEST2=1.0                                      D 590
      IF (VPRM(IX,IY).GT.0.09) TEST2=1.0                                D 600
      IF (REC(IX,IY).NE.0.0) TEST2=1.0                                  D 610
      IF (THCK(IX+1,IY+1).EQ.0.0.OR.THCK(IX+1,IY-1).EQ.0.0.OR.THCK(IX-1, D 620
     1IY+1).EQ.0.0.OR.THCK(IX-1,IY-1).EQ.0.0) TEST2=1.0                 D 630
      IF ((THCK(IX,IY+1).EQ.0.0.OR.THCK(IX,IY-1).EQ.0.0.OR.THCK(IX+1,IY) D 640
     1.EQ.0.0.OR.THCK(IX-1,IY).EQ.0.0).AND.NPTPND.GT.5) TEST2=1.0       D 650
      CNODE=C1*(1.0-F1)                                                 D 660
      IF (TEST.LT.98.0.OR.TEST2.GT.0.0) GO TO 70                        D 670
      SUMC=CONC(IX+1,IY)+CONC(IX-1,IY)+CONC(IX,IY+1)+CONC(IX,IY-1)      D 680
      IF (NCHK.EQ.4) GO TO 60                                           D 690
      SUMC=SUMC+CONC(IX+1,IY+1)+CONC(IX+1,IY-1)+CONC(IX-1,IY+1)+CONC(IX- D 700
     11,IY-1)                                                           D 710
   60 AVC=SUMC/NCHK                                                     D 720
      IF (AVC.GT.C1) METH=2                                             D 730
                                                                        D 740
         ---PUT 4 PARTICLES ON CELL DIAGONALS---                        D 750
   70 DO 140 IT=1,2                                                     D 760
```

FORTRAN IV program listing—Continued

```
      EVET=(-1.0)**IT                                              D 770
      DO 140 IS=1,2                                                D 780
      EVES=(-1.0)**IS                                              D 790
      PART(1,IND)=IX+F1*EVET                                       D 800
      PART(2,IND)=IY+F1*EVES                                       D 810
      PART(2,IND)=-PART(2,IND)                                     D 820
      PART(3,IND)=C1                                               D 830
      IF (TEST.LT.98.0.OR.TEST2.GT.0.0) GO TO 130                  D 840
      IXD=IX+EVET                                                  D 850
      IYD=IY+EVES                                                  D 860
      KR=KR+1                                                      D 870
      IPT(KR)=IND                                                  D 880
      IF (METH.EQ.2) GO TO 80                                      D 890
      PART(3,IND)=CNODE+CONC(IXD,IYD)*F1                           D 900
      GO TO 90                                                     D 910
   80 PART(3,IND)=2.0*C1*CONC(IXD,IYD)/(C1+CONC(IXD,IYD))          D 920
   90 IF (C1-CONC(IXD,IYD)) 100,110,120                            D 930
  100 RP(KR)=CONC(IXD,IYD)-PART(3,IND)                             D 940
      RN(KR)=C1-PART(3,IND)                                        D 950
      GO TO 130                                                    D 960
  110 RP(KR)=0.0                                                   D 970
      RN(KR)=0.0                                                   D 980
      GO TO 130                                                    D 990
  120 RP(KR)=C1-PART(3,IND)                                        D1000
      RN(KR)=CONC(IXD,IYD)-PART(3,IND)                             D1010
  130 IND=IND+1                                                    D1020
  140 CONTINUE                                                     D1030
                                                                   D1040
      IF (NPTPND.EQ.5.OR.NPTPND.EQ.9) GO TO 150                    D1050
      GO TO 160                                                    D1060
         ---PUT ONE PARTICLE AT CENTER OF CELL---                  D1070
  150 PART(1,IND)=-IX                                              D1080
      PART(2,IND)=-IY                                              D1090
      PART(3,IND)=C1                                               D1100
      IND=IND+1                                                    D1110
```

```
C     ---PLACE NORTH, SOUTH, EAST, AND WEST PARTICLES---                D1120
  160 IF (NPTPND.LT.8) GO TO 290                                        D1130
      CNODE=C1*(1.0-F2)                                                 D1140
      DO 280 IT=1,2                                                     D1150
      EVET=(-1.0)**IT                                                   D1160
      PART(1,IND)=IX+F2*EVET                                            D1170
      PART(2,IND)=-IY                                                   D1180
      PART(3,IND)=C1                                                    D1190
      IF (TEST.LT.98.0.OR.TEST2.GT.0.0) GO TO 220                       D1200
      IXD=IX+EVET                                                       D1210
      KR=KR+1                                                           D1220
      IPT(KR)=IND                                                       D1230
      IF (METH.EQ.2) GO TO 170                                          D1240
      PART(3,IND)=CNODE+CONC(IXD,IY)*F2                                 D1250
      GO TO 180                                                         D1260
  170 PART(3,IND)=2.0*C1*CONC(IXD,IY)/(C1+CONC(IXD,IY))                 D1270
  180 IF (C1-CONC(IXD,IY)) 190,200,210                                  D1280
  190 RP(KR)=CONC(IXD,IY)-PART(3,IND)                                   D1290
      RN(KR)=C1-PART(3,IND)                                             D1300
      GO TO 220                                                         D1310
  200 RP(KR)=0.0                                                        D1320
      RN(KR)=0.0                                                        D1330
      GO TO 220                                                         D1340
  210 RP(KR)=C1-PART(3,IND)                                             D1350
      RN(KR)=CONC(IXD,IY)-PART(3,IND)                                   D1360
  220 IND=IND+1                                                         D1370
      PART(1,IND)=IX                                                    D1380
```

FORTRAN IV program listing—Continued

```
      PART(2,IND)=IY+F2*EVET                                            D1390
      PART(2,IND)=-PART(2,IND)                                          D1400
      PART(3,IND)=C1                                                    D1410
      IF (TEST.LT.98.0.OR.TEST2.GT.0.0) GO TO 280                       D1420
      IYD=IY+EVET                                                       D1430
      KR=KR+1                                                           D1440
      IPT(KR)=IND                                                       D1450
      IF (METH.EQ.2) GO TO 230                                          D1460
      PART(3,IND)=CNODE+CONC(IX,IYD)*F2                                 D1470
      GO TO 240                                                         D1480
  230 PART(3,IND)=2.0*C1*CONC(IX,IYD)/(C1+CONC(IX,IYD))                 D1490
  240 IF (C1-CONC(IX,IYD)) 250,260,270                                  D1500
  250 RP(KR)=CONC(IX,IYD)-PART(3,IND)                                   D1510
      RN(KR)=C1-PART(3,IND)                                             D1520
      GO TO 280                                                         D1530
  260 RP(KR)=0.0                                                        D1540
      RN(KR)=0.0                                                        D1550
      GO TO 280                                                         D1560
  270 RP(KR)=C1-PART(3,IND)                                             D1570
      RN(KR)=CONC(IX,IYD)-PART(3,IND)                                   D1580
  280 IND=IND+1                                                         D1590
                                                                        D1600
  290 IF (TEST.LT.98.0.OR.TEST2.GT.0.0) GO TO 410                       D1610
      SUMPT=0.0                                                         D1620
         ---COMPUTE CONC. GRADIENT WITHIN CELL---                       D1630
      DO 300 KPT=1,NCHK                                                 D1640
      IK=IPT(KPT)                                                       D1650
  300 SUMPT=PART(3,IK)+SUMPT                                            D1660
      CBAR=SUMPT/NCHK                                                   D1670
         ---CHECK MASS BALANCE WITHIN CELL AND ADJUST PT. CONCS.---     D1680
      SUMPT=0.0                                                         D1690
      IF (CBAR-C1) 310,410,330                                          D1700
```

```
310 CRCT=1.0-(CBAR/C1)                                    D1710
    IF (METH.EQ.1) CRCT=CBAR/C1                           D1720
    DO 320 KPT=1,NCHK                                     D1730
    IK=IPT(KPT)                                           D1740
    PART(3,IK)=PART(3,IK)+RP(KPT)*CRCT                    D1750
320 SUMPT=SUMPT+PART(3,IK)                                D1760
    CBARN=SUMPT/NCHK                                      D1770
    GO TO 350                                             D1780
330 CRCT=1.0-(C1/CBAR)                                    D1790
    IF (METH.EQ.1) CRCT=C1/CBAR                           D1800
    DO 340 KPT=1,NCHK                                     D1810
    IK=IPT(KPT)                                           D1820
    PART(3,IK)=PART(3,IK)+RN(KPT)*CRCT                    D1830
340 SUMPT=SUMPT+PART(3,IK)                                D1840
    CBARN=SUMPT/NCHK                                      D1850
350 IF (CBARN.EQ.C1) GO TO 410                            D1860
       ---CORRECT FOR OVERCOMPENSATION---                 D1870
    CRCT=C1/CBARN                                         D1880
    DO 380 KPT=1,NCHK                                     D1890
    IK=IPT(KPT)                                           D1900
    PART(3,IK)=PART(3,IK)*CRCT                            D1910
       ---CHECK CONSTRAINTS---                            D1920
    IF (PART(3,IK)-C1) 360,380,370                        D1930
360 CLIM=C1-RP(KPT)+RN(KPT)                               D1940
    IF (PART(3,IK).LT.CLIM) GO TO 390                     D1950
    GO TO 380                                             D1960
370 CLIM=C1+RP(KPT)-RN(KPT)                               D1970
    IF (PART(3,IK).GT.CLIM) GO TO 390                     D1980
380 CONTINUE                                              D1990
    GO TO 410                                             D2000
```

FORTRAN IV program listing—Continued

```
  390 TEST2=1.0                                                          D2010
      DO 400 KPT=1,NCHK                                                  D2020
      IK=IPT(KPT)                                                        D2030
  400 PART(3,IK)=C1                                                      D2040
  410 CONTINUE                                                           D2050
      NP=IND                                                             D2060
      IF (INT.EQ.0) CALL CHMOT                                           D2070
C     ****************************************************************   D2080
      RETURN                                                             D2090
C     ****************************************************************   D2100
      END                                                                D2110-
      SUBROUTINE VELO                                                    E  10
      DOUBLE PRECISION DMIN1,DEXP,DLOG,DABS                              E  20
      REAL *8TMRX,VPRM,HI,HR,HC,HK,WT,REC,RECH,TIM,AOPT,TITLE            E  30
      REAL *8XDEL,YDEL,S,AREA,SUMT,RHO,PARAM,TEST,TOL,PINT,HMIN,PYR      E  40
      REAL *8RATE,SLEAK,DIV                                              E  50
      COMMON /PRMI/ NTIM,NPMP,NPNT,NITP,N,NX,NY,NP,NREC,INT,NNX,NNY,NUMO E  60
     1BS,NMOV,IMCV,NPMAX,ITMAX,NZCRIT,IPRNT,NPTPND,NPNTMV,NPNTVL,NPNTD,N E  70
     2PNCHV,NPDELC                                                       E  80
      COMMON /PRMK/ NODEID(20,20),NPCELL(20,20),LIMBO(500),IXOBS(5),IYOB E  90
     1S(5)                                                               E 100
      COMMON /HEDA/ THCK(20,20),PERM(20,20),TMWL(5,50),TMOBS(50),ANFCTR  E 110
      COMMON /HEDB/ TMRX(20,20,2),VPRM(20,20),HI(20,20),HR(20,20),HC(20, E 120
     120),HK(20,20),WT(20,20),REC(20,20),RECH(20,20),TIM(100),AOPT(20),T E 130
     2ITLE(10),XDEL,YDEL,S,AREA,SUMT,RHO,PARAM,TEST,TOL,PINT,HMIN,PYR    E 140
      COMMON /XINV/ DXINV,DYINV,ARINV,PORINV                             E 150
      COMMON /CHMA/ PART(3,3200),CONC(20,20),TMCN(5,50),VX(20,20),VY(20, E 160
     120),CONINT(20,20),CNRECH(20,20),POROS,SUMTCH,BETA,TIMV,STORM,STORM E 170
     2I,CMSIN,CMSOUT,FLMIN,FLMOT,SUMIO,CELDIS,DLTRAT,CSTORM              E 180
      COMMON /CHMC/ SUMC(20,20),VXBDY(20,20),VYBDY(20,20)                E 190
      COMMON /DIFUS/ DISP(20,20,4)                                       E 200
C     ****************************************************************   E 210
```

```
C     ---COMPUTE VELOCITIES AND STORE---                           E 220
      VMAX=1.0E-10                                                 E 230
      VMAY=1.0E-10                                                 E 240
      VMXBD=1.0E-10                                                E 250
      VMYBD=1.0E-10                                                E 260
      TMV=TIM(N)                                                   E 270
      LIM=0                                                        E 280
C                                                                  E 290
      DO 20 IX=1,NX                                                E 300
      DO 20 IY=1,NY                                                E 310
      DO 10 IZ=1,4                                                 E 320
   10 DISP(IX,IY,IZ)=0.0                                           E 330
C                                                                  E 340
      IF (THCK(IX,IY).EQ.0.0) GO TO 20                             E 350
      RATE=REC(IX,IY)/AREA                                         E 360
      SLEAK=(HK(IX,IY)-WT(IX,IY))*VPRM(IX,IY)                      E 370
      DIV=RATE+SLEAK+RECH(IX,IY)                                   E 380
C                                                                  E 390
C         ---VELOCITIES AT NODES---                                E 400
C             ---X-DIRECTION---                                    E 410
      GRDX=(HK(IX-1,IY)-HK(IX+1,IY))*DXINV*0.50                    E 420
      IF (THCK(IX-1,IY).EQ.0.0) GRDX=(HK(IX,IY)-HK(IX+1,IY))*DXINV E 430
      IF (THCK(IX+1,IY).EQ.0.0) GRDX=(HK(IX-1,IY)-HK(IX,IY))*DXINV E 440
      IF (THCK(IX-1,IY).EQ.0.0.AND.THCK(IX+1,IY).EQ.0.0) GRDX=0.0  E 450
      VX(IX,IY)=PERM(IX,IY)*GRDX*PORINV                            E 460
      ABVX=ABS(VX(IX,IY))                                          E 470
      IF (ABVX.GT.VMAX) VMAX=ABVX                                  E 480
C             ---Y-DIRECTION---                                    E 490
      GRDY=(HK(IX,IY-1)-HK(IX,IY+1))*DYINV*0.50                    E 500
      IF (THCK(IX,IY-1).EQ.0.0) GRDY=(HK(IX,IY)-HK(IX,IY+1))*DYINV E 510
```

FORTRAN IV program listing—Continued

```
      IF (THCK(IX,IY+1).EQ.0.0) GRDY=(HK(IX,IY-1)-HK(IX,IY))*DYINV        E 520
      IF (THCK(IX,IY-1).EQ.0.0.AND.THCK(IX,IY+1).EQ.0.0) GRDY=0.0         E 530
      VY(IX,IY)=PERM(IX,IY)*GRDY*PORINV*ANFCTR                            E 540
      ABVY=ABS(VY(IX,IY))                                                 E 550
      IF (ABVY.GT.VMAY) VMAY=ABVY                                         E 560
C                                                                         E 570
C        ---VELOCITIES AT CELL BOUNDARIES---                              E 580
      GRDX=(HK(IX,IY)-HK(IX+1,IY))*DXINV                                  E 590
      PERMX=2.0*PERM(IX,IY)*PERM(IX+1,IY)/(PERM(IX,IY)+PERM(IX+1,IY))     E 600
      VXBDY(IX,IY)=PERMX*GRDX*PORINV                                      E 610
      GRDY=(HK(IX,IY)-HK(IX,IY+1))*DYINV                                  E 620
      PERMY=2.0*PERM(IX,IY)*PERM(IX,IY+1)/(PERM(IX,IY)+PERM(IX,IY+1))     E 630
      VYBDY(IX,IY)=PERMY*GRDY*PORINV*ANFCTR                               E 640
      ABVX=ABS(VXBDY(IX,IY))                                              E 650
      ABVY=ABS(VYBDY(IX,IY))                                              E 660
      IF (ABVX.GT.VMXBD) VMXBD=ABVX                                       E 670
      IF (ABVY.GT.VMYBD) VMYBD=ABVY                                       E 680
C                                                                         E 690
      IF (DIV.GE.0.0) GO TO 20                                            E 700
      TDIV=(POROS*THCK(IX,IY))/DABS(DIV)                                  E 710
      IF (TDIV.LT.TMV) TMV=TDIV                                           E 720
   20 CONTINUE                                                            E 730
C     *****************************************************************   E 740
C     ---PRINT VELOCITIES---                                              E 750
      IF (NPNTVL.EQ.0) GO TO 80                                           E 760
      IF (NPNTVL.EQ.2) GO TO 30                                           E 770
      IF (NPNTVL.EQ.1.AND.N.EQ.1) GO TO 30                                E 780
      GO TO 80                                                            E 790
   30 WRITE (6,320)                                                       E 800
      WRITE (6,330)                                                       E 810
      DO 40 IY=1,NY                                                       E 820
   40 WRITE (6,350) (VX(IX,IY),IX=1,NX)                                   E 830
      WRITE (6,340)                                                       E 840
      DO 50 IY=1,NY                                                       E 850
```

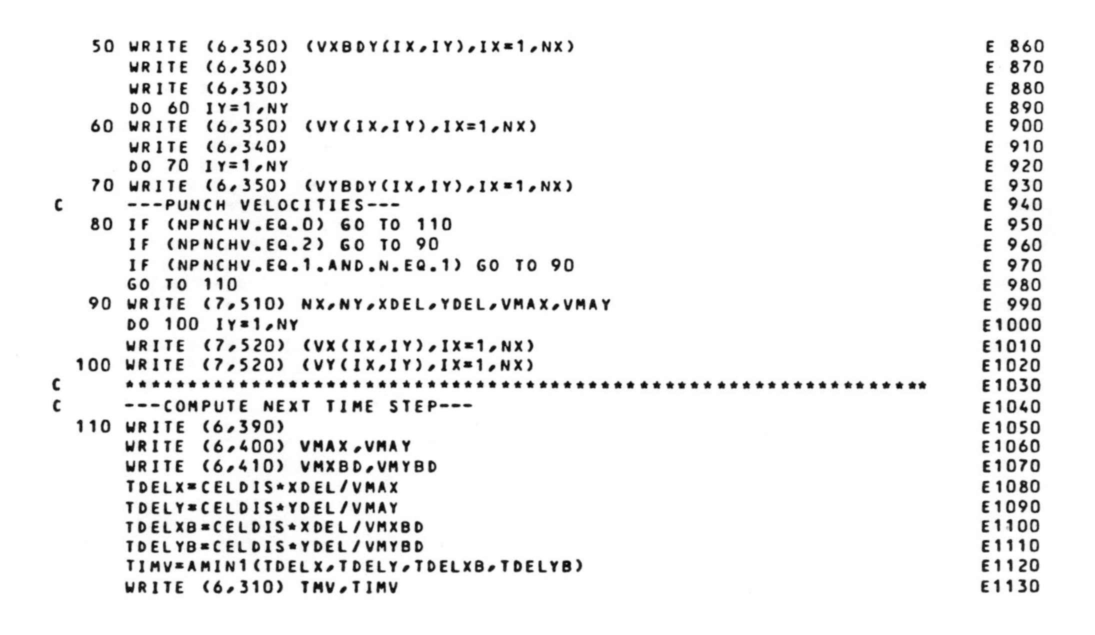

```
   50 WRITE (6,350) (VXBDY(IX,IY),IX=1,NX)                              E 860
      WRITE (6,360)                                                     E 870
      WRITE (6,330)                                                     E 880
      DO 60 IY=1,NY                                                     E 890
   60 WRITE (6,350) (VY(IX,IY),IX=1,NX)                                 E 900
      WRITE (6,340)                                                     E 910
      DO 70 IY=1,NY                                                     E 920
   70 WRITE (6,350) (VYBDY(IX,IY),IX=1,NX)                              E 930
C     ---PUNCH VELOCITIES---                                            E 940
   80 IF (NPNCHV.EQ.0) GO TO 110                                        E 950
      IF (NPNCHV.EQ.2) GO TO 90                                         E 960
      IF (NPNCHV.EQ.1.AND.N.EQ.1) GO TO 90                              E 970
      GO TO 110                                                         E 980
   90 WRITE (7,510) NX,NY,XDEL,YDEL,VMAX,VMAY                           E 990
      DO 100 IY=1,NY                                                    E1000
      WRITE (7,520) (VX(IX,IY),IX=1,NX)                                 E1010
  100 WRITE (7,520) (VY(IX,IY),IX=1,NX)                                 E1020
C     ******************************************************************E1030
C     ---COMPUTE NEXT TIME STEP---                                      E1040
  110 WRITE (6,390)                                                     E1050
      WRITE (6,400) VMAX,VMAY                                           E1060
      WRITE (6,410) VMXBD,VMYBD                                         E1070
      TDELX=CELDIS*XDEL/VMAX                                            E1080
      TDELY=CELDIS*YDEL/VMAY                                            E1090
      TDELXB=CELDIS*XDEL/VMXBD                                          E1100
      TDELYB=CELDIS*YDEL/VMYBD                                          E1110
      TIMV=AMIN1(TDELX,TDELY,TDELXB,TDELYB)                             E1120
      WRITE (6,310) TMV,TIMV                                            E1130
```

FORTRAN IV program listing—Continued

```
      IF (TMV.LT.TIMV) GO TO 120                                        E1140
      LIM=-1                                                            E1150
      GO TO 130                                                         E1160
  120 TIMV=TMV                                                          E1170
      LIM=1                                                             E1180
  130 NTIMV=TIM(N)/TIMV                                                 E1190
      NMOV=NTIMV+1                                                      E1200
      WRITE (6,420) TIMV,NTIMV,NMOV                                     E1210
      TIMV=TIM(N)/NMOV                                                  E1220
      WRITE (6,370) TIM(N)                                              E1230
      WRITE (6,380) TIMV                                                E1240
C                                                                       E1250
      IF (BETA.EQ.0.0) GO TO 200                                        E1260
C     ***************************************************************** E1270
C     ---COMPUTE DISPERSION COEFFICIENTS---                             E1280
      ALPHA=BETA                                                        E1290
      ALNG=ALPHA                                                        E1300
      TRAN=DLTRAT*ALPHA                                                 E1310
      XX2=XDEL*XDEL                                                     E1320
      YY2=YDEL*YDEL                                                     E1330
      XY2=4.0*XDEL*YDEL                                                 E1340
      DO 150 IX=2,NNX                                                   E1350
      DO 150 IY=2,NNY                                                   E1360
      IF (THCK(IX,IY).EQ.0.0) GO TO 150                                 E1370
      VXE=VXBDY(IX,IY)                                                  E1380
      VYS=VYBDY(IX,IY)                                                  E1390
      IF (THCK(IX+1,IY).EQ.0.0) GO TO 140                               E1400
C        ---FORWARD COEFFICIENTS: X-DIRECTION---                        E1410
      VYE=(VYBDY(IX,IY-1)+VYBDY(IX+1,IY-1)+VYS+VYBDY(IX+1,IY))/4.0      E1420
      VXE2=VXE*VXE                                                      E1430
      VYE2=VYE*VYE                                                      E1440
      VMGE=SQRT(VXE2+VYE2)                                              E1450
      IF (VMGE.LT.1.0E-20) GO TO 140                                    E1460
      DALN=ALNG*VMGE                                                    E1470
```

```
      DTRN=TRAN*VMGE                                                    E1480
      VMGE2=VMGE*VMGE                                                   E1490
C           ---XX COEFFICIENT---                                        E1500
      DISP(IX,IY,1)=(DALN*VXE2+DTRN*VYE2)/(VMGE2*XX2)                   E1510
C           ---XY COEFFICIENT---                                        E1520
      DISP(IX,IY,3)=(DALN-DTRN)*VXE*VYE/(VMGE2*XY2)                     E1530
C        ---FORWARD COEFFICIENTS: Y-DIRECTION---                        E1540
  140 IF (THCK(IX,IY+1).EQ.0.0) GO TO 150                               E1550
      VXS=(VXBDY(IX-1,IY)+VXE+VXBDY(IX-1,IY+1)+VXBDY(IX,IY+1))/4.0      E1560
      VYS2=VYS*VYS                                                      E1570
      VXS2=VXS*VXS                                                      E1580
      VMGS=SQRT(VXS2+VYS2)                                              E1590
      IF (VMGS.LT.1.0E-20) GO TO 150                                    E1600
      DALN=ALNG*VMGS                                                    E1610
      DTRN=TRAN*VMGS                                                    E1620
      VMGS2=VMGS*VMGS                                                   E1630
C           ---YY COEFFICIENT---                                        E1640
      DISP(IX,IY,2)=(DALN*VYS2+DTRN*VXS2)/(VMGS2*YY2)                   E1650
C           ---YX COEFFICIENT---                                        E1660
      DISP(IX,IY,4)=(DALN-DTRN)*VXS*VYS/(VMGS2*XY2)                     E1670
  150 CONTINUE                                                          E1680
C     ****************************************************************  E1690
C     ---ADJUST CROSS-PRODUCT TERMS FOR ZERO THICKNESS---               E1700
      DO 160 IX=2,NNX                                                   E1710
      DO 160 IY=2,NNY                                                   E1720
      IF (THCK(IX,IY+1).EQ.0.0.OR.THCK(IX+1,IY+1).EQ.0.0.OR.THCK(IX,IY-1 E1730
     1).EQ.0.0.OR.THCK(IX+1,IY-1).EQ.0.0) DISP(IX,IY,3)=0.0             E1740
      IF (THCK(IX+1,IY).EQ.0.0.OR.THCK(IX+1,IY+1).EQ.0.0.OR.THCK(IX-1,IY E1750
```

FORTRAN IV program listing—Continued

```
     1).EQ.0.0.OR.THCK(IX-1,IY+1).EQ.0.0) DISP(IX,IY,4)=0.0             E1760
  160 CONTINUE                                                           E1770
C     ****************************************************************** E1780
C     ---CHECK FOR STABILITY OF EXPLICIT METHOD---                       E1790
      TIMDIS=0.0                                                         E1800
      DO 170 IX=2,NNX                                                    E1810
      DO 170 IY=2,NNY                                                    E1820
      TDCO=DISP(IX,IY,1)+DISP(IX,IY,2)                                   E1830
  170 IF (TDCO.GT.TIMDIS) TIMDIS=TDCO                                    E1840
      TIMDC=0.5/TIMDIS                                                   E1850
      WRITE (6,440) TIMDC                                                E1860
      NTIMD=TIM(N)/TIMDC                                                 E1870
      NDISP=NTIMD+1                                                      E1880
      IF (NDISP.LE.NMOV) GO TO 180                                       E1890
      NMOV=NDISP                                                         E1900
      TIMV=TIM(N)/NMOV                                                   E1910
      LIM=0                                                              E1920
  180 WRITE (6,430) TIMV,NTIMD,NMOV                                      E1930
C     ****************************************************************** E1940
C     ---ADJUST DISP. EQUATION COEFFICIENTS FOR SATURATED THICKNESS---   E1950
      DO 190 IX=2,NNX                                                    E1960
      DO 190 IY=2,NNY                                                    E1970
      BAVX=0.5*(THCK(IX,IY)+THCK(IX+1,IY))                               E1980
      BAVY=0.5*(THCK(IX,IY)+THCK(IX,IY+1))                               E1990
      DISP(IX,IY,1)=DISP(IX,IY,1)*BAVX                                   E2000
      DISP(IX,IY,2)=DISP(IX,IY,2)*BAVY                                   E2010
      DISP(IX,IY,3)=DISP(IX,IY,3)*BAVX                                   E2020
  190 DISP(IX,IY,4)=DISP(IX,IY,4)*BAVY                                   E2030
C     ****************************************************************** E2040
  200 IF (LIM) 210,220,230                                               E2050
  210 WRITE (6,530)                                                      E2060
      GO TO 240                                                          E2070
  220 WRITE (6,540)                                                      E2080
      GO TO 240                                                          E2090
  230 WRITE (6,550)                                                      E2100
```

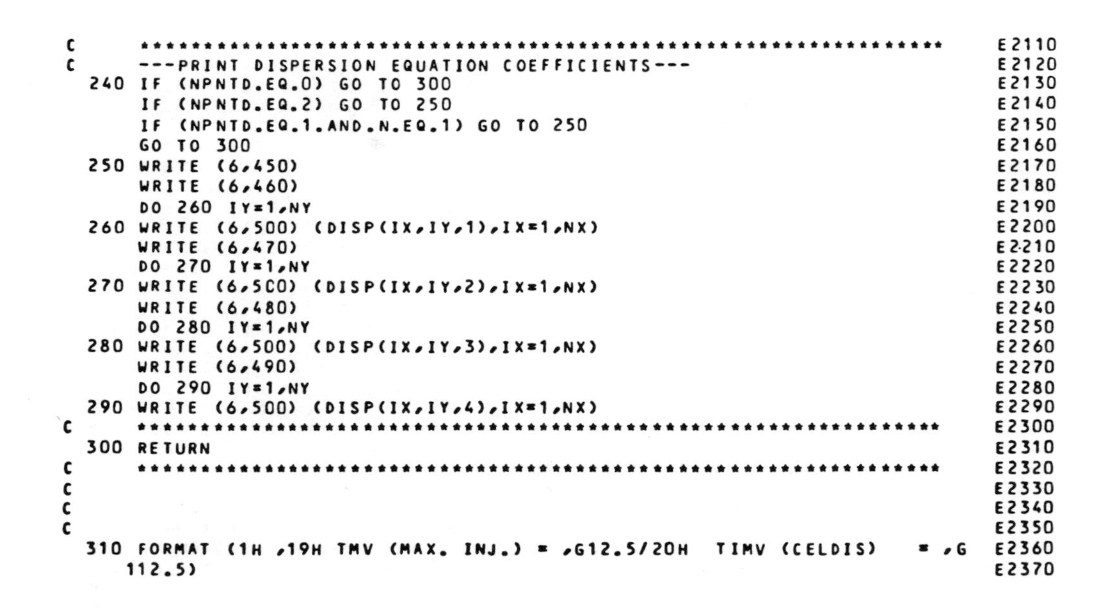

```
C     ******************************************************************   E2110
C     ---PRINT DISPERSION EQUATION COEFFICIENTS---                         E2120
  240 IF (NPNTD.EQ.0) GO TO 300                                            E2130
      IF (NPNTD.EQ.2) GO TO 250                                            E2140
      IF (NPNTD.EQ.1.AND.N.EQ.1) GO TO 250                                 E2150
      GO TO 300                                                            E2160
  250 WRITE (6,450)                                                        E2170
      WRITE (6,460)                                                        E2180
      DO 260 IY=1,NY                                                       E2190
  260 WRITE (6,500) (DISP(IX,IY,1),IX=1,NX)                                E2200
      WRITE (6,470)                                                        E2210
      DO 270 IY=1,NY                                                       E2220
  270 WRITE (6,500) (DISP(IX,IY,2),IX=1,NX)                                E2230
      WRITE (6,480)                                                        E2240
      DO 280 IY=1,NY                                                       E2250
  280 WRITE (6,500) (DISP(IX,IY,3),IX=1,NX)                                E2260
      WRITE (6,490)                                                        E2270
      DO 290 IY=1,NY                                                       E2280
  290 WRITE (6,500) (DISP(IX,IY,4),IX=1,NX)                                E2290
C     ******************************************************************   E2300
  300 RETURN                                                               E2310
C     ******************************************************************   E2320
C                                                                          E2330
C                                                                          E2340
C                                                                          E2350
  310 FORMAT (1H ,19H TMV (MAX. INJ.) = ,G12.5/20H  TIMV (CELDIS)   = ,G   E2360
     112.5)                                                                E2370
```

FORTRAN IV program listing—Continued

```
320 FORMAT (1H1,12HX VELOCITIES)                                          E2380
330 FORMAT (1H ,25X,8HAT NODES/)                                          E2390
340 FORMAT (1H0,25X,13HON BOUNDARIES/)                                    E2400
350 FORMAT (1H ,10G12.3)                                                  E2410
360 FORMAT (1H1,12HY VELOCITIES)                                          E2420
370 FORMAT (3H    ,11HTIM (N)   = ,1G12.5)                                E2430
380 FORMAT (3H    ,11HTIMEVELO = ,1G12.5)                                 E2440
390 FORMAT (1H1,10X,29HSTABILITY CRITERIA --- M.O.C.//)                   E2450
400 FORMAT (1H0,8H VMAX = ,1PE9.2,5X,7HVMAY = ,1PE9.2)                    E2460
410 FORMAT (1H ,8H VMXBD= ,1PE9.2,5X,7HVMYBD= ,1PE9.2)                    E2470
420 FORMAT (1H0,8H TIMV = ,1PE9.2,5X,8HNTIMV = ,I5,5X,7HNMOV = ,I5/)      E2480
430 FORMAT (1H0,8H TIMV = ,1PE9.2,5X,8HNTIMD = ,I5,5X,7HNMOV = ,I5)       E2490
440 FORMAT (3H    ,11HTIMEDISP = ,1E12.5)                                 E2500
450 FORMAT (1H1,32HDISPERSION EQUATION COEFFICIENTS,10X,25H=(D-IJ)*(B)    E2510
   1/(GRID FACTOR))                                                       E2520
460 FORMAT (1H ,35X,14HXX COEFFICIENT/)                                   E2530
470 FORMAT (1H ,35X,14HYY COEFFICIENT/)                                   E2540
480 FORMAT (1H ,35X,14HXY COEFFICIENT/)                                   E2550
490 FORMAT (1H ,35X,14HYX COEFFICIENT/)                                   E2560
500 FORMAT (1H ,1P10E8.1)                                                 E2570
510 FORMAT (2I4,2F10.1,2F10.7)                                            E2580
520 FORMAT (8F10.7)                                                       E2590
530 FORMAT (1H0,10X,42HTHE LIMITING STABILITY CRITERION IS CELDIS)        E2600
540 FORMAT (1HC,10X,40HTHE LIMITING STABILITY CRITERION IS BETA)          E2610
550 FORMAT (1H0,10X,58HTHE LIMITING STABILITY CRITERION IS MAXIMUM INJ    E2620
   1ECTION RATE)                                                          E2630
    END                                                                   E2640-
    SUBROUTINE MOVE                                                       F   10
    REAL *8TMRX,VPRM,HI,HR,HC,HK,WT,REC,RECH,TIM,AOPT,TITLE               F   20
    REAL *8XDEL,YDEL,S,AREA,SUMT,RHO,PARAM,TEST,TOL,PINT,HMIN,PYR          F   30
    COMMON /PRMI/ NTIM,NPMP,NPNT,NITP,N,NX,NY,NP,NREC,INT,NNX,NNY,NUMO    F   40
   1BS,NMOV,IMOV,NPMAX,ITMAX,NZCRIT,IPRNT,NPTPND,NPNTMV,NPNTVL,NPNTD,N    F   50
   2PNCHV,NPDELC                                                          F   60
    COMMON /PRMK/ NODEID(20,20),NPCELL(20,20),LIMBO(500),IXOBS(5),IYOB    F   70
```

```
      1S(5)                                                             F  80
       COMMON /HEDA/ THCK(20,20),PERM(20,20),TMWL(5,50),TMOBS(50),ANFCTR  F  90
       COMMON /HEDB/ TMRX(20,20,2),VPRM(20,20),HI(20,20),HR(20,20),HC(20, F 100
      120),HK(20,20),WT(20,20),REC(20,20),RECH(20,20),TIM(100),AOPT(20),T F 110
      2ITLE(10),XDEL,YDEL,S,AREA,SUMT,RHO,PARAM,TEST,TOL,PINT,HMIN,PYR    F 120
       COMMON /XINV/ DXINV,DYINV,ARINV,PORINV                             F 130
       COMMON /CHMA/ PART(3,3200),CONC(20,20),TMCN(5,50),VX(20,20),VY(20, F 140
      120),CONINT(20,20),CNRECH(20,20),POROS,SUMTCH,BETA,TIMV,STORM,STORM F 150
      2I,CMSIN,CMSOUT,FLMIN,FLMOT,SUMIO,CELDIS,DLTRAT,CSTORM              F 160
       COMMON /CHMC/ SUMC(20,20),VXBDY(20,20),VYBDY(20,20)                F 170
       DIMENSION XNEW(4), YNEW(4), DIST(4)                                F 180
C     ****************************************************************  F 190
      WRITE (6,680) NMOV                                                  F 200
      SUMTCH=SUMT-TIM(N)                                                  F 210
      F1=0.249                                                            F 220
      IF (NPTPND.EQ.5) F1=0.299                                           F 230
      IF (NPTPND.EQ.9) F1=0.333                                           F 240
      CONST1=TIMV*DXINV                                                   F 250
      CONST2=TIMV*DYINV                                                   F 260
C     ---MOVE PARTICLES 'NMOV' TIMES---                                   F 270
      DO 650 IMOV=1,NMOV                                                  F 280
   10 NPTM=NP                                                             F 290
C         ---MOVE EACH PARTICLE---                                        F 300
      DO 590 IN=1,NP                                                      F 310
      IF (PART(1,IN).EQ.0.0) GO TO 590                                    F 320
      KFLAG=0                                                             F 330
C     ****************************************************************  F 340
C             ---COMPUTE OLD LOCATION---                                  F 350
```

FORTRAN IV program listing—Continued

```
      JFLAG=1                                                            F 360
      IF (PART(1,IN).GT.0.0) GO TO 20                                    F 370
      JFLAG=-1                                                           F 380
      PART(1,IN)=-PART(1,IN)                                             F 390
   20 XOLD=PART(1,IN)                                                    F 400
      IX=XOLD+0.5                                                        F 410
      IFLAG=1                                                            F 420
      IF (PART(2,IN).GE.0.0) GO TO 30                                    F 430
      IFLAG=-1                                                           F 440
      PART(2,IN)=-PART(2,IN)                                             F 450
   30 YOLD=PART(2,IN)                                                    F 460
      IY=YOLD+0.5                                                        F 470
      IF (THCK(IX,IY).EQ.0.0) GO TO 560                                  F 480
C     ****************************************************************   F 490
C           ---COMPUTE NEW LOCATION AND LOCATE CLOSEST NODE---           F 500
C           ---LOCATE NORTHWEST CORNER---                                F 510
      IVX=XOLD                                                           F 520
      IVY=YOLD                                                           F 530
      IXE=IVX+1                                                          F 540
      IYS=IVY+1                                                          F 550
C     ****************************************************************   F 560
C        ---LOCATE QUADRANT, VEL. AT 4 CORNERS, CHECK FOR BOUNDARIES---  F 570
      CELDX=XOLD-IX                                                      F 580
      CELDY=YOLD-IY                                                      F 590
      IF (CELDX.EQ.0.0.AND.CELDY.EQ.0.0) GO TO 280                       F 600
      IF (CELDX.GE.0.0.OR.CELDY.GE.0.0) GO TO 70                         F 610
C              ---PT. IN NW QUADRANT---                                  F 620
      VXNW=VXBDY(IVX,IVY)                                                F 630
      VXNE=VX(IXE,IVY)                                                   F 640
      VXSW=VXBDY(IVX,IYS)                                                F 650
      VXSE=VX(IXE,IYS)                                                   F 660
      VYNW=VYBDY(IVX,IVY)                                                F 670
      VYNE=VYBDY(IXE,IVY)                                                F 680
      VYSW=VY(IVX,IYS)                                                   F 690
      VYSE=VY(IXE,IYS)                                                   F 700
```

```
      IF (THCK(IVX,IVY).EQ.0.0) GO TO 50                                F 710
      IF (REC(IXE,IVY).EQ.0.0.AND.VPRM(IXE,IVY).LT.0.09) GO TO 40       F 720
      VXNE=VXNW                                                         F 730
   40 IF (REC(IVX,IYS).EQ.0.0.AND.VPRM(IVX,IYS).LT.0.09) GO TO 50       F 740
      VYSW=VYNW                                                         F 750
   50 IF (REC(IXE,IYS).EQ.0.0.AND.VPRM(IXE,IYS).LT.0.09) GO TO 270      F 760
      IF (THCK(IVX,IYS).EQ.0.0) GO TO 60                                F 770
      VXSE=VXSW                                                         F 780
   60 IF (THCK(IXE,IVY).EQ.0.0) GO TO 270                               F 790
      VYSE=VYNE                                                         F 800
      GO TO 270                                                         F 810
C                                                                       F 820
   70 IF (CELDX.LE.0.0.OR.CELDY.GE.0.0) GO TO 130                       F 830
C              ---PT. IN NE QUADRANT---                                 F 840
   80 VXNW=VX(IVX,IVY)                                                  F 850
      VXNE=VXBDY(IVX,IVY)                                               F 860
      VXSW=VX(IVX,IYS)                                                  F 870
      VXSE=VXBDY(IVX,IYS)                                               F 880
      VYNW=VYBDY(IVX,IVY)                                               F 890
      VYNE=VYBDY(IXE,IVY)                                               F 900
      VYSW=VY(IVX,IYS)                                                  F 910
      VYSE=VY(IXE,IYS)                                                  F 920
      IF (CELDX.EQ.0.0) GO TO 120                                       F 930
      IF (THCK(IXE,IVY).EQ.0.0) GO TO 100                               F 940
      IF (REC(IVX,IVY).EQ.0.0.AND.VPRM(IVX,IVY).LT.0.09) GO TO 90       F 950
      VXNW=VXNE                                                         F 960
   90 IF (REC(IXE,IYS).EQ.0.0.AND.VPRM(IXE,IYS).LT.0.09) GO TO 100      F 970
```

FORTRAN IV program listing—Continued

```
      VYSE=VYNE                                                   F 980
  100 IF (REC(IVX,IYS).EQ.0.0.AND.VPRM(IVX,IYS).LT.0.09) GO TO 270 F 990
      IF (THCK(IXE,IYS).EQ.0.0) GO TO 110                          F1000
      VXSW=VXSE                                                    F1010
  110 IF (THCK(IVX,IVY).EQ.0.0) GO TO 270                          F1020
      VYSW=VYNW                                                    F1030
      GO TO 270                                                    F1040
  120 IF (REC(IVX,IYS).EQ.0.0.AND.VPRM(IVX,IYS).LE.0.09) GO TO 270 F1050
      IF (THCK(IVX,IVY).EQ.0.0) GO TO 270                          F1060
      VYSW=VYNW                                                    F1070
      GO TO 270                                                    F1080
C                                                                  F1090
  130 IF (CELDY.LE.0.0.OR.CELDX.GE.0.0) GO TO 190                  F1100
C                ---PT. IN SW QUADRANT---                          F1110
  140 VXNW=VXBDY(IVX,IVY)                                          F1120
      VXNE=VX(IXE,IVY)                                             F1130
      VXSW=VXBDY(IVX,IYS)                                          F1140
      VXSE=VX(IXE,IYS)                                             F1150
      VYNW=VY(IVX,IVY)                                             F1160
      VYNE=VY(IXE,IVY)                                             F1170
      VYSW=VYBDY(IVX,IVY)                                          F1180
      VYSE=VYBDY(IXE,IVY)                                          F1190
      IF (CELDY.EQ.0.0) GO TO 180                                  F1200
      IF (THCK(IVX,IYS).EQ.0.0) GO TO 160                          F1210
      IF (REC(IVX,IVY).EQ.0.0.AND.VPRM(IVX,IVY).LT.0.09) GO TO 150 F1220
      VYNW=VYSW                                                    F1230
  150 IF (REC(IXE,IYS).EQ.0.0.AND.VPRM(IXE,IYS).LT.0.09) GO TO 160 F1240
      VXSE=VXSW                                                    F1250
  160 IF (REC(IXE,IVY).EQ.0.0.AND.VPRM(IXE,IVY).LT.0.09) GO TO 270 F1260
      IF (THCK(IVX,IVY).EQ.0.0) GO TO 170                          F1270
      VXNE=VXNW                                                    F1280
  170 IF (THCK(IXE,IYS).EQ.0.0) GO TO 270                          F1290
      VYNE=VYSE                                                    F1300
      GO TO 270                                                    F1310
```

```
  180 IF (REC(IXE,IVY).EQ.0.0.AND.VPRM(IXE,IVY).LE.0.09) GO TO 270        F1320
      IF (THCK(IVX,IVY).EQ.0.0) GO TO 270                                  F1330
      VXNE=VXNW                                                            F1340
      GO TO 270                                                            F1350
C                                                                          F1360
  190 IF (CELDY.LE.0.0.OR.CELDX.LE.0.0) GO TO 260                          F1370
C              ---PT. IN SE QUADRANT---                                    F1380
  200 VXNW=VX(IVX,IVY)                                                     F1390
      VXNE=VXBDY(IVX,IVY)                                                  F1400
      VXSW=VX(IVX,IYS)                                                     F1410
      VXSE=VXBDY(IVX,IYS)                                                  F1420
      VYNW=VY(IVX,IVY)                                                     F1430
      VYNE=VY(IXE,IVY)                                                     F1440
      VYSW=VYBDY(IVX,IVY)                                                  F1450
      VYSE=VYBDY(IXE,IVY)                                                  F1460
      IF (CELDY.EQ.0.0) GO TO 240                                          F1470
      IF (CELDX.EQ.0.0) GO TO 250                                          F1480
      IF (THCK(IXE,IYS).EQ.0.0) GO TO 220                                  F1490
      IF (REC(IXE,IVY).EQ.0.0.AND.VPRM(IXE,IVY).LT.0.09) GO TO 210        F1500
      VYNE=VYSE                                                            F1510
  210 IF (REC(IVX,IYS).EQ.0.0.AND.VPRM(IVX,IYS).LT.0.09) GO TO 220        F1520
      VXSW=VXSE                                                            F1530
  220 IF (REC(IVX,IVY).EQ.0.0.AND.VPRM(IVX,IVY).LT.0.09) GO TO 270        F1540
      IF (THCK(IXE,IVY).EQ.0.0) GO TO 230                                  F1550
      VXNW=VXNE                                                            F1560
  230 IF (THCK(IVX,IYS).EQ.0.0) GO TO 270                                  F1570
      VYNW=VYSW                                                            F1580
      GO TO 270                                                            F1590
```

FORTRAN IV program listing—Continued

```
  240 IF (REC(IVX,IVY).EQ.0.0.AND.VPRM(IVX,IVY).LE.0.09) GO TO 270        F1600
      IF (THCK(IXE,IVY).EQ.0.0) GO TO 270                                  F1610
      VXNW=VXNE                                                            F1620
      GO TO 270                                                            F1630
  250 IF (REC(IVX,IVY).EQ.0.0.AND.VPRM(IVX,IVY).LE.0.09) GO TO 270        F1640
      IF (THCK(IVX,IYS).EQ.0.0) GO TO 270                                  F1650
      VYNW=VYSW                                                            F1660
      GO TO 270                                                            F1670
C                                                                          F1680
  260 IF (CELDX.EQ.0.0.AND.CELDY.LT.0.0) GO TO 80                          F1690
      IF (CELDX.LT.0.0.AND.CELDY.EQ.0.0) GO TO 140                         F1700
      IF (CELDX.GT.0.0.AND.CELDY.EQ.0.0) GO TO 200                         F1710
      IF (CELDX.EQ.0.0.AND.CELDY.GT.0.0) GO TO 200                         F1720
      WRITE (6,690) IN,IX,IY                                               F1730
  270 CONTINUE                                                             F1740
C     *****************************************************************   F1750
C        ---BILINEAR INTERPOLATION---                                      F1760
      CELXD=XOLD-IVX                                                       F1770
      CELDXH=AMOD(CELXD,0.5)                                               F1780
      CELDX=CELDXH*2.0                                                     F1790
      CELDY=YOLD-IVY                                                       F1800
C     *****************************************************************   F1810
C           ---X VELOCITY---                                               F1820
      VXN=VXNW*(1.0-CELDX)+VXNE*CELDX                                      F1830
      IF (THCK(IVX,IVY).EQ.0.0.OR.THCK(IXE,IVY).EQ.0.0) VXN=VXNW+VXNE      F1840
      VXS=VXSW*(1.0-CELDX)+VXSE*CELDX                                      F1850
      IF (THCK(IVX,IYS).EQ.0.0.OR.THCK(IXE,IYS).EQ.0.0) VXS=VXSW+VXSE      F1860
      XVEL=VXN*(1.0-CELDY)+VXS*CELDY                                       F1870
      IF (THCK(IVX,IVY).EQ.0.0.AND.THCK(IXE,IVY).EQ.0.0) XVEL=VXS          F1880
      IF (THCK(IVX,IYS).EQ.0.0.AND.THCK(IXE,IYS).EQ.0.0) XVEL=VXN          F1890
C           ---Y VELOCITY---                                               F1900
      CELDYH=AMOD(CELDY,0.5)                                               F1910
      CELDY=CELDYH*2.0                                                     F1920
      VYW=VYNW*(1.0-CELDY)+VYSW*CELDY                                      F1930
```

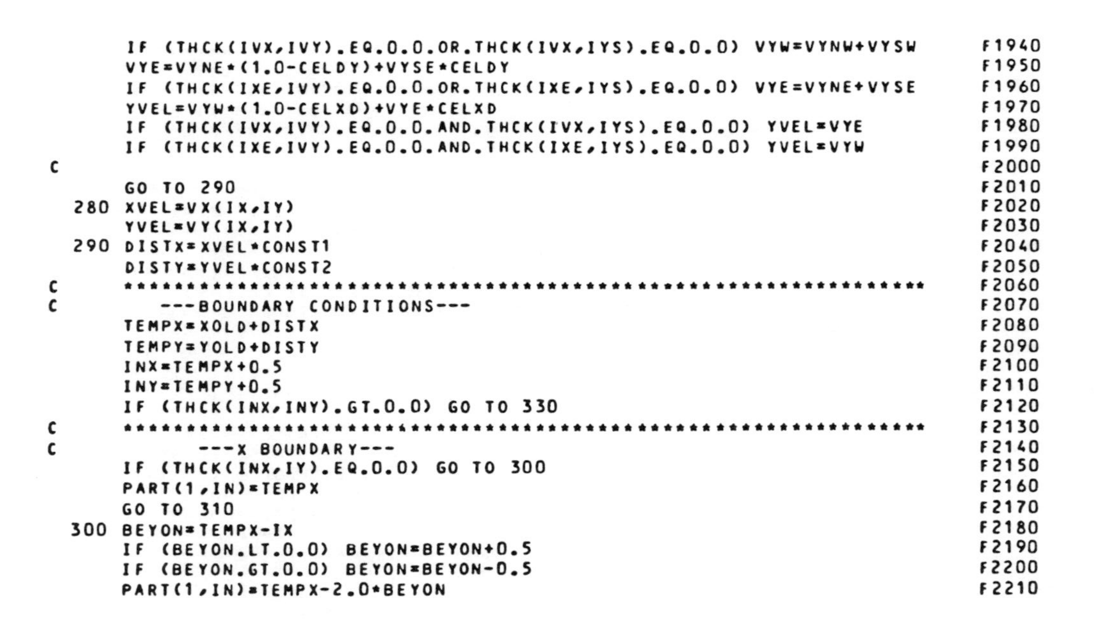

```
      IF (THCK(IVX,IVY).EQ.0.0.OR.THCK(IVX,IYS).EQ.0.0) VYW=VYNW+VYSW      F1940
      VYE=VYNE*(1.0-CELDY)+VYSE*CELDY                                      F1950
      IF (THCK(IXE,IVY).EQ.0.0.OR.THCK(IXE,IYS).EQ.0.0) VYE=VYNE+VYSE      F1960
      YVEL=VYW*(1.0-CELXD)+VYE*CELXD                                       F1970
      IF (THCK(IVX,IVY).EQ.0.0.AND.THCK(IVX,IYS).EQ.0.0) YVEL=VYE          F1980
      IF (THCK(IXE,IVY).EQ.0.0.AND.THCK(IXE,IYS).EQ.0.0) YVEL=VYW          F1990
C                                                                          F2000
      GO TO 290                                                            F2010
  280 XVEL=VX(IX,IY)                                                       F2020
      YVEL=VY(IX,IY)                                                       F2030
  290 DISTX=XVEL*CONST1                                                    F2040
      DISTY=YVEL*CONST2                                                    F2050
C     ******************************************************************   F2060
C        ---BOUNDARY CONDITIONS---                                         F2070
      TEMPX=XOLD+DISTX                                                     F2080
      TEMPY=YOLD+DISTY                                                     F2090
      INX=TEMPX+0.5                                                        F2100
      INY=TEMPY+0.5                                                        F2110
      IF (THCK(INX,INY).GT.0.0) GO TO 330                                  F2120
C     ******************************************************************   F2130
C           ---X BOUNDARY---                                               F2140
      IF (THCK(INX,IY).EQ.0.0) GO TO 300                                   F2150
      PART(1,IN)=TEMPX                                                     F2160
      GO TO 310                                                            F2170
  300 BEYON=TEMPX-IX                                                       F2180
      IF (BEYON.LT.0.0) BEYON=BEYON+0.5                                    F2190
      IF (BEYON.GT.0.0) BEYON=BEYON-0.5                                    F2200
      PART(1,IN)=TEMPX-2.0*BEYON                                           F2210
```

FORTRAN IV program listing—Continued

```
      INX=PART(1,IN)+0.5                                                F2220
      TEMPX=PART(1,IN)                                                  F2230
C     ************************************************************      F2240
C           ---Y BOUNDARY---                                            F2250
  310 IF (THCK(INX,INY).EQ.0.0) GO TO 320                               F2260
      PART(2,IN)=TEMPY                                                  F2270
      GO TO 340                                                         F2280
C     ************************************************************      F2290
  320 BEYON=TEMPY-IY                                                    F2300
      IF (BEYON.LT.0.0) BEYON=BEYON+0.5                                 F2310
      IF (BEYON.GT.0.0) BEYON=BEYON-0.5                                 F2320
      PART(2,IN)=TEMPY-2.0*BEYON                                        F2330
      INY=PART(2,IN)+0.5                                                F2340
      TEMPY=PART(2,IN)                                                  F2350
      GO TO 340                                                         F2360
  330 PART(1,IN)=TEMPX                                                  F2370
      PART(2,IN)=TEMPY                                                  F2380
  340 CONTINUE                                                          F2390
C     ************************************************************      F2400
C         ---SUM CONCENTRATIONS AND COUNT PARTICLES---                  F2410
      SUMC(INX,INY)=SUMC(INX,INY)+PART(3,IN)                            F2420
      NPCELL(INX,INY)=NPCELL(INX,INY)+1                                 F2430
C     ************************************************************      F2440
C         ---CHECK FOR CHANGE IN CELL LOCATION---                       F2450
      IF (IX.EQ.INX.AND.IY.EQ.INY) GO TO 580                            F2460
C           ---CHECK FOR CONST.-HEAD BDY. OR SOURCE AT OLD LOCATION---  F2470
      IF (REC(IX,IY).LT.0.0) GO TO 350                                  F2480
      IF (REC(IX,IY).GT.0.0) GO TO 360                                  F2490
      IF (VPRM(IX,IY).LT.0.09) GO TO 540                                F2500
      IF (WT(IX,IY).GT.HK(IX,IY)) GO TO 350                             F2510
      IF (WT(IX,IY).LT.HK(IX,IY)) GO TO 360                             F2520
      GO TO 540                                                         F2530
C     ************************************************************      F2540
```

```
C              ---CREATE NEW PARTICLES AT BOUNDARIES---                  F2550
  350 IF (IFLAG.GT.0) GO TO 550                                           F2560
      KFLAG=1                                                             F2570
  360 DO 370 IL=1,500                                                     F2580
      IF (LIMBO(IL).EQ.0) GO TO 370                                       F2590
      IP=LIMBO(IL)                                                        F2600
      IF (IP.LT.IN) GO TO 380                                             F2610
  370 CONTINUE                                                            F2620
C     *****************************************************************  F2630
C              ---GENERATE NEW PARTICLE---                                F2640
      IF (NPTM.EQ.NPMAX) GO TO 600                                        F2650
      NPTM=NPTM+1                                                         F2660
      IP=NPTM                                                             F2670
      GO TO 390                                                           F2680
  380 LIMBO(IL)=0                                                         F2690
C                                                                         F2700
  390 IF (KFLAG.EQ.0) GO TO 520                                           F2710
      IF (THCK(IX+1,IY).EQ.0.0.OR.THCK(IX-1,IY).EQ.0.0.OR.THCK(IX,IY+1).  F2720
     1EQ.0.0.OR.THCK(IX,IY-1).EQ.0.0) GO TO 520                           F2730
      IF (THCK(IX+1,IY+1).EQ.0.0.OR.THCK(IX+1,IY-1).EQ.0.0.OR.THCK(IX-1,  F2740
     1IY+1).EQ.0.0.OR.THCK(IX-1,IY-1).EQ.0.0) GO TO 520                   F2750
C                 ---IF CENTER SOURCE---                                  F2760
      IF (JFLAG.LT.0) GO TO 500                                           F2770
      JJ=4                                                                F2780
      AN=TEMPY-YOLD                                                       F2790
      AD=TEMPX-XOLD                                                       F2800
      DISTMV=SQRT((AD*AD)+(AN*AN))                                        F2810
      IF (AD.EQ.0.0) GO TO 410                                            F2820
      SLOPE=AN/AD                                                         F2830
```

FORTRAN IV program listing—Continued

```
      BI=YOLD-SLOPE*XOLD                                                F2840
      XC1=IX-F1                                                         F2850
      XC2=IX+F1                                                         F2860
      YC1=IY-F1                                                         F2870
      YC2=IY+F1                                                         F2880
C               ---COMPUTE NEW COORDINATES AND VERIFY---                F2890
      DO 400  IK=1,4                                                    F2900
      YNEW(IK)=0.0                                                      F2910
      XNEW(IK)=0.0                                                      F2920
  400 DIST(IK)=0.0                                                      F2930
      YNEW(1)=(SLOPE*XC1)+BI                                            F2940
      XNEW(1)=XC1                                                       F2950
      YNEW(2)=(SLOPE*XC2)+BI                                            F2960
      XNEW(2)=XC2                                                       F2970
      IF (SLOPE.EQ.0.0) GO TO 420                                       F2980
      YNEW(3)=YC1                                                       F2990
      XNEW(3)=(YC1-BI)/SLOPE                                            F3000
      YNEW(4)=YC2                                                       F3010
      XNEW(4)=(YC2-BI)/SLOPE                                            F3020
      GO TO 430                                                         F3030
  410 YNEW(1)=IY-F1                                                     F3040
      XNEW(1)=XOLD                                                      F3050
      YNEW(2)=IY+F1                                                     F3060
      XNEW(2)=XOLD                                                      F3070
  420 JJ=2                                                              F3080
  430 DO 440 II=1,JJ                                                    F3090
  440 DIST(II)=SQRT((XNEW(II)-TEMPX)**2+(YNEW(II)-TEMPY)**2)*1.00001    F3100
      IACC=0                                                            F3110
      DISTCK=2.0                                                        F3120
      DO 460 IG=1,JJ                                                    F3130
      IF (DIST(IG).GE.DISTMV.AND.DIST(IG).LT.DISTCK) GO TO 450          F3140
      GO TO 460                                                         F3150
  450 IXC=XNEW(IG)+0.50                                                 F3160
      IYC=YNEW(IG)+0.50                                                 F3170
```

```
      IF (IXC.NE.IX.OR.IYC.NE.IY) GO TO 460                            F3180
      IACC=IG                                                          F3190
      DISTCK=DIST(IG)                                                  F3200
  460 CONTINUE                                                         F3210
      IF (IACC.LT.1.OR.IACC.GT.4) GO TO 510                            F3220
      IF (XNEW(IACC).EQ.XC1.OR.XNEW(IACC).EQ.XC2) GO TO 470            F3230
      IF (YNEW(IACC).EQ.YC1.OR.YNEW(IACC).EQ.YC2) GO TO 480            F3240
      GO TO 510                                                        F3250
  470 IF (YNEW(IACC).LT.YC1) YNEW(IACC)=YC1                            F3260
      IF (YNEW(IACC).GT.YC2) YNEW(IACC)=YC2                            F3270
      GO TO 490                                                        F3280
  480 IF (XNEW(IACC).LT.XC1) XNEW(IACC)=XC1                            F3290
      IF (XNEW(IACC).GT.XC2) XNEW(IACC)=XC2                            F3300
  490 PART(1,IP)=XNEW(IACC)                                            F3310
      PART(2,IP)=YNEW(IACC)                                            F3320
      GO TO 530                                                        F3330
  500 PART(1,IP)=-IX                                                   F3340
      PART(2,IP)=IY                                                    F3350
      GO TO 530                                                        F3360
  510 PART(1,IP)=XOLD                                                  F3370
      PART(2,IP)=YOLD                                                  F3380
      GO TO 530                                                        F3390
C                ---IF EDGE SOURCE OR SINK---                          F3400
C                    ---X POSITION---                                  F3410
  520 DLX=INX-IX                                                       F3420
      PART(1,IP)=TEMPX-DLX                                             F3430
C                    ---Y POSITION---                                  F3440
      DLY=INY-IY                                                       F3450
```

FORTRAN IV program listing—Continued

```
      PART(2,IP)=TEMPY-DLY                                              F3460
      IF (KFLAG.GT.0) GO TO 530                                         F3470
C              ---IF SINK---                                            F3480
      SUMC(IX,IY)=SUMC(IX,IY)+CONC(IX,IY)                               F3490
      NPCELL(IX,IY)=NPCELL(IX,IY)+1                                     F3500
C                                                                       F3510
  530 PART(2,IP)=-PART(2,IP)                                            F3520
      PART(3,IP)=CONC(IX,IY)                                            F3530
      IF (REC(IX,IY).EQ.0.0) GO TO 540                                  F3540
C     *****************************************************************  F3550
C           ---CHECK FOR DISCHARGE BOUNDARY AT NEW LOCATION---          F3560
  540 IFLAG=1.0                                                         F3570
  550 IF (VPRM(INX,INY).GT.0.09.AND.WT(INX,INY).LT.HK(INX,INY)) GO TO 56 F3580
     10                                                                 F3590
      IF (REC(INX,INY).GT.0.0) GO TO 560                                F3600
      GO TO 590                                                         F3610
C     *****************************************************************  F3620
C           ---PUT PT. IN LIMBO---                                      F3630
  560 PART(1,IN)=0.0                                                    F3640
      PART(2,IN)=0.0                                                    F3650
      PART(3,IN)=0.0                                                    F3660
      DO 570 ID=1,500                                                   F3670
      IF (LIMBO(ID).GT.0) GO TO 570                                     F3680
      LIMBO(ID)=IN                                                      F3690
      GO TO 590                                                         F3700
  570 CONTINUE                                                          F3710
C                                                                       F3720
  580 IF (IFLAG.LT.0) PART(2,IN)=-TEMPY                                 F3730
      IF (JFLAG.LT.0) PART(1,IN)=-TEMPX                                 F3740
  590 CONTINUE                                                          F3750
C     ---END OF LOOP---                                                 F3760
C     *****************************************************************  F3770
```

```
      GO TO 620                                                         F3780
C         ---RESTART MOVE IF PT. LIMIT EXCEEDED---                      F3790
  600 WRITE (6,700) IMOV,IN                                             F3800
      TEST=100.0                                                        F3810
      CALL GENPT                                                        F3820
      DO 610 IX=1,NX                                                    F3830
      DO 610 IY=1,NY                                                    F3840
      SUMC(IX,IY)=0.0                                                   F3850
  610 NPCELL(IX,IY)=0                                                   F3860
      TEST=0.0                                                          F3870
      GO TO 10                                                          F3880
C     *****************************************************************F3890
  620 SUMTCH=SUMTCH+TIMV                                                F3900
C         ---ADJUST NUMBER OF PARTICLES---                              F3910
      NP=NPTM                                                           F3920
      WRITE (6,670) NP,IMOV                                             F3930
C     *****************************************************************F3940
      CALL CNCON                                                        F3950
C     *****************************************************************F3960
C         ---STORE OBS. WELL DATA FOR STEADY FLOW PROBLEMS---           F3970
      IF (S.GT.0.0) GO TO 640                                           F3980
      IF (NUMOBS.LE.0) GO TO 640                                        F3990
      J=MOD(IMOV,50)                                                    F4000
      IF (J.EQ.0) J=50                                                  F4010
      TMOBS(J)=SUMTCH                                                   F4020
      DO 630 I=1,NUMOBS                                                 F4030
      TMWL(I,J)=HK(IXOBS(I),IYOBS(I))                                   F4040
  630 TMCN(I,J)=CONC(IXOBS(I),IYOBS(I))                                 F4050
C         ---PRINT CHEMICAL OUTPUT---                                   F4060
  640 IF (IMOV.GE.NMOV) GO TO 660                                       F4070
```

FORTRAN IV program listing—Continued

```
  650 IF (MOD(IMOV,NPNTMV).EQ.0.OR.MOD(IMOV,50).EQ.0) CALL CHMOT          F4080
C     ******************************************************************  F4090
  660 RETURN                                                               F4100
C     ******************************************************************  F4110
C                                                                          F4120
C                                                                          F4130
C                                                                          F4140
C                                                                          F4150
  670 FORMAT (1HC,2X,2HNP,7X,2H= ,8X,I4,10X,11HIMOV      = ,8X,I4)         F4160
  680 FORMAT (1H0,10X,61HNO. OF PARTICLE MOVES REQUIRED TO COMPLETE THIS   F4170
     1 TIME STEP  = ,I4//)                                                 F4180
  690 FORMAT (1H0,5X,53H*** WARNING ***     QUADRANT NOT LOCATED FOR PT.   F4190
     1 NO. ,I5,11H , IN CELL ,2I4)                                         F4200
  700 FORMAT (1H0,5X,17H ***   NOTE   ***,10X,23HNPTM.EQ.NPMAX --- IMOV=   F4210
     1,I4,5X,8HPT. NO.=,I4,5X,10HCALL GENPT/)                              F4220-
      END                                                                  G   10
      SUBROUTINE CNCON                                                     G   20
      REAL *8TMRX,VPRM,HI,HR,HC,HK,WT,REC,RECH,TIM,AOPT,TITLE              G   30
      REAL *8XDEL,YDEL,S,AREA,SUMT,RHO,PARAM,TEST,TOL,PINT,HMIN,PYR        G   40
      REAL *8FLW                                                           G   50
      COMMON /PRMI/ NTIM,NPMP,NPNT,NITP,N,NX,NY,NP,NREC,INT,NNX,NNY,NUMO   G   60
     1BS,NMOV,IMOV,NPMAX,ITMAX,NZCRIT,IPRNT,NPTPND,NPNTMV,NPNTVL,NPNTD,N   G   70
     2PNCHV,NPDELC                                                         G   80
      COMMON /PRMK/ NODEID(20,20),NPCELL(20,20),LIMBO(500),IXOBS(5),IYOB   G   90
     1S(5)                                                                 G  100
      COMMON /HEDA/ THCK(20,20),PERM(20,20),TMWL(5,50),TMOBS(50),ANFCTR    G  110
      COMMON /HEDB/ TMRX(20,20,2),VPRM(20,20),HI(20,20),HR(20,20),HC(20,   G  120
     120),HK(20,20),WT(20,20),REC(20,20),RECH(20,20),TIM(100),AOPT(20),T   G  130
     2ITLE(10),XDEL,YDEL,S,AREA,SUMT,RHO,PARAM,TEST,TOL,PINT,HMIN,PYR      G  140
      COMMON /XINV/ DXINV,DYINV,ARINV,PORINV                               G  150
      COMMON /CHMA/ PART(3,3200),CONC(20,20),TMCN(5,50),VX(20,20),VY(20,   G  160
     120),CONINT(20,20),CNRECH(20,20),POROS,SUMTCH,BETA,TIMV,STORM,STORM   G  170
     2I,CMSIN,CMSOUT,FLMIN,FLMOT,SUMIO,CELDIS,DLTRAT,CSTORM                G  180
      COMMON /DIFUS/ DISP(20,20,4)                                         G  190
      COMMON /CHMC/ SUMC(20,20),VXBDY(20,20),VYBDY(20,20)                  G  200
      DIMENSION CNCNC(20,20), CNOLD(20,20)
```

```
C     *****************************************************************        G 210
      ITEST=0                                                                  G 220
      DO 10 IX=1,NX                                                            G 230
      DO 10 IY=1,NY                                                            G 240
      CNOLD(IX,IY)=CONC(IX,IY)                                                 G 250
   10 CNCNC(IX,IY)=0.0                                                         G 260
      APC=0.0                                                                  G 270
      NZERO=0                                                                  G 280
      TVA=AREA*TIMV                                                            G 290
      ARPOR=AREA*POROS                                                         G 300
C     *****************************************************************        G 310
C     ---CONC. CHANGE FOR 0.5*TIMV DUE TO:                                     G 320
C           RECHARGE, PUMPING, LEAKAGE, DIVERGENCE OF VELOCITY...              G 330
      CONST=0.5*TIMV                                                           G 340
   20 DO 60 IX=1,NX                                                            G 350
      DO 60 IY=1,NY                                                            G 360
      IF (THCK(IX,IY).EQ.0.0) GO TO 60                                         G 370
      EQFCT1=CONST/THCK(IX,IY)                                                 G 380
      EQFCT2=EQFCT1/POROS                                                      G 390
      C1=CONC(IX,IY)                                                           G 400
      CLKCN=0.0                                                                G 410
      SLEAK=(HK(IX,IY)-WT(IX,IY))*VPRM(IX,IY)                                  G 420
      IF (SLEAK.LT.0.0) CLKCN=CNRECH(IX,IY)                                    G 430
      IF (SLEAK.GT.0.0) CLKCN=C1                                               G 440
      CNREC=C1                                                                 G 450
      RATE=REC(IX,IY)*ARINV                                                    G 460
      IF (RATE.LT.0.0) CNREC=CNRECH(IX,IY)                                     G 470
```

FORTRAN IV program listing—Continued

```
      DIV=RATE+SLEAK+RECH(IX,IY)                                        G 480
      IF (S.EQ.0.0) GO TO 30                                            G 490
      DERH=(HK(IX,IY)-HR(IX,IY))/TIM(N)                                 G 500
      DIV=DIV+S*DERH                                                    G 510
      IF (S.LT.0.005) GO TO 30                                          G 520
C     ...NOTE: ABOVE STATEMENT ASSUMES THAT S=0.005 SEPARATES CONFINED  G 530
C              FROM UNCONFINED CONDITIONS; THIS CRITERION SHOULD BE     G 540
C              CHANGED IF FIELD CONDITIONS ARE DIFFERENT.               G 550
      DELC=EQFCT2*(C1*(DIV-POROS*DERH)-RATE*CNREC-SLEAK*CLKCN-RECH(IX,IY G 560
     1)*CNRECH(IX,IY))                                                  G 570
      GO TO 40                                                          G 580
   30 DELC=EQFCT2*(C1*DIV-RATE*CNREC-SLEAK*CLKCN-RECH(IX,IY)*CNRECH(IX,I G 590
     1Y))                                                               G 600
   40 CNCNC(IX,IY)=CNCNC(IX,IY)+DELC                                    G 610
C     ---CONC. CHANGE DUE TO DISPERSION FOR 0.5*TIMV---                 G 620
C        ---DISPERSION WITH TENSOR COEFFICIENTS---                      G 630
      IF (BETA.EQ.0.0) GO TO 50                                         G 640
      X1=DISP(IX,IY,1)*(CONC(IX+1,IY)-C1)                               G 650
      X2=DISP(IX-1,IY,1)*(CONC(IX-1,IY)-C1)                             G 660
      Y1=DISP(IX,IY,2)*(CONC(IX,IY+1)-C1)                               G 670
      Y2=DISP(IX,IY-1,2)*(CONC(IX,IY-1)-C1)                             G 680
      XX1=DISP(IX,IY,3)*(CONC(IX,IY+1)+CONC(IX+1,IY+1)-CONC(IX,IY-1)-CON G 690
     1C(IX+1,IY-1))                                                     G 700
      XX2=DISP(IX-1,IY,3)*(CONC(IX,IY+1)+CONC(IX-1,IY+1)-CONC(IX,IY-1)-C G 710
     1ONC(IX-1,IY-1))                                                   G 720
      YY1=DISP(IX,IY,4)*(CONC(IX+1,IY)+CONC(IX+1,IY+1)-CONC(IX-1,IY)-CON G 730
     1C(IX-1,IY+1))                                                     G 740
      YY2=DISP(IX,IY-1,4)*(CONC(IX+1,IY)+CONC(IX+1,IY-1)-CONC(IX-1,IY)-C G 750
     1ONC(IX-1,IY-1))                                                   G 760
   50 CNCNC(IX,IY)=CNCNC(IX,IY)+EQFCT1*(X1+X2+Y1+Y2+XX1-XX2+YY1-YY2)    G 770
   60 CONTINUE                                                          G 780
C     ****************************************************************  G 790
      ITEST=ITEST+1                                                     G 800
      IF (ITEST.EQ.1) GO TO 70                                          G 810
      GO TO 110                                                         G 820
```

```
C     ****************************************************************** G 830
C     ---CONC. CHANGE AT NODES DUE TO CONVECTION---                      G 840
   70 DO 90 IX=1,NX                                                      G 850
      DO 90 IY=1,NY                                                      G 860
      IF (THCK(IX,IY).EQ.0.0) GO TO 90                                   G 870
      APC=NPCELL(IX,IY)                                                  G 880
      IF (APC.GT.0.0) GO TO 80                                           G 890
      IF (REC(IX,IY).NE.0.0.OR.VPRM(IX,IY).GT.0.09) GO TO 90             G 900
      NZERO=NZERO+1                                                      G 910
      GO TO 90                                                           G 920
   80 CONC(IX,IY)=SUMC(IX,IY)/APC                                        G 930
   90 CONTINUE                                                           G 940
C          ---CHECK NUMBER OF CELLS VOID OF PTS.---                      G 950
      IF (NZERO.GT.0) WRITE (6,290) NZERO,IMOV                           G 960
      IF (NZERO.LE.NZCRIT) GO TO 20                                      G 970
      TEST=99.0                                                          G 980
      WRITE (6,300)                                                      G 990
      WRITE (6,320)                                                      G1000
      DO 100 IY=1,NY                                                     G1010
  100 WRITE (6,330) (NPCELL(IX,IY),IX=1,NX)                              G1020
      GO TO 20                                                           G1030
C     ****************************************************************** G1040
C     ---CHANGE CONCENTRATIONS AT NODES---                               G1050
  110 DO 130 IX=1,NX                                                     G1060
      DO 130 IY=1,NY                                                     G1070
      IF (THCK(IX,IY).EQ.0.0) GO TO 120                                  G1080
      CONC(IX,IY)=CONC(IX,IY)+CNCNC(IX,IY)                               G1090
```

FORTRAN IV program listing—Continued

```
      NPCELL(IX,IY)=0                                                  G1100
      SUMC(IX,IY)=0.0                                                  G1110
      IF (CONC(IX,IY).LE.0.0) GO TO 130                                G1120
      CNCPCT=CNCNC(IX,IY)/CONC(IX,IY)                                  G1130
      SUMC(IX,IY)=CNCPCT                                               G1140
      GO TO 130                                                        G1150
  120 IF (CONC(IX,IY).GT.0.0) WRITE (6,310) IX,IY,CONC(IX,IY)          G1160
      CONC(IX,IY)=0.0                                                  G1170
  130 CONTINUE                                                         G1180
C     ***************************************************************  G1190
C     ---CHANGE CONCENTRATION OF PARTICLES---                          G1200
      DO 180 IN=1,NP                                                   G1210
      IF (PART(1,IN).EQ.0.0) GO TO 180                                 G1220
      INX=ABS(PART(1,IN))+0.5                                          G1230
      INY=ABS(PART(2,IN))+0.5                                          G1240
C     ---UPDATE CONC. OF PTS. IN SINK/SOURCE CELLS---                  G1250
      IF (REC(INX,INY).NE.0.0) GO TO 140                               G1260
      IF (VPRM(INX,INY).LE.0.09) GO TO 150                             G1270
  140 PART(3,IN)=CONC(INX,INY)                                         G1280
      GO TO 180                                                        G1290
  150 IF (CNCNC(INX,INY).LT.0.0) GO TO 170                             G1300
  160 PART(3,IN)=PART(3,IN)+CNCNC(INX,INY)                             G1310
      GO TO 180                                                        G1320
  170 IF (CONC(INX,INY).LE.0.0) GO TO 160                              G1330
      IF (SUMC(INX,INY).LT.-1.0) GO TO 160                             G1340
      PART(3,IN)=PART(3,IN)+PART(3,IN)*SUMC(INX,INY)                   G1350
  180 CONTINUE                                                         G1360
      WRITE (6,280) TIM(N),TIMV,SUMTCH                                 G1370
C     ***************************************************************  G1380
C     ---COMPUTE MASS BALANCE FOR SOLUTE---                            G1390
      CSTORM=0.0                                                       G1400
      STORM=0.0                                                        G1410
      DO 270 IX=1,NX                                                   G1420
      DO 270 IY=1,NY                                                   G1430
      IF (THCK(IX,IY).EQ.0.0) GO TO 270                                G1440
```

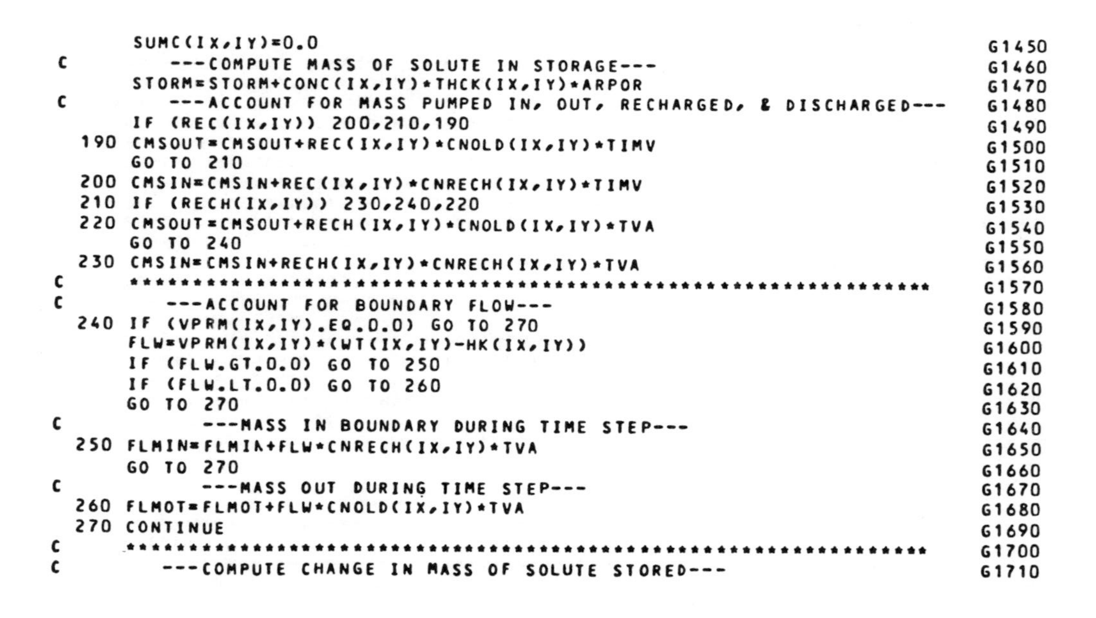

```
      SUMC(IX,IY)=0.0                                                   G1450
C         ---COMPUTE MASS OF SOLUTE IN STORAGE---                       G1460
      STORM=STORM+CONC(IX,IY)*THCK(IX,IY)*ARPOR                         G1470
C         ---ACCOUNT FOR MASS PUMPED IN, OUT, RECHARGED, & DISCHARGED---G1480
      IF (REC(IX,IY)) 200,210,190                                       G1490
  190 CMSOUT=CMSOUT+REC(IX,IY)*CNOLD(IX,IY)*TIMV                        G1500
      GO TO 210                                                         G1510
  200 CMSIN=CMSIN+REC(IX,IY)*CNRECH(IX,IY)*TIMV                         G1520
  210 IF (RECH(IX,IY)) 230,240,220                                      G1530
  220 CMSOUT=CMSOUT+RECH(IX,IY)*CNOLD(IX,IY)*TVA                        G1540
      GO TO 240                                                         G1550
  230 CMSIN=CMSIN+RECH(IX,IY)*CNRECH(IX,IY)*TVA                         G1560
C     ****************************************************************  G1570
C         ---ACCOUNT FOR BOUNDARY FLOW---                               G1580
  240 IF (VPRM(IX,IY).EQ.0.0) GO TO 270                                 G1590
      FLW=VPRM(IX,IY)*(WT(IX,IY)-HK(IX,IY))                             G1600
      IF (FLW.GT.0.0) GO TO 250                                         G1610
      IF (FLW.LT.0.0) GO TO 260                                         G1620
      GO TO 270                                                         G1630
C             ---MASS IN BOUNDARY DURING TIME STEP---                   G1640
  250 FLMIN=FLMIN+FLW*CNRECH(IX,IY)*TVA                                 G1650
      GO TO 270                                                         G1660
C             ---MASS OUT DURING TIME STEP---                           G1670
  260 FLMOT=FLMOT+FLW*CNOLD(IX,IY)*TVA                                  G1680
  270 CONTINUE                                                          G1690
C     ****************************************************************  G1700
C         ---COMPUTE CHANGE IN MASS OF SOLUTE STORED---                 G1710
```

FORTRAN IV program listing—Continued

```
      CSTORM=STORM-STORMI                                                 G1720
      SUMIO=FLMIN+FLMOT-CMSIN-CMSOUT                                      G1730
C     ****************************************************************    G1740
C     ---REGENERATE PARTICLES IF 'NZCRIT' EXCEEDED---                     G1750
      IF (TEST.GT.98.0) CALL GENPT                                        G1760
      TEST=0.0                                                            G1770
C     ****************************************************************    G1780
      RETURN                                                              G1790
C     ****************************************************************    G1800
C                                                                         G1810
C                                                                         G1820
C                                                                         G1830
  280 FORMAT (3H    ,11HTIM(N)    = ,1G12.5,10X,11HTIMV      = ,1G12.5,10X,  G1840
     19HSUMTCH = ,G12.5)                                                  G1850
  290 FORMAT (1H0,5X,40HNUMBER OF CELLS WITH ZERO PARTICLES   =   ,I4,5X,9  G1860
     1HIMOV . =   ,I4/)                                                   G1870
  300 FORMAT (1H0,5X,44H***    NZCRIT EXCEEDED  ---  CALL GENPT    ***/)   G1880
  310 FORMAT (1H ,5X,37H***CONC.GT.0.AND.THCK.EQ.0 AT NODE = ,2I4,4X,7HC  G1890
     1ONC = ,G10.4,4H ***)                                                G1900
  320 FORMAT (1H0,2X,6HNPCELL/)                                           G1910
  330 FORMAT (1H ,4X,20I3)                                                G1920
      END                                                                 G1930-
      SUBROUTINE OUTPT                                                    H  10
      REAL *8TMRX,VPRM,HI,HR,HC,HK,WT,REC,RECH,TIM,AOPT,TITLE             H  20
      REAL *8XDEL,YDEL,S,AREA,SUMT,RHO,PARAM,TEST,TOL,PINT,HMIN,PYR       H  30
      COMMON /PRMI/ NTIM,NPMP,NPNT,NITP,N,NX,NY,NP,NREC,INT,NNX,NNY,NUMO  H  40
     1BS,NMOV,IMOV,NPMAX,ITMAX,NZCRIT,IPRNT,NPTPND,NPNTMV,NPNTVL,NPNTD,N  H  50
     2PNCHV,NPDELC                                                        H  60
      COMMON /PRMK/ NODEID(20,20),NPCELL(20,20),LIMBO(500),IXOBS(5),IYOB  H  70
     1S(5)                                                                H  80
      COMMON /HEDA/ THCK(20,20),PERM(20,20),TMWL(5,50),TMOBS(50),ANFCTR   H  90
      COMMON /HEDB/ TMRX(20,20,2),VPRM(20,20),HI(20,20),HR(20,20),HC(20,  H 100
     120),HK(20,20),WT(20,20),REC(20,20),RECH(20,20),TIM(100),AOPT(20),T  H 110
     2ITLE(10),XDEL,YDEL,S,AREA,SUMT,RHO,PARAM,TEST,TOL,PINT,HMIN,PYR     H 120
```

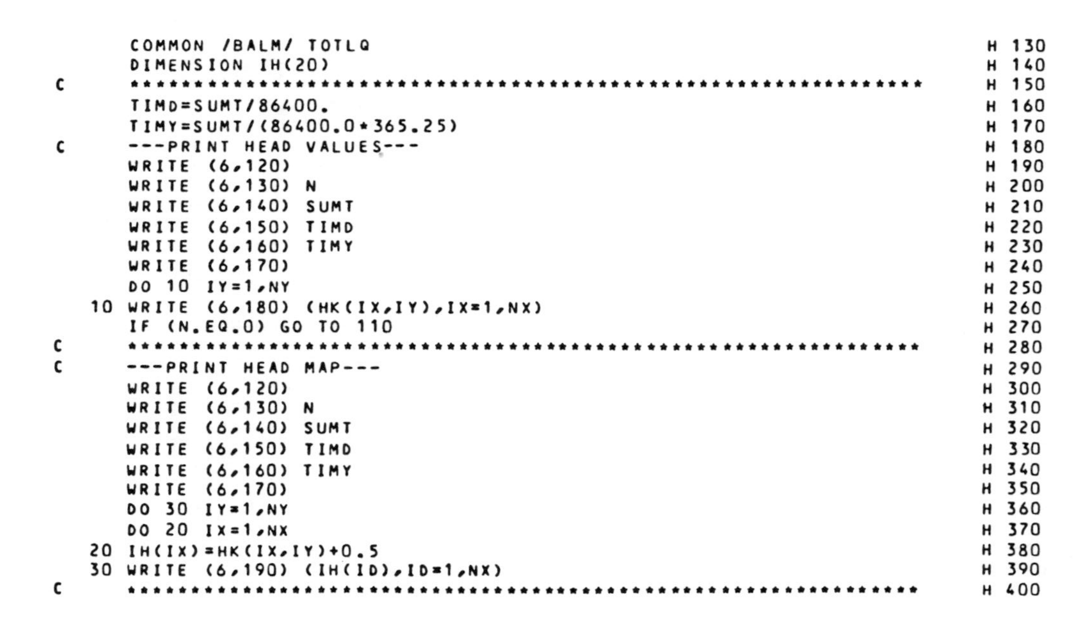

```
      COMMON /BALM/ TOTLQ                                               H 130
      DIMENSION IH(20)                                                  H 140
C     ******************************************************************H 150
      TIMD=SUMT/86400.                                                  H 160
      TIMY=SUMT/(86400.0*365.25)                                        H 170
C     ---PRINT HEAD VALUES---                                           H 180
      WRITE (6,120)                                                     H 190
      WRITE (6,130) N                                                   H 200
      WRITE (6,140) SUMT                                                H 210
      WRITE (6,150) TIMD                                                H 220
      WRITE (6,160) TIMY                                                H 230
      WRITE (6,170)                                                     H 240
      DO 10 IY=1,NY                                                     H 250
   10 WRITE (6,180) (HK(IX,IY),IX=1,NX)                                 H 260
      IF (N.EQ.0) GO TO 110                                             H 270
C     ******************************************************************H 280
C     ---PRINT HEAD MAP---                                              H 290
      WRITE (6,120)                                                     H 300
      WRITE (6,130) N                                                   H 310
      WRITE (6,140) SUMT                                                H 320
      WRITE (6,150) TIMD                                                H 330
      WRITE (6,160) TIMY                                                H 340
      WRITE (6,170)                                                     H 350
      DO 30 IY=1,NY                                                     H 360
      DO 20 IX=1,NX                                                     H 370
   20 IH(IX)=HK(IX,IY)+0.5                                              H 380
   30 WRITE (6,190) (IH(ID),ID=1,NX)                                    H 390
C     ******************************************************************H 400
```

FORTRAN IV program listing—Continued

```
C        ---COMPUTE WATER BALANCE AND DRAWDOWN---                          H 410
         QSTR=0.0                                                          H 420
         PUMP=0.0                                                          H 430
         TPUM=0.0                                                          H 440
         QIN=0.0                                                           H 450
         QOUT=0.0                                                          H 460
         QNET=0.0                                                          H 470
         DELQ=0.0                                                          H 480
         JCK=0                                                             H 490
         PCTERR=0.0                                                        H 500
         WRITE (6,290)                                                     H 510
C                                                                          H 520
         DO 80 IY=1,NY                                                     H 530
         DO 70 IX=1,NX                                                     H 540
         IH(IX)=0.0                                                        H 550
         IF (THCK(IX,IY).EQ.0.0) GO TO 70                                  H 560
         TPUM=REC(IX,IY)+RECH(IX,IY)*AREA+TPUM                             H 570
         IF (VPRM(IX,IY).EQ.0.0) GO TO 60                                  H 580
         DELQ=VPRM(IX,IY)*AREA*(WT(IX,IY)-HK(IX,IY))                       H 590
         IF (DELQ.GT.0.0) GO TO 40                                         H 600
         QOUT=QOUT+DELQ                                                    H 610
         GO TO 50                                                          H 620
   40    QIN=QIN+DELQ                                                      H 630
   50    QNET=QNET+DELQ                                                    H 640
   60    DDRW=HI(IX,IY)-HK(IX,IY)                                          H 650
         IH(IX)=DDRW+0.5                                                   H 660
         QSTR=QSTR+DDRW*AREA*S                                             H 670
   70    CONTINUE                                                          H 680
C           ---PRINT DRAWDOWN MAP---                                       H 690
         WRITE (6,300) (IH(IX),IX=1,NX)                                    H 700
   80    CONTINUE                                                          H 710
         PUMP=TPUM*SUMT                                                    H 720
         DELS=-QSTR/SUMT                                                   H 730
         ERRMB=PUMP-TOTLQ-QSTR                                             H 740
```

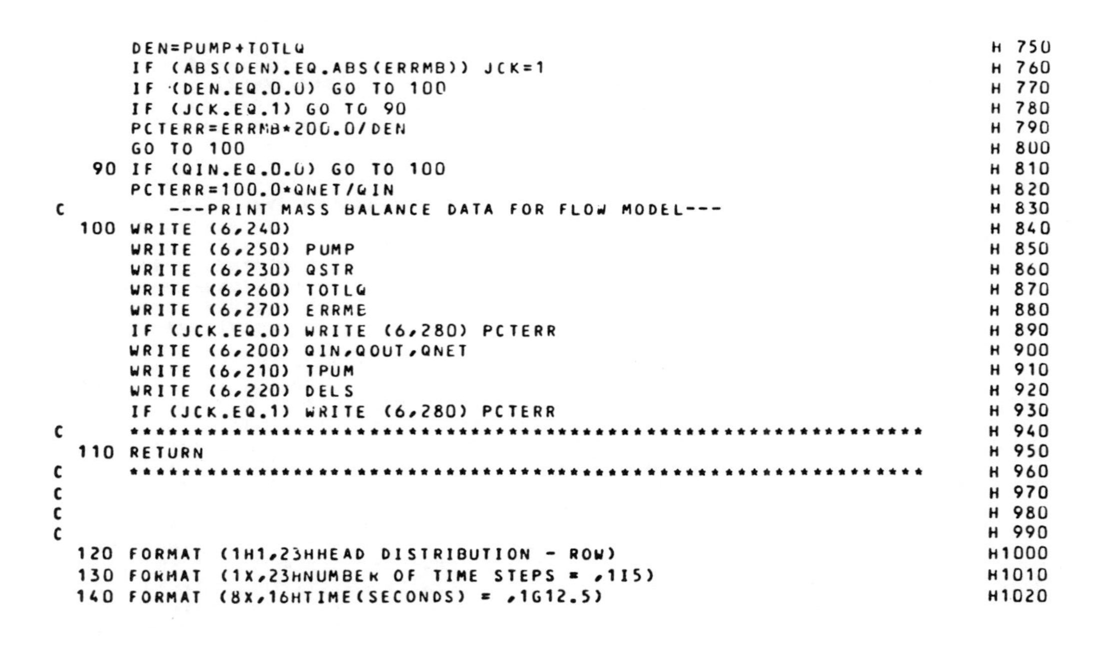

```
      DEN=PUMP+TOTLQ                                                    H 750
      IF (ABS(DEN).EQ.ABS(ERRMB)) JCK=1                                 H 760
      IF (DEN.EQ.0.0) GO TO 100                                         H 770
      IF (JCK.EQ.1) GO TO 90                                            H 780
      PCTERR=ERRMB*200.0/DEN                                            H 790
      GO TO 100                                                         H 800
   90 IF (QIN.EQ.0.0) GO TO 100                                         H 810
      PCTERR=100.0*QNET/QIN                                             H 820
C         ---PRINT MASS BALANCE DATA FOR FLOW MODEL---                  H 830
  100 WRITE (6,240)                                                     H 840
      WRITE (6,250) PUMP                                                H 850
      WRITE (6,230) QSTR                                                H 860
      WRITE (6,260) TOTLQ                                               H 870
      WRITE (6,270) ERRME                                               H 880
      IF (JCK.EQ.0) WRITE (6,280) PCTERR                                H 890
      WRITE (6,200) QIN,QOUT,QNET                                       H 900
      WRITE (6,210) TPUM                                                H 910
      WRITE (6,220) DELS                                                H 920
      IF (JCK.EQ.1) WRITE (6,280) PCTERR                                H 930
C     ****************************************************************  H 940
  110 RETURN                                                            H 950
C     ****************************************************************  H 960
C                                                                       H 970
C                                                                       H 980
C                                                                       H 990
  120 FORMAT (1H1,23HHEAD DISTRIBUTION - ROW)                           H1000
  130 FORMAT (1X,23HNUMBER OF TIME STEPS = ,I15)                        H1010
  140 FORMAT (8X,16HTIME(SECONDS) = ,1G12.5)                            H1020
```

FORTRAN IV program listing—Continued

```
150 FORMAT (8X,16HTIME(DAYS)     = ,1E12.5)                                H1030
160 FORMAT (8X,16HTIME(YEARS)    = ,1E12.5)                                H1040
170 FORMAT (1H )                                                           H1050
180 FORMAT (1H0,10F12.7/10F12.7)                                           H1060
190 FORMAT (1H0,20I4)                                                      H1070
200 FORMAT (1H0,2X,33HRATE MASS BALANCE -- (IN C.F.S.) //10X,8HQIN   =     H1080
   1 ,G12.5/10X,8HQOUT =   ,G12.5/10X,8HQNET =   ,G12.5/)                  H1090
210 FORMAT (1H ,17X,8HTPUM =   ,G12.5)                                     H1100
220 FORMAT (1H ,17X,8HDELS =   ,G12.5/)                                    H1110
230 FORMAT (4X,29HWATER RELEASE FROM STORAGE = ,1E12.5)                    H1120
240 FORMAT (1H0,2X,23HCUMULATIVE MASS BALANCE//)                           H1130
250 FORMAT (4X,29HCUMULATIVE  NET  PUMPAGE    = ,1E12.5)                   H1140
260 FORMAT (4X,29HCUMULATIVE  NET  LEAKAGE    = ,1E12.5)                   H1150
270 FORMAT (1H0,7X,25HMASS BALANCE RESIDUAL  = ,G12.5)                     H1160
280 FORMAT (1H ,7X,25HERROR  (AS PERCENT)    = ,G12.5/)                    H1170
290 FORMAT (1H1,8HDRAWDOWN)                                                H1180
300 FORMAT (3H   ,20I5)                                                    H1190
    END                                                                    H1200-
    SUBROUTINE CHMOT                                                       I  10
    REAL *8TMRX,VPRM,HI,HR,HC,HK,WT,REC,RECH,TIM,AOPT,TITLE                I  20
    REAL *8XDEL,YDEL,S,AREA,SUMT,RHO,PARAM,TEST,TOL,PINT,HMIN,PYR          I  30
    COMMON /PRMI/ NTIM,NPMP,NPNT,NITP,N,NX,NY,NP,NREC,INT,NNX,NNY,NUMO     I  40
   1BS,NMOV,IMOV,NPMAX,ITMAX,NZCRIT,IPRNT,NPTPND,NPNTMV,NPNTVL,NPNTD,N     I  50
   2PNCHV,NPDELC                                                           I  60
    COMMON /PRMK/ NODEID(20,20),NPCELL(20,20),LIMBO(500),IXOBS(5),IYOB     I  70
   1S(5)                                                                   I  80
    COMMON /HEDA/ THCK(20,20),PERM(20,20),TMWL(5,50),TMOBS(50),ANFCTR      I  90
    COMMON /HEDB/ TMRX(20,20,2),VPRM(20,20),HI(20,20),HR(20,20),HC(20,     I 100
   120),HK(20,20),WT(20,20),REC(20,20),RECH(20,20),TIM(100),AOPT(20),T     I 110
   2ITLE(10),XDEL,YDEL,S,AREA,SUMT,RHO,PARAM,TEST,TOL,PINT,HMIN,PYR        I 120
    COMMON /CHMA/ PART(3,3200),CONC(20,20),TMCN(5,50),VX(20,20),VY(20,     I 130
   120),CONINT(20,20),CNRECH(20,20),POROS,SUMTCH,BETA,TIMV,STORM,STORM     I 140
   2I,CMSIN,CMSOUT,FLMIN,FLMOT,SUMIO,CELDIS,DLTRAT,CSTORM                  I 150
    DIMENSION IC(20)                                                       I 160
```

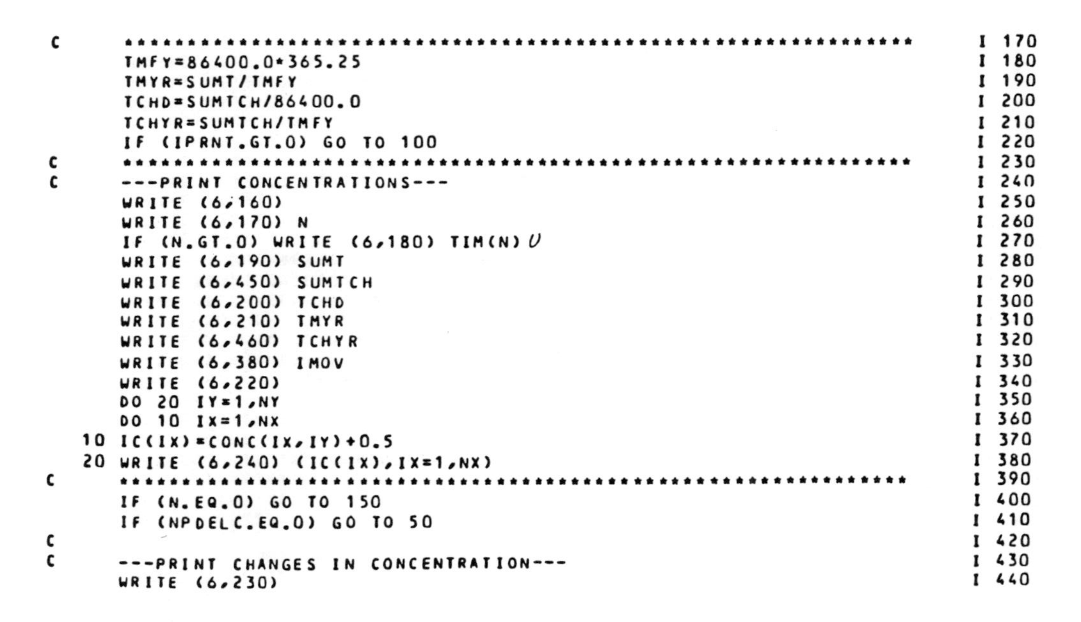

```
C     ***********************************************************************  I 170
      TMFY=86400.0*365.25                                                      I 180
      TMYR=SUMT/TMFY                                                           I 190
      TCHD=SUMTCH/86400.0                                                      I 200
      TCHYR=SUMTCH/TMFY                                                        I 210
      IF (IPRNT.GT.0) GO TO 100                                                I 220
C     ***********************************************************************  I 230
C     ---PRINT CONCENTRATIONS---                                               I 240
      WRITE (6,160)                                                            I 250
      WRITE (6,170) N                                                          I 260
      IF (N.GT.0) WRITE (6,180) TIM(N)                                         I 270
      WRITE (6,190) SUMT                                                       I 280
      WRITE (6,450) SUMTCH                                                     I 290
      WRITE (6,200) TCHD                                                       I 300
      WRITE (6,210) TMYR                                                       I 310
      WRITE (6,460) TCHYR                                                      I 320
      WRITE (6,380) IMOV                                                       I 330
      WRITE (6,220)                                                            I 340
      DO 20 IY=1,NY                                                            I 350
      DO 10 IX=1,NX                                                            I 360
   10 IC(IX)=CONC(IX,IY)+0.5                                                   I 370
   20 WRITE (6,240) (IC(IX),IX=1,NX)                                           I 380
C     ***********************************************************************  I 390
      IF (N.EQ.0) GO TO 150                                                    I 400
      IF (NPDELC.EQ.0) GO TO 50                                                I 410
C                                                                              I 420
C     ---PRINT CHANGES IN CONCENTRATION---                                     I 430
      WRITE (6,230)                                                            I 440
```

FORTRAN IV program listing—Continued

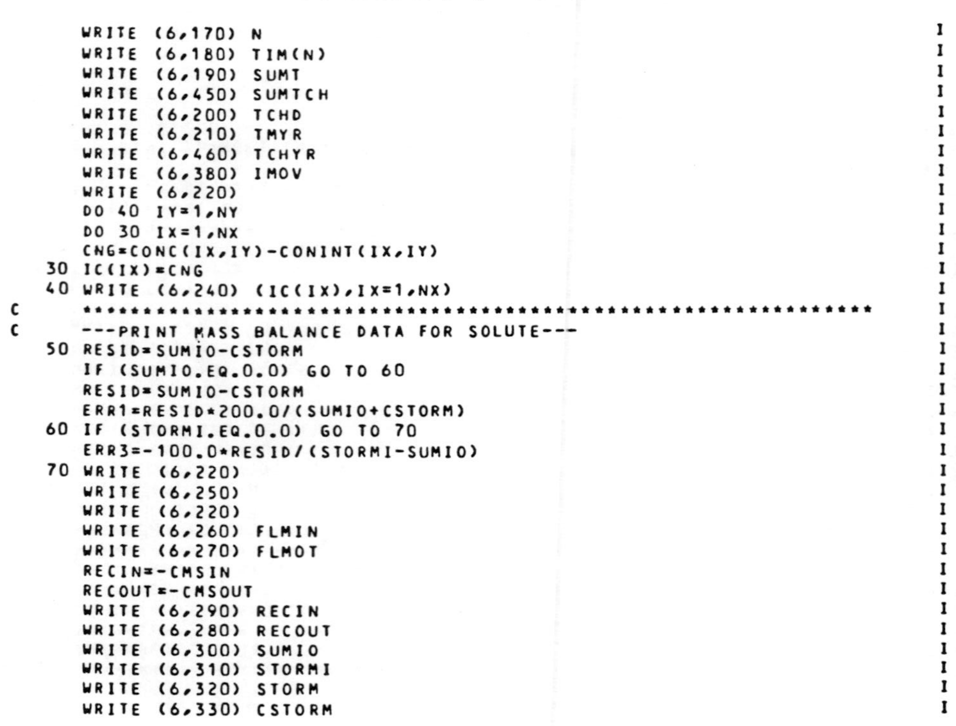

```
      WRITE (6,170) N                                                   I 450
      WRITE (6,180) TIM(N)                                              I 460
      WRITE (6,190) SUMT                                                I 470
      WRITE (6,450) SUMTCH                                              I 480
      WRITE (6,200) TCHD                                                I 490
      WRITE (6,210) TMYR                                                I 500
      WRITE (6,460) TCHYR                                               I 510
      WRITE (6,380) IMOV                                                I 520
      WRITE (6,220)                                                     I 530
      DO 40 IY=1,NY                                                     I 540
      DO 30 IX=1,NX                                                     I 550
      CNG=CONC(IX,IY)-CONINT(IX,IY)                                     I 560
   30 IC(IX)=CNG                                                        I 570
   40 WRITE (6,240) (IC(IX),IX=1,NX)                                    I 580
C     ***************************************************************** I 590
C     ---PRINT MASS BALANCE DATA FOR SOLUTE---                          I 600
   50 RESID=SUMIO-CSTORM                                                I 610
      IF (SUMIO.EQ.0.0) GO TO 60                                        I 620
      RESID=SUMIO-CSTORM                                                I 630
      ERR1=RESID*200.0/(SUMIO+CSTORM)                                   I 640
   60 IF (STORMI.EQ.0.0) GO TO 70                                       I 650
      ERR3=-100.0*RESID/(STORMI-SUMIO)                                  I 660
   70 WRITE (6,220)                                                     I 670
      WRITE (6,250)                                                     I 680
      WRITE (6,220)                                                     I 690
      WRITE (6,260) FLMIN                                               I 700
      WRITE (6,270) FLMOT                                               I 710
      RECIN=-CMSIN                                                      I 720
      RECOUT=-CMSOUT                                                    I 730
      WRITE (6,290) RECIN                                               I 740
      WRITE (6,280) RECOUT                                              I 750
      WRITE (6,300) SUMIO                                               I 760
      WRITE (6,310) STORMI                                              I 770
      WRITE (6,320) STORM                                               I 780
      WRITE (6,330) CSTORM                                              I 790
```

```
      IF (SUMIO.EQ.0.0) GO TO 80                                        I 800
      WRITE (6,340)                                                     I 810
      WRITE (6,350) RESID                                               I 820
      WRITE (6,360) ERR1                                                I 830
   80 IF (STORMI.EQ.0.0) GO TO 90                                       I 840
      WRITE (6,370)                                                     I 850
      WRITE (6,360) ERR3                                                I 860
C     ******************************************************************I 870
C     ---PRINT HYDROGRAPHS AFTER 50 STEPS OR END OF SIMULATION---       I 880
   90 IF (MOD(IMOV,50).EQ.0.AND.S.EQ.0.0) GO TO 100                     I 890
      IF (MOD(N,50).EQ.0.AND.S.GT.0.0) GO TO 100                        I 900
      GO TO 150                                                         I 910
  100 WRITE (6,390) TITLE                                               I 920
      IF (NUMOBS.LE.0) GO TO 150                                        I 930
      WRITE (6,400) INT                                                 I 940
      IF (S.GT.0.0) WRITE (6,410)                                       I 950
      IF (S.EQ.0.0) WRITE (6,420)                                       I 960
C        ---TABULATE HYDROGRAPH DATA---                                 I 970
      MOZ=0                                                             I 980
      IF (S.GT.0.0) GO TO 110                                           I 990
      NTO=NMOV                                                          I1000
      IF (NMOV.GT.50) NTO=MOD(IMOV,50)                                  I1010
      GO TO 120                                                         I1020
  110 NTO=NTIM                                                          I1030
      IF (NTIM.GT.50) NTO=MOD(N,50)                                     I1040
  120 IF (NTO.EQ.0) NTO=50                                              I1050
      DO 140 J=1,NUMOBS                                                 I1060
```

FORTRAN IV program listing—Continued

```
      TMYR=0.0                                                          I1070
      WRITE (6,430) J,IXOBS(J),IYOBS(J)                                 I1080
      WRITE (6,440) MOZ,WT(IXOBS(J),IYOBS(J)),CONINT(IXOBS(J),IYOBS(J)), I1090
     1TMYR                                                              I1100
      DO 130 M=1,NTO                                                    I1110
      TMYR=TMOBS(M)/TMFY                                                I1120
  130 WRITE (6,440) M,TMWL(J,M),TMCN(J,M),TMYR                          I1130
  140 CONTINUE                                                          I1140
C     ***************************************************************   I1150
  150 RETURN                                                            I1160
C     ***************************************************************   I1170
C                                                                       I1180
C                                                                       I1190
C                                                                       I1200
  160 FORMAT (1H1,13HCONCENTRATION/)                                    I1210
  170 FORMAT (1X,23HNUMBER OF TIME STEPS = ,1I5)                        I1220
  180 FORMAT (8X,16HDELTA T        = ,1G12.5)                           I1230
  190 FORMAT (8X,16HTIME(SECONDS) = ,1G12.5)                            I1240
  200 FORMAT (3X,21HCHEM.TIME(DAYS)      = ,1E12.5)                     I1250
  210 FORMAT (8X,16HTIME(YEARS)    = ,1E12.5)                           I1260
  220 FORMAT (1H )                                                      I1270
  230 FORMAT (1H1,23HCHANGE IN CONCENTRATION/)                          I1280
  240 FORMAT (1H0,20I5)                                                 I1290
  250 FORMAT (1H ,21HCHEMICAL MASS BALANCE)                             I1300
  260 FORMAT (8X,25HMASS IN BOUNDARIES       = ,1E12.5)                 I1310
  270 FORMAT (8X,25HMASS OUT BOUNDARIES      = ,1E12.5)                 I1320
  280 FORMAT (8X,25HMASS PUMPED OUT          = ,1E12.5)                 I1330
  290 FORMAT (8X,25HMASS PUMPED IN           = ,1E12.5)                 I1340
  300 FORMAT (8X,25HINFLOW MINUS OUTFLOW     = ,1E12.5)                 I1350
  310 FORMAT (8X,25HINITIAL MASS STORED      = ,1E12.5)                 I1360
  320 FORMAT (8X,25HPRESENT MASS STORED      = ,1E12.5)                 I1370
  330 FORMAT (8X,25HCHANGE MASS STORED       = ,1E12.5)                 I1380
  340 FORMAT (1H ,5X,53HCOMPARE RESIDUAL WITH NET FLUX AND MASS ACCUMULA I1390
     1TION:)                                                            I1400
```

```
350 FORMAT (8X,25HMASS BALANCE RESIDUAL  = ,1E12.5)                    I1410
360 FORMAT (8X,25HERROR  (AS PERCENT)    = ,1E12.5)                    I1420
370 FORMAT (1H ,5X,55HCOMPARE INITIAL MASS STORED WITH CHANGE IN MASS  I1430
   1STORED:)                                                           I1440
380 FORMAT (1X,23H NO. MOVES COMPLETED = ,1I5)                         I1450
390 FORMAT (1H1,10A8//)                                                I1460
400 FORMAT (1H0,5X,65HTIME VERSUS HEAD AND CONCENTRATION AT SELECTED O I1470
   1BSERVATION POINTS//15X,19HPUMPING PERIOD NO. ,I4////)              I1480
410 FORMAT (1H0,16X,19HTRANSIENT  SOLUTION////)                        I1490
420 FORMAT (1H0,15X,21HSTEADY-STATE SOLUTION////)                      I1500
430 FORMAT (1H0,20X,22HOBS.WELL NO.    X    Y,17X,1HN,6X,40HHEAD (FT)  I1510
   1    CONC.(MG/L)     TIME (YEARS)//24X,I3,9X,I2,3X,I2//)            I1520
440 FORMAT (1H ,58X,I2,6X,F7.1,8X,F7.1,8X,F7.2)                        I1530
450 FORMAT (1H ,2X,21HCHEM.TIME(SECONDS) = ,E12.5)                     I1540
460 FORMAT (1H ,2X,21HCHEM.TIME(YEARS)   = ,E12.5)                     I1550
    END                                                                I1560-
```

Definition of Selected Program Variables

Variable	Definition
AAQ	area of aquifer in model
ALNG	BETA
ANFCTR	anisotropy factor (ratio of T_{yy} to T_{xx})
AOPT	iteration parameters
AREA	area of one cell in finite-difference grid
BETA	longitudinal dispersivity of porous medium
CELDIS	maximum distance across one cell that a particle is permitted to move in one step (as fraction of width of cell)
CLKCN	concentration of leakage through confining layer or streambed
CMSIN	mass of solute recharged into aquifer
CMSOUT	mass of solute discharged from aquifer
CNCNC	change in concentration due to dispersion and sources
CNCPCT	change in concentration as percentage of concentration at node
CNOLD	concentration at node at end of previous time increment
CNREC	concentration of well withdrawal or injection
CNRECH	concentration in fluid source
CONC	concentration in aquifer at node
CONINT	concentration in aquifer at start of simulation
C1	CONC at node (IX,IY)
DALN	longitudinal dispersion coefficient
DDRW	drawdown
DELQ	volumetric rate of leakage across a confining layer or streambed
HR	head from row computation in subroutine ITERAT; elsewhere HR represents head from previous time step
IMOV	particle movement step number
INT	pumping period number
IPRNT	print control index for hydrographs
ITMAX	maximum permitted number of iterations
IXOBS	x-coordinate of observation point
IYOBS	y-coordinate of observation point
KOUNT	iteration number for ADIP
LIMBO	array for temporary storage of particles
N	time step number
NCA	number of aquifer nodes in model
NCODES	number of node identification codes
NITP	number of iteration parameters
NMOV	number of particle movements (or time increments) required to complete time step
NODEID	node identification code
NP	total number of active particles in grid
NPCELL	number of particles in a cell during time increment
NPMAX	maximum number of available particles
NPMP	number of pumping periods or simulation periods
NPNT	number of time steps between printouts
NPTPND	initial number of particles per node
NREC	number of pumping wells
NTIM	number of time steps
NUMOBS	number of observation wells

DELS	rate of change in ground-water storage
DERH	change in head with respect to time
DISP	dispersion equation coefficients
DISTX	distance particle moves in x-direction during time increment
DISTY	distance particle moves in y-direction during time increment
DLTRAT	ratio of transverse to longitudinal dispersivity
DTRN	transverse dispersion coefficient
FCTR	multiplication or conversion factor
FLMIN	solute mass entering modeled area during time step
FLMOT	solute mass leaving modeled area during time step
GRDX	hydraulic gradient in x-direction
GRDY	hydraulic gradient in y-direction
HC	head from column computation
HI	initial head in aquifer
HK	computed head at end of time step
HMIN	minimum iteration parameter
NX	number of nodes in x-direction
NY	number of nodes in y-direction
NZCRIT	maximum number of cells that can be void of particles
NZERO	number of cells that are void of particles at the end of a time increment
PARAM	iteration parameter for current iteration
PART	1. x-coordinate of particle; 2. y-coordinate of particle; 3. concentration of particle. Also note that the signs of coordinates are used as flags to store information on original location of particle.
PERM	hydraulic conductivity (in LT^{-1})
PINT	pumping period in years
POROS	effective porosity
PUMP	cumulative net pumpage
PYR	total duration of pumping period (in seconds)
QNET	net water flux (in L^3T^{-1})

Definition of selected program variables—Continued

Variable	Definition
QSTR	cumulative change in volume of water in storage
REC	point source or sink; negative for injection, positive for withdrawal (in L^3T^{-1})
RECH	diffuse recharge or discharge; negative for recharge, positive for discharge (in LT^{-1})
RN	range in concentration between regenerated particle and adjacent node having lower concentration
RP	range in concentration between regenerated particle and adjacent node having higher concentration
S	storage coefficient (or specific yield)
SLEAK	rate of leakage through confining layer or streambed
STORM	change in total solute mass in storage (by summation)
STORMI	initial mass of solute in storage
SUMC	summation of concentrations of all particles in a cell
SUMIO	change in total solute mass in storage (from inflows—outflows)
SUMT	total elapsed time (in seconds)
SUMTCH	cumulative elapsed time during particle moves (in seconds)
THCK	saturated thickness of aquifer
TIM	length of specific time step (in seconds)
TIMD	elapsed time in days
TMRX	transmissivity coefficients (harmonic means on cell boundaries; forward values are stored)
TMWL	computed heads at observation points
TOL	convergence criteria (ADIP)
TOTLQ	cumulative net leakage through confining layer or streambed
TRAN	transverse dispersivity of porous medium
VMAX	maximum value of VX
VMAY	maximum value of VY
VMGE	magnitude of velocity vector
VMXBD	maximum value of VXBDY
VMYBD	maximum value of VYBDY
VPRM	initially used to read transmissivity values at nodes; then after line B2270, VPRM equals leakance factor for confining layer or streambed (vertical hydraulic conductivity/thickness). If VPRM≥0.09, then the program assumes that the node is a constant-head boundary and is flagged for subsequent special treatment in calculating convective transport.
VX	velocity in x-direction at a node
VXBDY	velocity in x-direction on a boundary between nodes
VY	velocity in y-direction at a node
VYBDY	velocity in y-direction on a boundary between nodes

TIMY elapsed time in years
TIMV length of time increment for particle movement (in seconds)
TIMX time step multiplier for transient flow problems
TINIT size of initial time step for transient flow problems (in seconds)
TITLE problem description
TMCN computed concentrations at observation points
TMOBS elapsed times for observation point records
WT initial water-table or potentiometric elevation, or constant head in stream or source bed
XDEL grid spacing in x-direction
XOLD x-coordinate of particle at end of previous time increment
XVEL velocity of particle in x-direction
YDEL grid spacing in y-direction
YOLD y-coordinate of particle at end of previous time increment
YVEL velocity of particle in y-direction

INDEX